On Active Grounds

Environmental Humanities Series

Environmental thought pursues with renewed urgency the grand concerns of the humanities: who we think we are, how we relate to others, and how we live in the world. Scholarship in the environmental humanities explores these questions by crossing the lines that separate human from animal, social from material, and objects and bodies from techno-ecological networks. Humanistic accounts of political representation and ethical recognition are re-examined in consideration of other species. Social identities are studied in relation to conceptions of the natural, the animal, the bodily, place, space, landscape, risk, and technology, and in relation to the material distribution and contestation of environmental hazards and pleasures.

The Environmental Humanities Series features research that adopts and adapts the methods of the humanities to clarify the cultural meanings associated with environmental debate. The scope of the series is broad. Film, literature, television, Web-based media, visual art, and physical landscape—all are crucial sites for exploring how ecological relationships and identities are lived and imagined. The Environmental Humanities Series publishes scholarly monographs and essay collections in environmental cultural studies, including popular culture, film, media, and visual cultures; environmental literary criticism; cultural geography; environmental philosophy, ethics, and religious studies; and other cross-disciplinary research that probes what it means to be human, animal, and technological in an ecological world.

Gathering research and writing in environmental philosophy, ethics, cultural studies, and literature under a single umbrella, the series aims to make visible the contributions of humanities research to environmental studies, and to foster discussion that challenges and reconceptualizes the humanities.

On Active Grounds

Agency and
Time in the
Environmental
Humanities

Robert Boschman
and Mario Trono,
editors

WLU PRESS

**WILFRID LAURIER
UNIVERSITY PRESS**

LAURIER
Inspiring Lives.

This book has been published with the help of a grant from the Canadian Federation for the Humanities and Social Sciences, through the Awards to Scholarly Publications Program, using funds provided by the Social Sciences and Humanities Research Council of Canada. Wilfrid Laurier University Press acknowledges the support of the Canada Council for the Arts for our publishing program. We acknowledge the financial support of the Government of Canada through the Canada Book Fund for our publishing activities. This work was supported by the Research Support Fund.

Canada

Canada Council   Conseil des arts
for the Arts   du Canada

ONTARIO ARTS COUNCIL
CONSEIL DES ARTS DE L'ONTARIO
an Ontario government agency
un organisme du gouvernement de l'Ontario

Library and Archives Canada Cataloguing in Publication

On active grounds : agency and time in the environmental humanities / Robert Boschman and Mario Trono, editors.

(Environmental humanities)
Includes bibliographical references and index.
Issued in print and electronic formats.
ISBN 978-1-77112-339-6 (softcover).—ISBN 978-1-77112-341-9 (EPUB).—
ISBN 978-1-77112-340-2 (PDF)

1. Environmental sciences—Philosophy. 2. Humanities—Philosophy. I. Boschman, Robert Wayne, 1961–, editor II. Trono, Mario, [date], editor III. Series: Environmental humanities

GE40.O5 2019   179'.1   C2018-904626-0
                                                                      C2018-904627-9

Cover photo courtesy of Robert Boschman. Cover design by Sandra Friesen. Interior design by Angela Booth Malleau, designbooth.ca

This book is printed on FSC® certified paper and is certified Rainforest Alliance™ and Ancient Forest Friendly™. It contains post-consumer fibre, is processed chlorine free, and is manufactured using biogas energy.

Printed in Canada

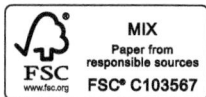

MIX
Paper from
responsible sources
FSC
www.fsc.org   FSC® C103567

CONTENTS

LIST OF IMAGES

PHOTO ESSAY

ACKNOWLEDGEMENTS

The chapters in this volume are part of the wider, dynamic, and committed community of Environmental Humanities scholars who function in adaptive and shifting configurations across disciplinary and international boundaries. We acknowledge that community's dedication and vitality and its roles, direct and indirect, in helping bring this book forward.

Two people in particular at Wilfrid Laurier University Press, Director Lisa Quinn and Senior Editor Siobhan McMenemy, have been instrumental in bringing *On Active Grounds* to press. The rest of the staff at WLUP have been equally committed in their efforts on behalf of this book, and we the editors gratefully acknowledge those efforts. The look and feel of this volume attests to their labours as much as anyone else's.

Our colleagues at Mount Royal University are responsible for the genial, supportive environment in which this project germinated and grew. From our department (English, Languages, and Cultures) to our Faculty (Arts) and the wider university community, we have been the beneficiaries of all kinds of support. We especially acknowledge and thank two successive Deans of the Faculty of Arts, Jeffrey Keshen and Jennifer Pettit, for their unremitting backing of *On Active Grounds*. Like Keshen and Pettit, Connie Van Der Byl, Director of the Institute for Environmental Sustainability (IES), has anchored us with funding support without which *OAG* would not exist, as has Michael Quinn, Associate Vice-President of Research. Quinn's Office for Research, Scholarship, and Community Engagement (ORSCE) maintains an open-door policy that has been essential to us.

Ecocritical Agency in Time

Mario Trono and Robert Boschman

What a diversity of time, and of events!
—Jan Zalasiewicz, 2012

The question of how to exercise agency on behalf of natural environments in the time we have left—as individuals, as a species—is as complex as the weave and stuff of existence itself.

And we are part of that complexity, with our embodied thoughts arising from quantum and other mysteries. In hoping to change a given state of affairs involving humans and the biosphere, we face the daunting processual density of matter and event in time; the intricacies of actors in their networks; and the histories of ideas that play out unpredictably in omnidirectional flows of affects and effects that create equivalencies to the complex non-linear systems of earth and atmosphere. The feeling that narrowing the scope of considerations would be irresponsible (given the interconnectedness of things) can slow an environmental critic down. But pragmatically delimiting one's investigations and conscientiously connecting them to advocacy does not automatically lead to quick or adequate (or sometimes even any) results. Basic education programs and awareness campaigns hit their limits, due to bio-political intricacies that are hard to see as a whole. Meaningful or lasting changes to behaviours and laws seem ever out of reach. Pre-existing social systems that carry built-in protections for some and not others adeptly co-opt and divert environmental initiatives. The need for sustenance and the seductions of amenity ensure that *akrasia* (acting against one's better judgment) remains part of the human condition. And there is fear of getting it wrong, of the

1

legacies of bad ideas once thought good—early-twentieth-century eugenics, hasty introductions of new species to habitats, certainty of an ever-regenerating planet, or absolutism as a necessary means to an end. Concentration during purposive research hours can be broken by communications in "real time" about another ecodisaster, dispersing one's energies further.

At times, though, bad news can galvanize an environmental humanities researcher. Passions return and paths of environmental right action seem clear enough. The goal of keeping drinking water clean and available, for example, is easy to embrace. But deciding on grounds for reforms and protections before vectoring personal and collective energies into political and other means to rebuild social-ecological systems to achieve specific objectives is quite another matter. Buoyed, however, by what seems our innate sense of possibility, we accept that there is much to do and get to it, examining environmental and other kinds of scientific information; studying political ecologies; reflecting on motivation and intention; considering moral and ethical dimensions; positioning oneself in a discourse relative to others; geo-locating oneself in relation to an aspect of materiality that concerns; and deciding how to proceed—all before finding that, as is the case with states of matter in time, things fall apart and that changing course is likely. The process is dynamic and can be exhilarating until environmental stressors and forcing agents, acting in time, bring on the possibility or fact of suffering for human and non-human species. The sense that we may already be too late[1] or that an avenging Gaia has found a way of dealing with us[2] hovers at the edge of consciousness. We may find ourselves again dispersed, reticent, perhaps not doing enough while consoling ourselves with arguments that thought *is* action or that what is required is more thought and less pseudo-action (see our section on agency below). All the while, points of no return approach, or have migrated from futurity into states of arrival, and many kinds of hope become essential. This felt work dynamic is experientially cyclical and endemic to Environmental Humanities (EH) studies, and it pulses in relation to agency and time, factors that give the work its sense of urgency.

The impetus for assembling this collection was our sense that in so much of EH scholarship we read, writers were engaging either directly or obliquely with questions of agency and time. The preoccupations are sometimes incidental, sometimes explicit, but always there. It struck us that EH critics, by the nature of their work, are required to adjust their senses of human and non-human agency, the potentials of matter, and the related influence of time in its varied lines and material substrates. EH researchers need to do several things simultaneously and continuously on these fronts because their research lands them in states of affairs and life processes already under way: "To begin to think life, we must begin in the middle with an activist sense of life at no remove: in the middling immediacy of its always 'going on'" (Massumi 1). This active at-onceness characterizes our interdiscipline. It is a default acknowledgement of the depth of biospheric relations based on "concepts of event, change, potential, and creativity that implicate matter in life and life in matter, and both in mentality" (Massumi 181). The active grounds to which our title refers are, essentially, grounds per se, since being is active. As William James puts it, "We are only as we are active" (1996, 161–62).[3]

The Humanities for the Environment 2018 Report—Ways to Here, Ways Forward, authored by members of the Trinity Centre for Environmental Humanities, highlights what it terms "environmental humanities in action" (Holm and Brennan 2018, 4). It is a prime example of EH scholarship concerned (in almost every paragraph in this case) with aspects of agency and time while only occasionally using either word. While we felt a collection devoted to analyzing how these twinned preoccupations are manifest would recognize an important metadiscourse, it was also clear that the preoccupations finally spoke for themselves within work focused primarily on specific and important aspects of cultures and environments. So we asked contributors to link their current work to agency and time while carrying on with their other work: to foreground the at-onceness of EH labours this introduction delineates. The photo essay at the heart of this collection expresses the double theme as well in memoir and photographs: "Agency and Time on Active Grounds: A Memoir of Bruno Latour and Gaia Global Circus." That piece is bracketed by four sections: Eco-Temporal Literacies, Timelines and Indigeneity, Animal Agents and Human / Non-Human Interactions, and Systems Change in Time. The remainder of this introduction discusses time and agency even though the act of addressing each in order militates against the fact of how *interrelated* the concepts are in what physicists call "spacetime." But English-language writing sequences from left to right and presents timelines as moving in the same direction. Sentences coordinate and subordinate in what seems a forward flow of reasoning towards ends, tilting usually towards teleologies that may or may not stand up to scrutiny. Things get taken one at a time, usually within linear notions of time, even as new theories of time suggest that doing so proves such hallmarks of writing to be incongruent with how physics tells us things really "go." But with these limits of the English language in peripheral view—and the assumption in hand that, since change is constant, "making a difference" means trying to influence what *kinds* of change emerge—we proceed nonetheless, motivated by the insight from non-linear systems analysis that a wide diversity of agents in a system results in richer emergent patterns and more possibilities.

AGENCY

This collection acknowledges that agency as a term has diverse manifestations and meanings, both on and off university campuses, in various EH disciplines, and certainly within what we can term the environmental post-humanities. As regards this last, in "Agency at the Time of the Anthropocene," Bruno Latour explains that even before the term "Anthropocene" became widespread, there was a dawning realization that a "complete subversion of the respective positions of subject and object" was under way and that a subject could no longer "act autonomously in front of an objective background, but [has] *to share agency with other subjects that have also lost their autonomy.* It is because we are now confronted with those subjects—or rather *quasi*-subjects—that we have to shift away from dreams of mastery as well as from the threat of being fully naturalized" (2014, 5). In Latour's view, humans now share the planetary stage and are moving towards a "common articulation that could allow speaking with and about former 'facts of nature' in a different way" (1). This view is in keeping with the Latourian notion of actants functioning in enacted alliances, and all of this is commensurate with several other complications of agency.

In *Bodily Natures: Science, Environment, and the Material Self*, Stacy Alaimo articulates her notion of trans-corporeality as follows (an idea informing Karla Armbruster's chapter in this volume):

> Potent ethical and political possibilities emerge from the literal contact zone between human corporeality and more-than-human nature. Imagining human corporeality as trans-corporeality, in which the human is always intermeshed with the more-than-human world, underlines the extent to which the substance of the human is ultimately inseparable from "the environment." It makes it difficult to pose nature as mere background ... for the exploits of the human since "nature" is always as close as one's own skin—perhaps even closer. (2)

This new materialist orientation finds a related strand in *Meeting the Universe Halfway: Quantum Physics and the Entanglement of Matter and Meaning*, in which Karen Barad develops a theory of agentive realism and the concept of intra-action (2007). The concept abandons the term "interaction" (which presupposes pre-existing embodied agents that move to interact with one another) to suggest that agency is not an inherent impulse or property but follows from a dynamism of forces that perpetually inter-constitute agents. This view meshes well with Jane Bennett's notion of distributive agency (see Mario Trono's chapter in this volume), which explains how agentive forces in assemblages are not discretely located. In *Vibrant Matter: a Political Ecology of Things* (2010), Bennett offers the related concept of vital materialism, which elaborates the "active powers issuing from non-subjects," as she argues that "the image of dead or thoroughly instrumentalized matter feeds human hubris and our earth-destroying fantasies of conquest and consumption" (ix). These perspectives and others have complicated the idea of agency for the better.

But just because these views have taken hold in various academic precincts does not mean that other conceptions of the term "agency" do not exist. They do, and they have an impact. Off campus on the wider cultural field, various professional discourses use this term in different ways. In business, the term "distributed agency" (Bennett uses instead the word "distributive") refers to an arrangement where an organization hires workers who labour at a distance from the main organizational entity. Also in business, the "agency dilemma" describes scenarios where a representative acts against a principal's best interests. Commercial law involves agents with authority to operationalize the rights and actions of others through contracts and legal processes. An agency is a basic structure of systemic organization, as in bureaus, departments, and firms. Intelligence agencies affect the course of geo-political events. In applied chemistry an agent refers to any substance, natural law, or force that can produce an effect. The term has a long semantic history of usages that are primarily instrumental in nature. Debates within and between academic disciplines over agency have grown in complexity and intensity in recent decades in philosophy, psychology, anthropology, and sociology, but also in disciplines like English, history, education, and comparative religion, or periodically in any field where political ecologies and the spatio-temporal life tra-jectory of one or many living things becomes a subject of study.

Psychology has developed agency theory to explain human obedi-ence. But more profoundly, it asks (along with philosophers and cogni-tive scientists) if the idea of a bounded self unjustifiably presupposes a stable consciousness the contents of which are reliably present *to* that consciousness. The implication for agency being that, as with a drunk elephant driver, we may not really be in charge, and actions taken may not play out as we consciously hope. Theories of identity (see the chapter by Morgan Zedalis and Sean Gould) abound and differ over which factors internal and external are constitutive of what gets called a "self," with clear implications for agency. The field has its predictive theory of reasoned action. And it warns against the actor–observer bias wherein one attributes one's actions to external causes, while believing others' arise from internal ones. Elsewhere in the humanities, compar-ative religionists examine various notions related to agency operative in human societies, including soul consciousness, predestination, good and evil, and free will in relation to various monotheisms. Anthropol-ogists speak of "agentivity" in relation to manifestations of agency that arise in response to social structures, with the specifics of social ontology (categories of being and their relations) kept responsibly in view. And in many disciplines, anthropocentrism gets bracketed in order to ask if non-human animals can be said to exercise some forms of agency.

Philosophers consider concepts of action, intentionality, causation, and mental states before moving to theories of agency. Ethics comes into play. Questions about how one can or cannot, should or should not act get asked, requiring operative definitions of individual, citizen, and subject. Are there coherent selves seeking freedom from oppressive forces? Or is selfdom an illusion shimmering atop a mere and messy conglomeration of bio-cultural tendencies adrift in random influences? If it can be demonstrated philosophically that autonomous selfhood is an untenable product of Enlightenment thought, what are the grounds for a clear and functional definition of legal personhood? Is agency a definable right? If liberalism is an exhausted ideal and the notion of a human subject is no longer stable, what does that mean for collective action, or for any theory of groups, if it cannot be established what it is a group consists of, especially in laws?

The very idea of "taking action" is a composite notion. It requires a definition of causation as both a process and a result that can be appealed to as a kind of proof or validation in cases where purposive action can be said to have occurred. But how can causation be adequately understood in a world of stunningly complex systems that persistently give rise to unintended consequences? If there is no verifiable result for an action, is the action then meaningless? Or altogether non-existent? What is the distinction between action and non-action if *both* can bring forth consequences and changed states of affairs? And what of the problematic distinction between thought and deed, between the state of being *in practice* as opposed to *in theory*? How do we reconcile the *feeling* that thought differs very much from action when it can be argued that the processes of mind are, themselves, a kind of environmental activity (Bem and Keijzer 1996)? Or, if one can differentiate thinking and doing in some ways, what if what is needed is more thought and less pseudo-action (Žižek 2016)?

What if Latour, Alaimo, Barad, and Bennett and others are correct in claiming that one of the key threats to the global environment is the very concept of human agency as it has been historically defined? How best to counter its role in proliferating populations, its scientific adventurism in the service of empire, and its corporate–industrial manifestations driven by voracious consumptions and facilitated now by globalization? We must also consider that not everyone agrees with post-humanist views of agency. Langdon Winner, for instance, decried actor-network theory's now often challenged claim that non-humans do carry agency within systems (1993). In any case, while our debates unfold, industrial corporate entities have spent or "wasted" no such time in examining the philosophical roots of their actions. They have instead launched themselves out along a wide range of agentive vectors to reach commercial goals and gain formidable social powers. After long ago securing in law the operational definitions that established their multinational character and status as legal persons, they continue to strategically expand their reach and functioning, propelling into the future a range of ever-adaptive (hence agentive in some sense) cultural-industrial systems. These require ongoing expansion to continue drawing returns on investments in order to afford significant gains in personal wealth for certain persons very well defined and protected within the neoliberalization of laws. Questions of collective agency arise when we consider that on the factory floors of Fordist modernity advances in automation, robotics, and programming began to attenuate the agencies of collective labour, with artificial intelligence poised today to continue this long-term trend. Legal scholars claim that "[h]ow the law chooses to treat machines without principals will be the central legal question that accompanies the introduction of truly autonomous machines" (Vladeck 124). These are machines acting in accordance with forms of agency that are subset to more complex forms of agency that have resulted in

> laws to restrain labor organization and restrict consumer movements; corporate participation on school and university boards; favorable tax laws for investors; corporate ownership and control of the media ... the legal defense of corporate, financial power to limit consumer information about the policies that affect them ... maintain[ed] conditions for unfettered markets and to obscure or state clean up disasters created by under-regulated financial and corporate activity; and state/bureaucratic delays to hold off action on global climate change. (2012, n.p.)

Traditional notions of agency, matter, and the status of non-human animals will likely not yield to recent trends in EH thought, not with so much capital at stake. Science fictionists have long imagined capital itself as a kind of AI (Canavan 2015), and the language used to describe corporate AI planning is usually rife with creationist dreams of machinic displacements of human agency and action. Description is often breathless and accords with how well corporations have strategically decentralized their core power centres, mimicking their own advanced technologies—in what is, intriguingly, not an endorsement of distributive agency but a strategic distribution of agency as traditionally understood. For example, the Artificial Intelligence Laboratory of the Massachusetts Institute of Technology writes: "The notions of central and peripheral systems evaporate—everything is both central and peripheral. Based on these principles we have built a very successful series of mobile robots which operate without supervision as Creatures in standard office environments" (Brooks 1991, 139). But they *do* operate under supervision, just of a vaster and more complex kind intent on lowering costs and maximizing profits: recent surveys of Japanese firms "examine the impact of AI-related technologies on business and employment ... [and] expect a positive impact on business but a negative impact on employment" (Morikawa 2016).

Accompanying a caste system of high- and low-tech workers will be increased levels of e-waste, especially in Third World countries. It remains to be seen if advances in biotechnology can ameliorate this and other kinds of environmental damages as they continue to occur in what may become instead a strained cycle of accelerated degradation marked by partial reclamation, powered by technological determinism and boosterism: "As an enabling technology, nanotechnology has the potential to permeate and improve *all* industrial sectors and influence their development toward sustainability" (Karn 2007, my emphasis). The corporate contexts for biotech development are not operationally programmed in the main by command functions involving principles of environmental right action. They are entities created and guided by articles of incorporation, internal bylaws, employment agreements, initial public offerings, stock exchanges or over-the-counter markets, asset transfer protocols, and shareholder agreements. What the embrace of nanotechnology might well entail is acceptance of a process where, from a global view, the structures of agencies responsible for damages to ecosystems depend on ruination as part of a systems-wide business model that involves the drawing forth of ingeniously envisioned but finally disingenuous ventures that leverage capital in hidden ways. Recent EH thinking regarding agency is proved correct, then, if human agency as traditionally imagined is revealed as carrying the seeds of its own destruction. The problem is a philosophical one where enterprise assumes that ecosystems are external to us and able to be acted upon with limited consequence. It is a thought problem Timothy Morton calls a "dark ecology," the form of which

> is that of noir film. The noir narrator begins investigating a supposedly external situation, from a supposedly neutral point of view, only to discover that she or he is implicated in it. The point of view of the narrator herself becomes stained with desire. There is no metaposition from which we can make ecological pronouncements. (2012, 16–17)

From what positions can the environmental humanities speak in such a situation? And given how film noir always ends for its characters, why? Many kinds of hope become essential.

Human consciousness appears designed to formulate dualisms it soon finds itself battling. While historically this habit of mind had dire results for some persons/environments but not others, its reach now extends to all of the biosphere and the species within it. If mind/body, self/other, culture/nature, and subject/object seemed problematic divisions before, they are more so now in the face of so much scientific and other kinds of evidence that things are not distinct but very much of one another. If our thoughts are nested in electro-chemical pockets that are environmentally/materially of a piece with the rock making up the granitic crust of the planet, then, for example, both we as human beings and the diegesis involving Sisyphus are both, in a sense, of rock. In this view, the Sisyphean classical tragedy and metaphor for the human condition is revealed as a comedy of philosophical errors. There is something freeing and revitalizing here, a welcome escape from the relentlessness of dualisms and the hierarchies they erect. There are objects, there are states of affairs, and there are events. There need not be absurd exercises in valuation and a related warring of oppositions that plays out destructively in biocultural contexts.

But how can a philosophical aversion to dualism play into agency in the context of an environmental ethic when so much of what needs to be done appears to be a matter of simple oppositions in terms of key conflicts and operative agents on each side: warming/cooling, exploitation/conservation, unregulated/regulated, private/public, individual/collective? Perhaps dualisms could remain present in thought but be prevented from functioning like genomes that guide all future formations. Their uses in bringing dynamic processes into theory could then be held on to while their abuses are held at bay. This is perhaps why Timothy Morton enjoins us to "hang out in what feels like dualism" (2009, 205). Bruno Latour imagines "a political philosophy in which *there are no longer two* major poles of attraction" while wondering "how to recruit an assembly, without continuing to worry about the ancient titles that sent some to sit in nature's ranks and others on society's benches" (2004, 59). His aim may be akin to Massumi's, when the latter argues that social "aspects … [need not be] treated as in contradiction or opposition, but as co-occurring dimensions of every event's relaying of formative potential" (12). One enduring dualism in the environmental humanities, however, involves the traditional categories of right and left, and in this Massumi is again welcome, as he dismisses the shuffling crutches of a tired "'ought' … [wherein] the appetite is for subordinating activity to statement, and statement to program. This way of doing 'politics' is always process poor." We should think less about how our language can make statements and more about

> [t]he degree to which a technique of existence avails itself of its *imaginative powers*: its ability to … catalyze what's to come, emergently, inventively, unpreprogrammed and reflective of no past model. This is the power of language to perform virtual events of foretracing … A 'magic' power to invoke relational realities into world-lining. (173)

These "imaginative powers" result in what Whitehead calls the "production of novelty" (21). We hope this collection functions as a produced novelty with its emphasis on the multi-view at-onceness of EH scholarship where agency and time shift unpredictably and actively between backgrounds, mid-grounds, and foregrounds.

TIME

The time of substance is one of the great topics and ongoing mysteries of contemporary physics and earth sciences. EH receives ever more fantastic dispatches and updates from the sciences about the nature of time: temporality is a variable in universes that may be multiple, may curve through six dimensions, or may not exist at all except as a measure of our ignorance. A pebble, writes the geologist Jan Zalasiewicz (2012), contains multiple histories that require multiple chronometers, and as Latour says, "Geologists need these chronometers, for they have to deal with colossal amounts of time ... a kind of soup of events" (2010, 206). Much of EH research draws on information arriving from various scientific fields the way environmental science itself does generally, from geology, biology, atmospheric science, and so on. EH sometimes writes back about how the cultures of the sciences, the slipperiness of language, and the philosophies of researchers can affect findings and how these get explained to the culture at large. Combined approaches have appeared, such as complex systems studies that trade in philosophical propositions and mathematical modelling, or animal studies which are zoologically informed as they examine how the non-human is conceived of through the lenses of psychology, sociology, and literary/filmic/visual art histories. What all investigations involve are events in time, and disciplines "take up time" in different ways. They are themselves events that intermingle in space and time with what they can only partly espy, given that any subject/perceiver is indistinct from an object/perceived. We have a general *sense* of the staggering number and kind of temporally ensconced material events, ranging from a cosmological one involving the visible universe's beginnings, or a geologic apprehension of planetary deep time, to the relativity of light in its speed or the possibility that some kinds of animal consciousness may carry a rudimentary, abstract sense of time. New research on the temporal dimensions of landscape ecology[4] can help us imagine an integrated scene of existence before us that is forever calling forth events of attention involving materiality, agency, time. Even when one adopts a necessarily circumscribed view and delineated topic, the multi-layered *depths* of locations and events in time flash into vertiginous view. Responsive and dynamic maps of potential relations must be layered atop one another, interconnections discovered, and changing states of affairs tracked, even as we wonder how to situate ourselves on these maps.

Literature has always aspired to respond in kind to the endless, emergent novelty of existence through aesthetic correspondence. Literary works surprise with their own patterns and other intricacies, conjuring the same awe, terror, and delight that follow from immersion in other registers involving the processual complexities of the life world. Writers well understand that time is not directly observable, only apparent in events. So they create simulations that are also their own happenings, what ifs and what thens, knowing their stories will factor into yet other events involving intentional readers immersed in language, reading at other points in space and time. Words vector far beyond their origins: "Go, my little book! Go, my little tragedy!" writes the character of Criseyde in Chaucer's *Troilus and Criseyde*, addressing her own text and imagining a kind of agency for it. Studies of time in literature reveal vast and many kinds of engagement with temporality, beginning with the time "travel" afforded by tenses in language. Story and discourse (or reader) time are considered distinct. There are encyclopaedic accountings of representations of time units (seconds, minutes, hours, days, seasons, years, and the many symbolic manifestations of these), along with structural modelling of same, in *topoi* and other instances. For Ricoeur, narration phenomenologically facilitates the human experience of time: "time becomes human to the extent that it is articulated through a narrative mode, and narrative attains its full meaning when it becomes a condition of temporal existence" (52). In cognitivist approaches, how readers construct meaning is the focus, and it is claimed that surprise arises out of a "generic interplay between times" (Sternberg 1992, 523). William James spoke of the specious present, those moments of duration where our most acute sense of time is felt (1890). Reading, as we know, can marvellously suspend this sense whenever we "lose" ourselves in a book.

This effect is not exclusive of course to literature and may come from immersion in any art, however defined. The Lakota conceive broadly of art, "Taku waste'la ka pi ag a xapi – whatever we make and admire" (Medicine). In the environmental humanities, creative academic work can sometimes provide transport too, creative not necessarily in the sense of writing style but in terms of the felt immediacy and intensity generated when aliveness engines and techniques of existence find or express novelty—not for its own sake but as part of a profound interplay with the universe. And there are stylistic experiments too. In *Dark Ecology: For a Logic of Future Coexistence* (2016), for example, Timothy Morton attempts a style at once poetic and theoretical, as if agreeing with Massumi that there is tremendous potential in "putting art and philosophy, theory and practice, on the same creative plane, in the same ripple pool. Art and philosophy, theory and practice, can themselves resonate and effectively fuse" (83). This is a promising principle for research and composition in the environmental humanities as it moves towards writing its many selves into social text more widely. We can see this principle in action in, for instance, the rich collaboration between philosopher Bruno Latour, director Frédérique Aït-Touati, and playwright Pierre Daubigny, resulting in the Alberta production of Gaïa Global Circus memorialized in this book by Robert Boschman. Its range of disciplines makes EH ideally suited to doing so, especially if its current non-dualistic thinking and recombinatory impulses remain strong. This is fortunate because, as contract law puts it, time is of the essence. Environmental tipping points suggest themselves, then loom, as we try to imagine how to reconnect to natural environments we somehow convinced ourselves were external: "We have not just one of time's arrows but two: the first, modernist, which goes toward detachment; the second, nonmodern, which goes toward reattachment" (Latour 2009, 194).

In *Philosophy and Temporality from Kant to Critical Theory,* Espen Hammer explains that modern temporality is oriented towards the future, towards imagined points of achieved satisfaction. The present is devalued, reduced to only a stage for means operating towards an end. Intervening time (a progression of unsatiated presents) is devalued as well. If end points are all that matter, then rationality insists that "speed is here the crucial factor. If the means are employed effectively, they will make possible a relatively swift execution of the action and shorten as much as possible the time until the end is achieved" (51). The relentless push forward, Hammer maintains, enshrines efficiency a high value and insists that time be accurately measured. And indeed, we see a horological march from Swiss mechanic to Japanese quartz to American atomic clocks. This mechanistic and forward-moving view of time reaches its apotheosis and juggernaut expression in capitalism neoliberally expressed. The problem being that, as Paul Virilio says in *The Original Accident,* our accelerated cultures offer insufficient protection "from *excess in virtual speed,* from what unexpectedly happens to 'substance,' meaning to what lies *beneath* the engineer's awareness as producer" (13). Accelerated behaviours become cultural imperatives that become visible in human health problems, at which point commercial entities stand ready to ensure we are built for speed—Big Pharma offers drugs with fast-acting agents while airlines move to offering high-speed wifi on flights to passengers frustrated that commercial air travel has not become faster since the 1960s (the reason is fuel cost), although corporate jets somehow manage to become increasingly supersonic. Rumours of commuters one day being able to fly continue to circulate (Halvorson).

In an article on biotemporality, nootemporality,[5] and comparative views of time in behavior, Richelle and colleagues argue "it is obvious that time in living systems on the one hand and time in human and infrahuman behaviour and mind on the other appear as two distinct fields" and that we should be especially concerned with "the emergence of the concept of time in the child" (75). Contemporary expectations among youths as regards instantaneity and audiovisual stimulation is without historical precedent, and this has been accompanied by a contiguous rise among North American young people in often unattributed levels of stress and anxiety (Clinton). In a culture where meaning is divorced from power and power decentralizes and distributes itself to escape responsibility (Laïdi), electronically naturalized market messaging compels us towards time horizons where obtainment of material satisfactions awaits before they and we crash en route, in "accidents" that are actually standardized outcomes in globalized systems programs that carry risks for some but not others. As Donald Trump said in a 2006 audiobook from Trump University about a predicted US housing market crash (which did actually follow in 2008), "I sort of hope that happens because then people like me would go in and buy" (Diamond 2016). "That's called business, by the way," he later said of such activity during presidential debates (Nussbaum 2016) in a supreme instance of default bias. For worker-consumers, wages could sometimes purchase (on what was believed to be a road to end-point prosperity) enough free time that one could afford to "kill" it once in a while. But the temporally geared and techno-semiotic reach of both product placement and work demands into even the most private minutes of the lived Western day (into bathroom stalls and bedrooms, and soon, already, flesh itself, as biosphere becomes technosphere) has attenuated what free time remained.

Oddly, techno-capitalism and EH research both require novelty to carry on, the former to create the next "best thing" and the latter to express new agentive pathways in convincing ways. Perhaps there is leverage, then, in that EH can encourage radically new inscriptions in other conceptions and habits given that scholars have such a range of cultural-historical resources from which to assemble novelty. As the Swiss prog rock band Clepsydra sings, "Time hitch-hikers join their memories, sound from all other times."[6]

ACTIVE GROUNDS AND THIS BOOK

That the environmental humanities is an emergent interdiscipline focused both implicitly and explicitly on time and agency can be seen demonstrated in our volume, wherein scholar-teachers from various disciplines disclose in their operations a protean, post-secondary response to environmental issues and crises. Agency and time, as central facets of this collection, are clearly latticed throughout. They help define a pivot in the history of the academy as it actively engages critical environmental change. In "Eco-Temporal Literacies," the first section of the volume, the first sentence of Paul Huebener's chapter, "'The clock's wound up': Critical Reading Practices in the Time of Social Acceleration and Ecological Collapse," sets the tone for the collection: "The environmental crisis is, in many ways, a crisis of time." A literary scholar and teacher at Athabasca University in Alberta, Huebener brings into his own interdisciplinary practice the work of sociologist Hartmut Rosa, who "argues that social acceleration is incontrovertibly real and carries profound consequences." Huebener writes:

> Most worrying of all, Rosa argues that the cycle of acceleration and its "unbridled onward rush into the abyss" is essentially unstoppable and will most likely culminate "in either the collapse of the ecosystem or in the ultimate breakdown of the modern social order and its values under the pressure of growing acceleration pathologies and the power of the enemies these foster." (36)

Much like Randy and Kent Schroeder writing in Chapter 10 of this volume, Huebener zeroes in on non-linear events that can be read in everything from advertising to poetry, creating openings for the resolutely pluralist nature of the environmental humanities. Unlike Rosa, however, he also uncovers important linkages between time and agency and, through these linkages, some reason for hope. If time is experienced as "the constructed result of many different social and cultural decisions that exist in negotiation with the complex and nonlinear temporalities of the ecosphere itself" (39), then the environmental crises are not (yet) unstoppable. Their trajectories may still be altered. A significant component of agency for Huebener is what he calls "critical temporal literacy, skills that are vital for productive analyses of acceleration and other temporal dimensions of ecological crisis."

Like Huebener, ecocinema critic Mario Trono argues for enhanced critical literacy, albeit in terms of film viewing, one that "calls interpreters into vocational being." In Trono's "A Better Distribution Deal: Ecocinematic Viewing and Montagist Reply," all films, not only those with an overtly environmental purview, constitute rich fields for ecocritical inquiry and active response. It is essential that this be the case—that "'[e]cocinematic' does not refer here to the features of ecocinema, a genre of film, but to an embodied and contextualizing way of seeing moving images, a process-based operation that can take place at almost any point of cultural reception" (60)—because in the Anthropocene era viewer engagement with film can and should be active. Trono's proposed two processes, ecocinematic viewing and montagist reply, seek to rehabituate viewers as both interpreters and producers of image texts, whose experience, knowledge, intuition, and memory are agentive and can be rapidly redirected through video responses disseminated online. Trono's call to *view* and *reply* is a refusal to cede filmic experience to the consumerist aims of studios. Using *Avatar* as an example, he points out how "a popular film's primary power is evident in the consumptions it calls forth and relies on to succeed commercially. And consumption itself ... is what strains environments" (63). In video response lies the heightened power of agency, itself a response to commercialized renderings of time as a major constituent of cinematic experience; with its ellipses, compressions, and truncations, commercial cinema offers a fetishized spectacle of immediacy, "*not* the immersion in task and time that the non-filmic world so often demands" (68). He thus calls for a detailed reevaluation of film viewing that requires not only creative action (both online and off) but also a fresh, fourfold temporal awareness that includes running time, body time, story time, and deep time. He argues that filmic experience can include *choosing* what we "install in our experience," hence acting on Bruno Latour's call in *Politics of Nature* "to force ourselves to slow down" (2004, 3). Over a decade ago now, Latour wrote, "It is all this temporal machinery, this time factory, this clock, this time-clock, that political ecology has to attack in full awareness of what it is doing. It has to modify the mechanism that generates the difference between the past and the future; it has to suspend the tick-tock that gave the temporality of the moderns its rhythm" (189).

In Chapter 3, "*Allô, ici la terre*: Agency in Ecological Music Composition, Performance, and Listening," eco-musicologist Sabine Feisst creates a fascinating complement to Trono's core argument regarding receptive agency as she documents the work of four performance artists focused on active environmental listening. Luc Ferrari, Leah Barclay, David Rothenberg, and Hildegard Westerkamp Ferrari (1929–2005) were interested in recording and reconceptualizing what Ferrari called "the voices of others," using what Feisst describes as "edited acoustic snapshots of sounds from natural and built environments." A participant in the Stockholm Conference on the Human Environment (1972), Ferrari worked with a wide array of notable figures in the sciences, philosophy, conservation, music, and journalism. Leah Barclay (b. 1985) "builds on Ferrari's work" and sees herself as an "agent of change." Her award-winning *Sound Mirrors* (2010) constitutes a recording of global river biospheres in an effort to reveal the agency of non-human voices; Barclay works closely with scientists and Indigenous peoples. Feisst also documents the work of musician David Rothenberg (b. 1962), whose interspecies communication and compositions include non-humans such as the Humpback whale and raise serious and controversial questions regarding *zoomusicology*: for one, "whether animals have music and use it just like humans in functional and aesthetic ways" (97). Feisst also notes Rothenberg's collaborative engagement with "whale time," while also critiquing his assumptions and practices. The work of Hildegard Westerkamp (b. 1946) rounds out Feisst's chapter and emphasizes the importance of critical agency. Westerkamp "pioneered soundwalking, which she defined as 'any excursion whose main purpose is listening to the environment ... Wherever we go we will give our ears priority. They have been neglected by us for a long time'" (100). Soundwalkers listen actively to the non-human world rather than filtering it from their experience. Active listening, they contend, positions them to "face the monster," one of the focal points of Robert Boschman's succeeding chapter, "The Environmental Vampire: Terror, Time, and Territory after 9/11."

Boschman rereads the figure of the vampire after 9/11 as one of categorical change, centrally involving time, terror, territory, and agency. Citing Nina Auerbach's argument that every age has its own vampire and that the vampiric characters of the 1990s were exhausted, Boschman demonstrates that after 9/11 a renewed and altered figure—what he calls the Environmental Vampire—appears in graphic-novel form and from there in a rapidly growing array of novels and films. "Conventionally," writes Boschman, "the vampire has evoked horror concerning the hominid Other as characterized by its ability to exist in a nocturnal state and to change form continually, but the Environmental Vampire is an even far more disruptive figure that represents pressures on the boundaries of the human in new ways" (108). Citing the 1996 essay by Donna Haraway, "Universal Donors in a Vampire Culture," Boschman argues that this transformed character extends her "pre-9/11 categories of race, population, and genome" to include territory, which he theorizes in terms of the work of political philosopher Stuart Elden while also analyzing the *30 Days of Night* textual universe that began in 2002 with the graphic novel by the same title. Distinguishing between *terror* and *horror* in terms of how each operates temporally and relating both terms to *catastrophe* (an inherently temporal concept), Boschman shows how "if the 9/11 date is what *30 Days of Night* begins with, ... Global Climate Change is what it points to by 2011" (126). The protagonist of the inaugural *30 Days of Night* graphic novel, Eben Olemaun, demonstrates agentive change by way of both his office as police chief of Barrow, Alaska, and his Inuit heritage. The tension between Olemaun's neo-colonial position and his indigeneity, as he fights to defend his oil-pipeline community from a sudden invasion of vampires on the very eve of a month of circumpolar darkness, enables him to become the Other in order to defeat the Other by injecting himself with vampire blood. Olemaun's decision "is both activist and biopolitical in that it involves his own transformation from human to non-human, from diurnal to nocturnal, with the object of redefining the threat that causes terror" (124) and jeopardizes human existence. A hybrid everyman, Eben Olemaun resets the "hypermodernism" that causes people to "have so much difficulty taking seriously that they are of this earth" (Latour et al. 2015, 11).[7] With his dynamic character as Inuit lawman, "Eben Olemaun eventually reshapes the failed Westernized quest for territorial sovereignty and security into a new ecological paradigm" (124).

This volume's second section, "Timelines and Indigeneity," assumes that reading the Indigenous as a component of time and agency in the environmental humanities is both crucial and necessary. Indigenous Studies and EH are partners in the reconstitution of environmental concerns within the academy and, indeed, across the global North and South. Time and indigeneity are taken up by two chapters that speak to critical temporal and agentive issues faced by Indigenous cultures in North America specifically. And these issues are almost always tied to land, water, and environmental stewardship. Geneviève Susemihl's chapter on Blackfoot Crossing and Head-Smashed-In Buffalo Jump, the latter a UNESCO World Heritage Site located in southern Alberta, Canada, speaks to the interdisciplinary and community engagement challenges around governance of an ancient human site sacred to the Blackfoot nations of North America. Susemihl takes readers into the on-the-ground complexities of understanding and acknowledges a place that speaks powerfully to tourists from around the world at the same time as it represents local Indigenous communities and their sense of the past, even while they have little or no ownership of the place itself. Communal agency comes into play here as a component of collection and exhibition, as academics argue for representation that takes in the nuances and hidden niches of the deep human past. Heritage becomes a tangled concept, used differently by governments, academies, and communities, not to mention the global tourism industry. The attachment of the UNESCO World Heritage Site moniker, prestigious though it is, takes the tangle of understanding to another level entirely. "Our collective memory," writes Susemihl, "is formed ... by historic environments, which contain an infinity of ancient and recent stories, written in stone, brick, or wood or inscribed on the landscape, and these become the focus of community identity and pride" (142). That the Blackfoot have limited governance over Head-Smashed-In points to an inevitable conflict, especially between the governments of Canada and the Blackfoot Nation, whose relationship to the site is, in Susemihl's words, "a long and intimate" one (146).

Like the Blackfoot of Alberta, Canada, the Navajo of the Four Corners region of the United States have been subject to forces that would separate them from their own lands, ancestral lands on which they continue to live despite being contaminated by the mining of uranium and coal. Lea Rekow's personal odyssey of documenting the histories of uranium and coal mining and dumping in this area, and the impacts of these histories on the Navajo peoples who have lived here for centuries, itself exemplifies a kind of activist, agentive engagement. Rekow's work in this region, both on the ground and from the air, is a testament to courage given her willingness to take personal health risks. She describes her experience in Shiprock, New Mexico, of "being caught between the disposal cell and the evaporation pond in a particularly intense dust storm. It wasn't the first I had been caught in" (184). Since the lethal effects of uranium contamination in the human body can take decades to appear, Rekow "continue[s] to experience anxiety about the potential health impacts" (184). Her photos of the Navajo landscape and peoples themselves speak eloquently throughout her chapter, and her title, "Mapping the Mining Legacy of Navajo Nation," attests to time and agency. Her work not only records the tragic histories of uranium and coal exploitation in the Four Corners area, where New Mexico meets Colorado, Utah, and Arizona, and where the San Juan River flows, but also "depicts the policy advocacy and environmental reclamation work tirelessly being undertaken by Indigenous and regional stakeholders themselves to confront these injustices, mitigate the damages, and demand accountability" (164). This is activist work at its most effective, and Rekow pulls no punches in her descriptions of what has transpired in this area, citing the extremely deleterious effects on the Navajo people and lands. She recounts witnessing how "a family's home had to be abandoned because of the high levels of radiation being emitted from the rock walls" (178) and how the replacement home also had to be vacated when it too was found to be radioactive. The ironies Rekow sees are numerous: many Navajo are still without electricity, despite the coal plants next to their places of residence; and "the motto for the Central Arizona Project is 'your water, your future'" (172) even though watersheds are badly polluted from decades of dumping.

Rekow's narrated, agentive immersion in a region, and in the discourses that determine the area, inspired Robert Boschman to provide this collection with its thematic centrepiece. It is a hybrid text that comprises memoir, photographic essay, and relevant theorization to describe the visit to Alberta in the fall of 2016 by the French philosopher Bruno Latour and Gaïa Global Circus, a troupe of environmental actors led by director Frédérique Aït-Touati. In terms of its critical energies and ideas, "Agency and Time on Active Grounds: A Memoir of Bruno Latour and Gaïa Global Circus" conceptually dramatizes the volume's preceding chapters and anticipates the six that follow it. Boschman recounts his trip with Latour to the Alberta Badlands, an area near Drumheller where extensive coal mining and oil extraction change the landscape at the same time as palaeontologists patiently excavate dinosaur fossils below the Cretaceous–Paleogene Boundary (referred to in popular nomenclature as the K-T Boundary). Boschman took Latour and colleague Olivier Vallet to see this boundary for themselves, how it conspicuously marks the region's horizontal strata with "shocked minerals" evincing the sudden end of the Cretaceous in geological time.

If polluted environments such as those that Lea Rekow describes are to become less frequent, it may take a significant reframing of what it means to be human and what the power of human agency could entail. In the chapter that begins the section titled "Animal Agents and Human–Non-Human Interactions," Karla Armbruster accounts for the epistemic stance required to pollute the biosphere in the first place. In "The Gaze of Predators and the Redefinition of the Human," citing the works of Sy Montgomery, John Berger, Val Plumwood, and wildlife biologist Joel Berger, among others, Armbruster argues that the Western human attitude towards air, water, and earth stems in large part from the default human position that being edible is unacceptable. To be edible is to be part of an ecosystem—and neoliberal humans certainly cannot stomach such a prospect, considering themselves exempt from the natural world. As Armbruster puts it, "Any animal who rips through this illusion [of being edible] by attacking a human for food is almost always quickly eliminated, having committed the ultimate crime against our self-proclaimed position atop the web of predator–prey interactions" (209–10). Like Robert Boschman does in Chapter 4, Armbruster here employs the term *horror* to describe the Western human response to being eaten by other animals—and how this particular horror re-emerges frequently in film, written texts, and dreams, suggesting long and evolved histories and habits. Armbruster's objective is both fascinating and crucial, as she "explores the possibility that the intensity of our feelings about becoming prey can propel us beyond horror ... to a point where we can instead use the idea to help us recalibrate our understanding of the human role in the biosphere" (211). The needed sense of being a part of surroundings that matter both to ourselves and to others would then begin with the body, itself an environment. Going to noted examples of humans encountering their own edibility, such as Val Plumwood, Timothy Treadwell, and the people of the Sundarbans, Armbruster demonstrates in detailed portraits how such individuals and communities have recalibrated their cosmologies.

Regarding time, Karla Armbruster asks us to recall our evolutionary past to better our species' relations in the future. Her statement, "[T]he exchange of looks between humans and animals was at the heart of the significance they once held for us," also links her chapter to the succeeding essay by Pamela Banting, "Anim-oils: Wild Animals in Petro-Cultural Landscapes." The relationship in time between humans and animals is central to Banting's chapter, and like Armbruster she sees the agentive nature of this relationship as, first, a bodily one, asking the question, "Might narratives of encounter between us and other animals illuminate our own corporeality?" (238). Along with Armbruster, Banting points out how current Western culture constructs human identity as exempt from "ecological relationship." Through her analyses of texts by Eden Robinson, Farley Mowat, and Sid Marty, Banting argues that "[c]ulture is essentially ecological relationship. Culture is the result of trans-species flows" (242). Moreover, in the terms of edibility tackled by Armbruster, Banting delineates how "culture begins with metabolism: food, eating, energy" (242). The fat of the animal body—and here Banting focuses primarily on the bear—is interconnected with the "fat" of petroleum, itself the distillation of organic bodies from deep planetary history. Bear grease and engine oil are more alike than we might care to think. "The bear's body is fuelled in both present and future tenses by a thick layer of insulating and nourishing fat. While obviously body fat is in no way the same substance as crude oil or bitumen, as forms of energy and fuel for propulsion through space and time their functions and histories overlap" (247). Banting's acute descriptions—of how human automobility (a powerful form of agency) has estranged us from the bear, and of how driving itself constitutes a statement of power and identity, literally gorging on time and space—are startling. "We drive; therefore we are human. Therefore, we are not for the eating. When we drive, we deny or disavow our basic mammalian nature, the speed and energy limitations of our legs and feet, and our place in historical, sociopolitical, and eco-systems" (252). Yet when we are on foot and meet the bear, we encounter "our double"(253), one who, as in the controversial instance of Timothy Treadwell, may also eat and reconstitute our bodies as bear.

If Pamela Banting "examin[es] oil's role in encounters between humans and other species" (not only the bear but also the whale and other animals valued for their "oil"), Morgan Zedalis and Sean Gould close this book's third section with the results of their primary research among Idaho ranchers concerning their relationship to reintroduced gray wolves. "Reacting to Wolves: The Historical Construction of Identity and Value" demonstrates a fresh and interdisciplinary approach to a particularly troubled historical relationship between humans and wolves. Their work as philosophers engaged in primary research in the field is based on painstaking, documented interviews with Idaho ranchers. All interviewees opposed the wolf reintroduction that began with a handful of animals in 1995. Zedalis and Gould, like Sabine Feisst, emphasize the importance of listening in a period fraught with environmental conflicts. As EH scholars, they demonstrate how to meet with others on their own ground in order to identify paths forward, in this instance regarding the cultural complexities of the rancher–wolf relationship. "Reacting to Wolves" takes place, in a real sense, *on active grounds*. The interview results are given in an interdisciplinary context that "draws from environmental anthropology, environmental philosophy, and moral psychology" (264). Individual and community agency play core roles in this chapter; Zedalis and Gould argue that "the reactions to wolves can be understood as serving an instrumental purpose of establishing each individual's tie to the broader ranching community and the identity to which they aspire" (265). Social identity plays a primary role and must be taken into account whenever economic compensation is deployed: "economic approaches assume that value exists as a fixed, antecedent variable tied to various demographics which can be accommodated through trading one value with others" (267). Zedalis and Gould indicate that their methodology could be valuable for assessing species reintroductions worldwide, especially where monetary compensation is used. If human agencies and practices are going to become more resilient, then the approach of Zedalis and Gould may serve as a model for understanding the social forces that impact identity and "historical perceptions of the land and wildlife" (266–267) in a specific region. In this case, ranchers' resentment towards Defenders of Wildlife and that group's role in both gray wolf reintroduction and rancher compensation for domesticated animal losses needs to be understood historically as well.

We curated this volume's final three chapters under the rubric "Systems Change in Time" in order to emphasize their respective, practical policy formulations for current and future action. We take as axiomatic the claim that EH research of all kinds should resonate in close relation to policy initiatives. Under this heading, agency and time can be read as being of acute importance within the environmental humanities. We have calibrated these final chapters as examples of the kinds of concrete steps EH can posit for a sustainable world. Randy Schroeder and Kent Schroeder's Chapter 10, "Declarations of Interdependence: Unexpected Human–Animal Conflict and Bhutanese Non-Linear Policy" functions in tandem with Mishka Lysack's "Effective Environmental Action in Canada: the German *Energiewende* as a Model of Public Agency" to set forth specific environmental policy analyses of specific locations and histories: Bhutan and Germany, respectively.

Lysack's analysis of policy initiatives in Germany (as well as throughout parts of the European Union) points obviously to an opportunity for Canada's federal government to enact similar objectives. Germany's detailed *Energiewende* and *Klimaschutz* program, he argues, could be adapted for "the mobilization of Canadians as stakeholders across sectors" (310). And Lysack's work is important temporally, in a multi-generational sense, as he argues for the empowerment of citizens as public agents for change at a time when, increasingly, people of all ages are mobilizing. Citing numerous on-the-ground examples of working emissions reductions and renewable energy projects, Lysack argues that "[c]ommunicating research will also continue to be crucial in order to enable education and community engagement, both of which can transform energy infrastructures" (318).

Randy Schroeder and Kent Schroeder similarly bring another country's successful environmental policy history forward to Canadian readers. They construe agency and time as "the iridescence of interdependence" (288). In describing Bhutan's Gross National Happiness

(GNH) model, built on the concept of ecological interdependence, they dispel romantic and/or facile assumptions regarding both happiness and interdependence. Their chapter is a nuanced reading of both concepts. Distinct from "the pursuit of happiness" articulated in the American Declaration of Independence, Bhutan's early 1970s Gross National Happiness policy "is ultimately about the achievement of the 'full and innate potential' of being human" (292) and is a development strategy that recognizes the relationships between humans and non-humans as well as between individuals and communities. Interdependence is viewed and acted on as a complex set of conditions and relationships subject to antagonistic pleiotropy, an evolutionary concept that acts as "an active figure for interdependence [that] is far more interdependent than the most enthusiastic Western cheerleader for interdependence could ever admit" (290). Schroeder and Schroeder describe how Bhutan has demonstrated resilience at the policy level over the last forty years by viewing the culture–nature nexus as inherently mutable. They also focus on a specific problem, human wildlife conflict (HWC), and how Bhutan has adopted "a policy pivot" in response to this emergent situation stemming from its initiation of GNH. Like Zedalis and Gould in "Reacting to Wolves," Schroeder and Schroeder give an account of HWC in terms of one government's policy approach that "links conservation and rural livelihoods as two interdependent components" (295).

According to Nancy C. Doubleday, whose chapter, "Culture as Vector: (Re)Locating Agency in Social-Ecological Systems Change," closes out this collection,

> [i]f we wish for the capacity to develop alternative scenarios for the future that support equitable social-ecological resilience, and seek to avoid "unthinkable catastrophes" ... then arguably, we are at a point where we need to advance our conceptual, technological, and communicative approaches and deepen collaborative abilities across scales. (328)

Doubleday's chapter takes into account many of the previous chapters even as it pointedly deals with "wicked problems" (329) that plague both the environment and agentive, policy vectors that people set forth. As the Hope Chair in Peace and Health, Department of Philosophy, at

McMaster University, she argues for the possibilities that arise when we "[understand] culture as *action* as well as *object*, thus as a *culture vector* possessing a capacity for agency within social-ecological systems (SESs)" (331). Taking culture immediately out of its usual, exhausted binary relationship with nature invites readers to see the term again in its sense as a verb involving *conservation, transmission, definition, observation,* and *connection.* Interdisciplinarity and communication step forward as agentive components in acting on new opportunities and new syntheses and what she calls "the ever-present challenge of re-conceptualizing the human project" (331). Doubleday's chapter cogently addresses the issues of time and agency that are core to this book, affirms this volume's interdisciplinary character, and functions both here and at the end of this volume as a conclusion: "There is a growing social consensus in support of recognizing ... interdisciplinarity in order to flourish for two reasons: first to provide context for wicked problems across scales, both temporal and spatial; and second as a source of community formation and validation necessary to support the development of agency in individuals and groups" (329).

Notes

1 See Fenner (2010) on extinction.
2 See Lovelock (2007).
3 We are indebted to Massumi (2011) for drawing out attention to James's remark.
4 See Bissonette and Storch (2007).
5 "Noosphere" was a term postulated by the French idealist philosopher Teilhard de Chardin and refers to an evolutionary stage of humans entirely made up of the conscious mind enmeshed in interpersonal relations.
6 From "Fading Clouds of Time" on the band's 1991 album *Hologram.*
7 Writing in the immediate aftermath of the November 2015 terrorist attack in Paris, which preceded COP21 Paris by two weeks, Latour—himself a Parisian—describes "our zeitgeist" as "on the one hand, religious wars for the colonization of a non-existent afterworld; on the other, economic wars for the colonization of territories that are equally insubstantial. Two situations deeply linked by some strange politico-religious urge to 'replace' the world as it is with another world that is transcendent as an ideal yet totally deadly when violently inserted into the world of below. Two forms of criminal intoxication by the beyond that make it impossible for ordinary humans to get back to earth, to become earthly; as if it were impossible to render religion, politics, science, and economics secular again, material, mundane" (2016, 11).

References

Alaimo, Stacy. 2010. *Bodily Natures: Science, Environment, and the Material Self.* Bloomington: Indiana University Press.

Barad, Karen. 2007. *Meeting the Universe Halfway: Quantum Physics and the Entanglement of Matter and Meaning.* Durham: Duke University Press.

Bem, Sacha, and Fed Keijzer. 1996. "Recent Changes in the Concept of Cognition." *Theory and Psychology* 6(3): 449–69.

Bennett, Jane. 2010. *Vibrant Matter: A Political Ecology of Things.* Durham: Duke University Press.

Bissonette, John A., and Ilse Storch, eds. 2007. *Temporal Dimensions of Landscape Ecology: Wildlife Responses to Variable Resources.* New York: Springer.

Brooks, Rodney A. "Intelligence without Representation." *Artificial Intelligence* 47(1–3): 139–59.

Canavan, Gerry. 2015. "Capital as Artificial Intelligence." *Journal of American Studies* 49(4): 685–709.

Clinton, Jean. 2013. Interview: "Young Minds: Stress, Anxiety Plaguing Canadian Youth." *Global News.* http://globalnews.ca/news/530141/young-minds-stress-anxiety-plaguing-canadian-youth. Accessed 13 October 2016.

Connolly, William E. 2012. "Steps toward an Ecology of Late Capitalism." *Theory and Event* 15(1).

Diamond, Jeremy. 2016. "Donald Trump in 2006: I 'sort of hope' real estate market tanks." *CNN Politics,* 19 May. http://www.cnn.com/2016/05/19/politics/donald-trump-2006-hopes-real-estate-market-crashes.

Fenner, Frank. 2010. "Frank Fenner Sees No Hope for Humans." *The Australian,* 16 June.

Halvorson, Bengt. "Commuter Drones: Uber Hopes to Transcend Gridlock with, Yes, Flying Cars." *Car and Driver.* http://blog.caranddriver.com/commuter-drones-uber-hopes-to-transcend-gridlock-with-yes-flying-cars. Accessed 27 August 2016.

Hammer, Espen. 2011. *Philosophy and Temporality from Kant to Critical Theory.* Cambridge: Cambridge University Press.

Holm, Poul, and Ruth Brennan. "Humanities for the Environment 2018 Report—Ways to Here, Ways Forward." *Humanities* 7(1): 3.

James, William. 1996. *Essays in Radical Empiricism.* Lincoln: University of Nebraska Press.

James, William. 1890. *The Principles of Psychology.* New York: Henry Holt.

Karn, Barbara Parish. 2007. "Green Nanotechnology: A Path to Sustainability?" *Bridges* 4. Office of Science and Technology Austria—Washington. http://ostaustria.org/bridges-magazine/volume-14-july-12-2007/item/2294-green-nanotechnology-a-path-to-sustainability. Accessed 5 September 2016.

Laïdi, Zaki. 1998. *A World without Meaning: The Crisis of Meaning in International Politics*. Translated by June Burnham and Jenny Coulon. London: Routledge.

Latour, Bruno. 2004. *Politics of Nature: How to Bring the Sciences into Democracy*. Cambridge, MA: Harvard University Press.

———. 2014. "Agency at the Time of the Anthropocene." *New Literary History* 45(1). Baltimore: Johns Hopkins University Press.

———. 2016. "Introduction: Let's Touch Base!" *Reset Modernity!* Edited by Bruno Latour, with C. Leclercq. Cambridge, MA: MIT Press.

Latour, Bruno, Naisargi N. Dave, Mary L. Gray, Cymene Howe, Tom Boellstorff, Rudolf Gaudio, Martin Manalansan, and David Valentine. 2015. "A Question from Bruno Latour." *Cultural Anthropology*, 21 July 2015. https://culanth.org/fieldsights/703-a-question-from-bruno-latour.

Lovelock, James. 2007. *The Revenge of Gaia: Earth's Climate Crisis and the Fate of Humanity*. New York: Penguin.

Massumi, Brian. 2011. *Semblance and Event: Activist Philosophy and the Occurrent Arts*. Cambridge, MA: MIT Press.

Medicine, Bea. 1999. "Lakota Views of 'Art' and Artistic Expression." Lecture and Exhibition Essay. Vancouver: Grunt Gallery.

Morikawa, Masayuki. 2016. "Artificial Intelligence and Employment." Center for Economic Policy Research. http://voxeu.org/article/artificial-intelligence-and-employment. Accessed 21 August 2016.

Morton, Timothy. 2009. *Ecology without Nature: Rethinking Environmental Aesthetics*. Cambridge, MA: Harvard University Press.

———. 2012. *The Ecological Thought*. Cambridge, MA: Harvard University Press.

———. 2016. *Dark Ecology: For a Logic of Future Coexistence*. New York: Columbia University Press.

Nussbaum, Matthew. 2016. "Trump cheering housing collapse: 'That's called business.'" *Politico*, 26 September. http://www.politico.com/story/2016/09/trump-housing collapse-228708.

Richelle, Marc, Helga Lejeune, Jean-Jacques Perikel, and Patrik Fery. 1985. "From Biotemporality to Nootemporality: Toward an Integrative and Comparative View of Time in Behavior." In *Time, Mind, and Behavior*. New York: Springer Verlag

Ricoeur, Paul. 1984. *Time and Narrative*. Chicago: University of Chicago Press.

Virilio, Paul. 2006. *The Original Accident*. Cambridge, MA: Polity Press.

Vladeck, D.C. 2014. "Machines without Principles: Liability Rules and Artificial Intelligence." *Washington Law Review* 89(1): 117–50.

Whitehead. Alfred North. [1929]1978. *Process and Reality: An Essay in Cosmology*. New York: Free Press.

Winner, Langdon. 1993. "Upon Opening the Black Box and Finding It Empty: Social Constructivism and the Philosophy of Technology." *Science, Technology, and Human Values* 18(3): 362–78.

Zalasiewicz, Jan. 2012. *The Planet in a Pebble: A Journey into Earth's Deep History*. New York: Oxford University Press.
Žižek, Slavoj. *"Don't Act. Just Think."* YouTube. Accessed 15 September 2016.

I.
ECO-TEMPORAL
LITERACIES

Reading a landscape: An organized social scene for touring humans viewing geologic epochs. Off Highway 570, south of Drumheller, Alberta. The K–T Layer, evidence of the dinosaur extinction event, can be seen here here. (Photo courtesy Robert Boschman)

CHAPTER 1

"The clock's wound up": Critical Reading Practices in the Time of Social Acceleration and Ecological Collapse

Paul Huebener

The environmental crisis is, in many ways, a crisis of time: from the apparent inability of societies to act with a clear sense of future responsibility, to the tensions that exist between accelerating resource use and the need for persistent ecosystems, to the anxious realization that time, for many humans and other members of the ecosphere, will soon run out, if it hasn't already. The concept of the Anthropocene, the epoch in which humanity acts as the driving force of planetary change, has gained rapid currency over the past decade, spurring scholars from many disciplines to try to bridge the gaps between scholarship and environmental activism and politics. On this front, one of the tasks to which the humanities appear especially well suited is the assessment and reconfiguring of social assumptions about time. As Mario Trono and Robert Boschman write in *Found in Alberta: Environmental Themes for the Anthropocene*, "the environmental humanities can help call for an operationally safe conception of time" (Trono and Boschman 2014, 3). Commenting specifically on the temporality of representations of climate change, Anita Girvan writes that "the work of de-naturalizing given perspectives on time emerges as a significant intervention in the cultural politics of climate change" (Girvan 2014, 348). In his book *Slow Violence and the Environmentalism of the Poor*, meanwhile, Rob Nixon calls for us to understand long, slow-moving environmental disasters and displacements as literal forms of violence and to consider how we might more effectively "convert into image and narrative the disasters that are slow moving and long in the making" (Nixon 2011, 3).

Because "time" and "the environment" both represent complex clusters of experiences and concepts, critical examinations of the cultural politics of time and the environment must consider a wide range of concerns, from the relationship between the pace of climate change and the pace of policy change, to the massively long-term impacts that will result from the actions that we take across the span of just a few decades, to the painful experiences of duration for victims of environmental racism, to the assumptions of perpetual economic growth and "progress" through time.

Much of the above work emphasizes long-term or even epochal time spans, and while problems that unfold over such large scales of time indeed deserve our close and sustained attention, my focus here is somewhat different. In this chapter I examine some of the ways in which time itself, especially in the sense of acceleration, operates socially as a form of power in relation to the environment. I suggest that an everyday form of critical temporal literacy is a vital type of agency if we are to understand and respond adequately to the particular conceptualizations of time that exacerbate unsustainable social behaviours. A politically aware critical approach to the ways in which cultural assumptions about time are connected to the problems of sustainability—a practice we might call *ecocritical time studies*—can help us to read culture with a thoughtful and transformative awareness of the implications of temporal power and to understand the need for temporal justice alongside social and environmental justice. In demonstrating this approach, I first offer a response to the phenomenon of social acceleration. I then examine two case studies—one drawn from an everyday cultural encounter and one from a literary text—to illustrate how critical temporal literacy allows us to read, and reshape, the cultural manipulation of time as it operates in social acceleration and human–ecological relations.

RESPONDING TO SOCIAL ACCELERATION

There is a widespread sense in modern societies that life is speeding up. But even while this sense of acceleration may profoundly affect everyday experience as well as broad social and environmental phenomena, it tends to remain a largely unsubstantiated suspicion. In his 2013 book *Social Acceleration: A New Theory of Modernity*, sociologist

Hartmut Rosa argues that social acceleration is incontrovertibly real and carries profound consequences. He identifies three distinct types of acceleration within modern societies, each of which feeds into and exacerbates the others. The first type is technical or technological acceleration (the intentional acceleration of particular processes such as transportation and computer processing); the second is the acceleration of social change (that is, "the acceleration of social changes that are not inherently goal directed," such as "the acceleration of changeover in jobs, political party preferences, intimate partners, ... occupational and family structures, artistic styles, etc."); and the third type is the acceleration of the pace of life ("an increase of episodes of action and/ or experience per unit of time that is linked with a scarcity of temporal resources") (Rosa 2013, 64). Through a systematic analysis of these structures of acceleration, Rosa finds not only that each type of acceleration demonstrably occurs within modern societies but also that the three processes fuel one another in a perpetually strengthening cycle of speed: technical acceleration causes an acceleration of social change, which in turn contributes towards the acceleration of the pace of life, which completes the cycle by necessitating further technical acceleration (156). While he argues convincingly that social acceleration is a very real phenomenon in the specific forms he describes, Rosa is careful to point out that not *everything* is speeding up. Things that cannot be sped up, or have avoided being sped up, include natural processes such as brain stimulus processing, the actual creation of fossil fuels, "the capacity of the ecosystems of the earth to process toxic substances and waste materials" (81), specific communities such as the Amish (83); dysfunctional decelerations such as traffic jams, unemployment, and economic recessions (84); and, to a certain extent, instances of intentional deceleration employed as ideology, such as the Slow Food movement or the little-known Union for the Slowing Down of Time (85–6). None of these, however, "embody a structural and/or cultural countertrend that could equal the acceleration dynamic of modernity" (90).

For Rosa, the consequences of social acceleration include not only the rise of a prevalent sense of time pressure and a loss of overall social stability and predictability, but also the erosion of the viability of democratic "self-steering" as relatively slow-moving political entities struggle to maintain relevance in a fast-moving and unpredictable world (159); the isolation of young people from old people as "the experiences, practices, and stock of knowledge of the parental generation increasingly become for the youth anachronistic and meaningless" (115); and even the breaking up of "stable, long-term individual identities" (148). Most worrying of all, Rosa argues that the cycle of acceleration and its "unbridled onward rush into an abyss" is essentially unstoppable and will most likely culminate "in either the collapse of the ecosystem or ... the ultimate breakdown of the modern social order and its values under the pressure of growing acceleration pathologies and the power of the enemies these foster" (322).

In articulating these forms of social acceleration systematically, Rosa's study marks a major contribution to knowledge. His dark comment, though, about the likely "collapse of the ecosystem" occurs only on the final page of the volume, and the collapse he envisions is not investigated in any detail. This apocalyptic conclusion, then, calls quite urgently to be filled in with greater complexity and for its lessons and limitations to be more fully illuminated. Perhaps most importantly, it is necessary to realize that despite the implication of Rosa's phrasing, the collapse of the ecosystem cannot accurately be envisioned as a single event. The many ecosystems and bioregions that constitute the ecosphere as a whole are communities of immense complexity, whose degradation occurs not in a single identifiable moment, but rather in the form of complex (and accelerating) clusters of non-linear events. Different ecosystems, portions of ecosystems, and populations collapse or shift in different ways at different moments—and these disruptions in turn cause non-linear disruptions to our own social, economic, imaginative, and subsistence activities. This is one of the key ways in which the environmental crisis is a temporal crisis. One of the strengths of Rosa's study is that by understanding modern societies as globally linked networks all engaged in accelerative processes, he is able to craft a larger comprehensive model of social acceleration as a dominant global force. The weakness that comes with this approach, though, is that the study does not have the chance to develop insights into the particularities of different regions, places, or populations, in either their human or more-than-human dimensions.[1]

When we understand the collapse of the ecosystem under the weight of the accelerative human exhaustion of the ecosphere not as a single event, but as a cluster of many events, important opportunities for agency and intervention arise. Over time, accelerating ecological disruptions gradually encourage, and indeed force, the transformation of major social and cultural practices that otherwise might have been characterized by seemingly unstoppable inertia. For example, the tendency of modern societies to rapidly "develop" land through expansive construction projects, even in low-lying areas near rivers and oceans, is one of many characteristic features of the acceleration society. However, as sea levels rise, and as oceanside and inland flooding events become more frequent and more severe, these clusters of events begin to describe a narrative of risk, disaster, and the need for change and precaution—changes that would not be possible if the accelerative rush "into an abyss" caused a singular moment of complete destruction. In 2011, for instance, the British Columbia Ministry of Environment released a report warning builders and developers to ensure that their plans accounted for a projected sea level rise of about one metre by 2100 and two metres by 2200 (BC Ministry of Environment 2011, 5).[2] While these time frames are not overly long by generational standards, in the context of commercial and residential developments the time frames represent a dramatic lengthening, and a slowing, of the status quo. The need to anticipate the ways in which ecosystems will change over the next century is a decelerative impulse in that it builds long-term thinking explicitly into present-day planning. Paradoxically, the increasing pace and unpredictability of climate change can inspire us to lengthen and slow the modern sensibility of hasty and immediate time.

As we know, there are many examples of the ways in which societies *fail* to react to accelerating signs of ecological disruptions; but as these disruptions become increasingly obvious, frequent, and intense, so too must human responses. Rosa's assertion, then, that the unsustainable practices tied to modern social acceleration cannot be stopped by anything short of wholesale social collapse is troubled by the fragmentary and protracted form that ecological and social collapses take. Much like Rosa, Fredric Jameson has observed that "it is easier to imagine the end of the world than the end of capitalism" (Jameson 2005, 199), a recognition of the fact that the apparently linear model of progress, development, and economic growth has colonized our imaginations to the point where alternative forms of existence seem impossible. However, when the "end of the world" turns out to happen gradually, through increasingly frequent disruptions and devastating events, the radically disrupted world becomes easier and easier to imagine, simply by observing everyday events, while the transformation of capitalism and the accelerative modern society may, in fits and starts, become a simple necessity, a manifestation of what Nancy Doubleday, in her contribution to this volume, calls "an emergent culture" (343). As Robert Boschman and Mario Trono point out in their introduction to this volume, the notion of acceleration may carry particular significance in modern culture (14), but time is nevertheless a complex collection of many different experiences illuminated through a wide range of disciplines, and it cannot be reduced to a single process (11–12). The end of the world has already occurred for many people and other species through the loss of homes, habitats, livelihoods, and lives, but the problems brought about by social acceleration contain enough temporal complexity and multiplicity that certain reactions and adaptations remain possible.

Another reason to conceptualize the collapse of ecosystems as multiple non-linear events is tied to the need to recognize forms of temporal agency within the natural world itself. In her work on temporality and nature, Michelle Bastian comments on the Western world's tendency to imagine human cultural changes as purposeful acts of linear progress that occur against the backdrop of a natural world conceived of as a relatively static store of resources. Because of this dominant perspective, Bastian writes, "Western assumptions about time have fostered notions of nature as timeless, and as without agency" (Bastian 2009, 102). She argues that the concept of agency is closely linked to conceptualizations of time and that "when one understands time *as agency*, seeing nature as without significant changes, without *time*, is to also see it as without agency" (102–3). Ultimately, Bastian argues that the long, slow changes visible within ecosystems and species can be understood, at least partially, as forms of invention, agency, and other-than-human temporalities. This perspective complicates the assumption of a static backdrop of natural resources and services and thus refuses to reduce ecosystems to the simplistic categories of either functioning or exhausted.

It is critically important, then, to understand social acceleration and other implementations of social time not as inevitable—their own kind of natural forces—but rather as the constructed result of many different social and cultural decisions that exist in negotiation with the complex and non-linear temporalities of the ecosphere itself. While Rosa's study is explicitly pessimistic about the possibilities of temporal resistance, it simultaneously sounds a note of optimism in the thought that there is something to be gained from documenting the apparent accelerative descent into the abyss in a thorough and rigorous way. By engaging thoughtfully with the implications of this type of study, we exercise the skills of critical temporal literacy, skills that are vital for productive analyses of acceleration and other temporal dimensions of ecological crisis. Only by articulating and understanding how time operates socially as an immense but unstable tool of power can we productively investigate the ways in which the environmental crisis is also a crisis in the way we conceptualize time, and adequately assess the possibilities for changing course.

THE PROBLEM OF UNSCHEDULING: READING EVERYDAY REPRESENTATIONS OF TIME AND ENVIRONMENT

There are countless examples of everyday encounters with cultural representations of time in the environment, and each of these encounters opens up possibilities for assessing cultural assumptions about ecological and social time. One such example is an advertisement, titled "Unscheduled," produced for Go RVing Canada in 2013, which has aired on television as well as online through YouTube (Go RVing Canada 2013). Go RVing is a coalition of recreational vehicle (RV) manufacturers and dealers as well as campground organizations. The purpose of the organization is to promote RV travel, and it does so in this ad through a series of time-based associations.

The 55-second ad begins with the members of a suburban white middle-class family scrambling in their laundry room, kitchen, and two-car driveway to catch up to the scheduling demands of their day. The series of shots within this first part of the ad is fast-paced, with frequent cuts between shaky, poorly framed camera angles creating the impression that the footage itself was shot at the same frantic pace the family is experiencing and that handheld cameras were necessary because the events happened in too much of a rush to allow for more carefully orchestrated cinematography. The dialogue between the mother, the father, and their four children is so hectic as to be largely unintelligible. "You're going to pick up Chelsea from ballet at 7:30," the mother says rapidly to her husband (who is checking his phone and consulting a note stuck on the fridge in the kitchen) while she pulls socks onto her daughter's feet in the laundry room. The husband points out that picking up Chelsea will create a scheduling conflict, and the rapid dialogue descends into a series of panicked interruptions, all spoken while the parents try to corral their boisterous children into the vehicles in the driveway. References are made to guitar lessons, martial arts lessons, math tutoring, and a school play, all while the various family members make and partly consume sandwiches ("Just shove it in your mouth, let's go," the father says), put on various clothes, find lost keys, break up a fight between at least two of the children, confirm addresses and appointment times, determine which family members must be present in each of the two vehicles in order to orchestrate the appointment pickups that will occur later in the day, and finally determine who will actually *fit* into each vehicle, causing a panic and a complete switch of both vehicles' occupants when the tail of Jack's dinosaur costume for the school play will not fit into the smaller car. Twenty-eight seconds have now elapsed.

At this moment, the calmly singing voices of a second family become audible, and the first family stops, mid-panic, to watch as a recreational vehicle coasts by at an entirely relaxed pace, while the occupants sing the French Canadian children's song "Alouette." Slow smiles break across the faces of the first family as they gaze upon the RV, and a calming piano composition begins to play, drowning out the sounds of the suburban street. The shot fades to black, then fades back

in to the first family, now seated around a secluded campfire at night, an RV framed prominently behind them. The family members are roasting wieners on sticks, playing guitar, and holding one another around the fire. The perfect clarity of the soothing piano music is accompanied by the barely audible lapping of gentle waves on the shore of a lake. A final shot appears, in which the camera has pulled back to show the campfire and the RV nestled beneath a large tree alongside the serene waters. The darkness of this nighttime shot contrasts the pure white word that fades into the centre of the screen—"Unschedule"—which finally gives way to the web address GoRVing.ca.

Several important factors are at work in this ad. We see, most starkly, that human culture is *fast*—fast to the point of chaos, panic, pointlessness, and perpetual anxiety. Founded on what David Landes calls "time discipline"—the internalized drive to be punctual that is tied to the ubiquitous presence of clocks (Landes 1983, 7)—modern culture in this ad is driven by speed, complexity, deadlines, schedules, and multitasking, and as a civilizational strategy it risks being self-defeating. Nature, by contrast, is slow; it is a place where the demands of scheduling and social acceleration have been lifted, leaving behind a pure experience of recreation, clarity, and communal gathering. The ad is selling RVs and camping by linking them conceptually with what some sociologists of time refer to as "event time," a cultural approach to time in which events transpire at their own usually relaxed pace, as opposed to a faster and more rigid clock-based time. As social psychologist Robert Levine concludes in his cross-cultural analysis of the pace

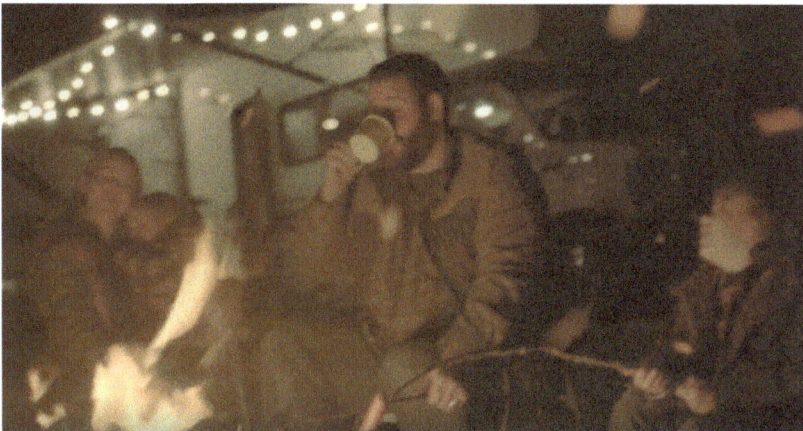

of life, "One of the most significant differences in the pace of life is whether people use the hour on the clock to schedule the beginning and ending of activities, or whether the activities are allowed to transpire according to their own spontaneous schedule. These two approaches are known, respectively, as living by clock time and living by event time" (Levine 1997, 82).

Those societies that primarily emphasize event time (typically in less industrialized nations) experience both the benefits and the costs of their temporal socialization; Levine points to survey data that indicate that people in "slower," event-time-driven places are much less likely to die from heart disease but are also less likely "to be satisfied with their lives" (155, 158). However, for the fast-paced, clock-driven societies of late modernity, the demands of accelerative scheduling and multitasking mean that event time is rarely practised in any sustained way. Because the complex realities of event time are largely unfamiliar to members of late modernity, it can therefore be represented in an idealized form as a "natural," pre-existing state of temporal perfection, as the antidote to the accelerative anxieties of daily life.

Indeed, the use of event time as a marketing strategy in the Go RVing ad closely echoes the similar use of event time in the well-known series of tourism advertisements run in recent years by Newfoundland and Labrador Tourism. Playing on the unique time zone that the province inhabits within the global standard time system, these ads characterize the region as a sanctuary from the time pressures of modern life. One such ad features slow-motion footage of people engaged in leisure activities—hiking, throwing stones into the water, playing music, buying a cake—while a soothing voice says, "When you're always a half hour ahead, you never feel the need to catch up" (Newfoundland and Labrador Tourism 2011). The province's website elaborates: "We've never been afraid to be ourselves or go at our own pace. And, although we have our very own time zone, we don't really measure time in seconds or half hours, but in moments and experiences. Here, you can set your watch a half hour ahead, or you can just leave it at home" (Newfoundland and Labrador Tourism 2011). This fetishization of event time is echoed not only in the "Unschedule" tagline of the Go RVing ad, but also on the Go RVing website, which asks, "Have you ever returned home from a holiday and felt like you needed a vacation from your vacation? That's what happens when you're stuck following someone else's timeline … A vacation in your RV comes with a blank itinerary. Life happens on your time" (Go RVing Canada n.d.). If the accelerative time discipline of late modernity is the disease, then event time, as sold through tourism and recreation agencies, especially those that emphasize enjoyment of the natural environment, is the cure. However, this representation of slow nature as the antidote to fast culture is problematic in several important ways.

A conventional ecocritical reading of the Go RVing ad would likely borrow from William Cronon's well-known observations about the paradoxes and forms of denial that are built into the romanticized vision of wilderness as a place to which we can "return" in order to seek refuge from civilization:

> [T]o the extent that we live in an urban-industrial civilization but at the same time pretend to ourselves that our real home is in the wilderness, to just that extent we give ourselves permission to evade responsibility for the lives we actually lead. We inhabit civilization while holding some part of ourselves—what we imagine to be the most precious part—aloof from its entanglements. We work our nine-to-five jobs in its institutions, we eat its food, we drive its cars (not least to reach the wilderness), we benefit from the intricate and all too invisible networks with which it shelters us, all the while pretending that these things are not an essential part of who we are. By imagining that our true home is in the wilderness, we forgive ourselves the homes we actually inhabit. (Cronon 1995)

Cronon perhaps overstates the degree to which urbanites might disown the "entanglements" of modern culture. Still, representations of nature such as those in "Unschedule" clearly confirm his central observation. In subscribing to the contradictory vision of wilderness travel as both a refuge and a return home, we risk failing to address the problem of how we might actually hope to make a sustainable living from the land, and we also evade responsibility for addressing the problems of unsustainability that exist in the homes and cities where we do live. This conceptual problem that Cronon identified over twenty years ago remains with us, and his insights can still illuminate contemporary representations of culture and the environment.

As a complement to this approach, a critical focus on time can illuminate the situation in several ways that, while borrowing directly from Cronon's insights, would not necessarily be apparent through a focus based primarily on understandings of place. One concern is that representations of nature like the one in the Go RVing ad conceal the fact that the natural environment itself does involve fast and urgent temporalities, from sudden storms and fleeing animals to our own encroaching hunger (hunger that is alleviated in the ad only because the family brought with them packages of processed food that they acquired through their accelerated, multitasking cultural labours). Another consequence is that this kind of representation tells us that all of the concerns of the social and political spheres—the problems of acceleration tied to overwork, urban sprawl, late modern consumer culture, even scheduling itself—no longer exist once we make the transition into nature. The ad attempts to depoliticize nature's time even while it strategically manipulates the supposed slowness of nature.

Sarah Sharma notes that attempts to resist the fast pace of social time often take the form of *spatial* strategies. A Slow Food practitioner, for instance, might go on a pilgrimage to a special restaurant, a farm, or a culinary region, while high-speed business travellers seek refuge in the enclosed space of airport sleeping pods. "Slowness," Sharma writes, "appears to be about getting away, maintaining *distance* from the temporal and the complex multiplicity of time" (2014, 111). Arguing that these kinds of "spatial solutions" actually avoid addressing the complex inequities associated with cultural temporalities, she concludes that "the cultural turn to slowness is a depoliticization of time" (111).

We may find it useful here to draw a distinction between, on the one hand, slowness as an *escape* from cultural time (the tactic of spatial distance that Sharma critiques) and, on the other hand, slowness as a form of *engagement* with cultural time. This latter sense of slowness as engagement seems necessary in order for us to recognize the genuine political efforts involved in, for instance, calls for guaranteed parental leave or for a four-day or three-day work week. Of course, the distinction between engagement and escape is often murky, which is one of the reasons that the slow movement sparks such passionate debates. In any case, Sharma's critique of slowness applies most usefully to slowness as escape, and indeed, the initiative in the Go RVing ad to escape fast culture does regretfully miss the opportunity to engage with the politics of fast culture in favour of firing up the combustion engine to briefly abscond from them.

Given the ways in which "spatial solutions" can serve to depoliticize time, it is especially significant that the spatial solution in the Go RVing ad involves a physical passage into nature, or at least into a somewhat manicured campsite that passes for wilderness. When nature is associated closely with deceleration—when nature is seen as the spatial home of a pristine, essential slowness—nature's time, too, becomes inevitably vacated of its complexities and the politics through which we relate to it. In this kind of advertisement, the simplistic representation of two different temporal worlds—fast culture and slow nature—closes down any conversation about temporal nuance, about the intertwined existence of cultural and ecological temporalities, and about the forms of power and responsibility that are tied to different forms of time. It stops us from acknowledging the ways in which many temporalities in the environment are linked powerfully and problematically with human activity. After all, the problems of social acceleration in industrial and late modern culture play a major role in disrupting the very stability of ecosystems that is so fetishized here; the fast pace of resource consumption that enables social acceleration (and that enables RVing, along with so many other activities) also disables sustainability. This representation forecloses insights into the ways that slower, more thoughtful forms of action and behaviour already exist within certain aspects of the everyday lives of citizens of late modernity, as well as the ways in which the more frenetic elements of the acceleration society might be addressed, carefully and self-consciously, from within. It discourages us from addressing the social problems of overwork, multitasking, time crunch, and systemic unsustainability by suggesting that we instead take a short holiday from these problems by renting a campground—an escape that not only adds yet another item to our to-do lists, but also mischaracterizes the natural world as a separate sphere of existence.

This analysis sounds pretty negative. Yet somehow, it is difficult to watch the ad without breaking into a smile. Despite all of the problems, there remains a compelling sense in which the advertisement reflects an experience of joy; it asks viewers to take delight in the sense of relief from the relentless pressures of the acceleration society. It gives us permission to sigh, to relax, to put down our to-do lists, even for a moment. As such, it holds out the alluring promise of enabling viewers to step back from the obsessions of everyday time discipline and to reevaluate the ways in which slowness and ecological awareness might disrupt the problematic assumptions of speed and apparent productivity. In a somewhat paradoxical sense, then, even while the ad's problematic representational strategies primarily foster a vision of slowness as escape, the ad also teases us with the possibility of slowness as engagement.

Ultimately, though, the tease is probably just a tease. The ad does frame the environment and its landscapes as a separate realm, as a getaway from culture rather than a shared component of the ecosphere that the acceleration society is rapidly destroying. The mild sense of countercultural initiative represented in the Go RVing ad remains compromised at a deeper level through its unwillingness to account for its contradictions. In this sense, the ad confirms and expands Mario Trono's observation, in his contribution to this volume, of the all-too-common scenario in which "positive environmental themes in popular films function industrially against their own messaging" (60). While the insights that could be gained through a thoughtful re-evaluation of social acceleration and ecological awareness are significant, the ad somewhat inevitably sacrifices these insights to the needs of selling the idea of consumer spending on campgrounds and RVs.

In identifying the possibilities and limitations of one particular representation—in this case an advertisement—the larger goal should be to develop and put into widespread practice the broader skills of critical temporal literacy. If we can identify the ways in which different representations of time are productive in their reframing of cultural assumptions, the ways in which they are limited and inadequate, or the ways in which they privilege one particular model of time while erasing other forms of time that have important implications for social power and sustainability, then we will be better prepared to read time critically as it operates systemically or culturally, and we will also have better insight into the ways in which problematic assumptions about time have infiltrated our own minds.

MAKING PROGRESS IN THE FOREST: DON MCKAY'S "WAITING FOR SHAY"

I have been suggesting that the ways in which we represent time in the environment can serve either to reinforce or to question cultural assumptions, to conceal or to illuminate the ideologies and politics tied to the complex interactions between human times and ecological times. In this final section, I want to draw attention to a poem that comments on the hazards and the thrill of the imposition of cultural models of time onto the natural world. The lesson I take from this poem has to do not only with the method by which it conjures and challenges the relationship between the Western model of progress and the temporalities of nature, but also with the way it illustrates another central point—that as we encounter the diverse forms of representation that surround us, we must recognize those moments when a text has something to teach us about natural and cultural time, and we must also embrace those moments when we need to become the teachers.[3]

Don McKay's poem "Waiting for Shay," from his 2006 volume *Strike/Slip*, considers the arrival of modern progressive time in the coastal forests of British Columbia. It sees this arrival as the central event of the broader transformation of the region from a wilderness that expresses temporality through the life cycles of leaves, moss, and bears, into an acculturated seat of linearity, rapid development, and the countdown towards extinction. As such, the poem asks us to see ourselves as temporal colonizers who are simultaneously in love with, blind to, and horrified by our own cultural ideologies of time.

The "powerful, dependable" Shay locomotive, McKay explains in a note, is a train engine that was used widely on Vancouver Island in the first half of the twentieth century to haul timber out of the forest along logging railways (McKay 2006, 75). His poem begins in the moment before the Shay locomotive arrives in the wilderness: "Summon it from sea mist: a dark shape / afloat on its barge in the half-air half- / water dreamtime of the coast" (19). This sense of mythical time and a vaguely realized place in which specific forms have yet to take shape could fit within various traditional accounts of creation; it invokes Australian Aboriginal cosmology as well as the moment in the Book of Genesis when God has created the earth and the light but has not yet given form to his creations or divided "the waters from the waters" (Gen 1:6). However, while we might expect the "dark shape" in the sea mist to take the form of some leviathan emerging from the deep, it turns out to be the Shay locomotive, a human-made object on a human-made barge, whose arrival interrupts the state of the ecosystem:

> What is coming, we might ask, as the mist
> lifts a flirtatious hem to give a glimpse
> of massive body, petrified
> rhinoceros or else some manic futurist's
> reclining nude. [...]
> Does the forest simply go on making moss
> and rot and whispering translations of translations, rain
> into leaf into berry into bear as Shay
> slides by on the tide? (19)

The process by which human presence, and in this case modern European colonization, lends places and times a particular character is at work here. The lifting of the mist, with which the surroundings begin to take shape, coincides precisely with the introduction of our presence—the word "we"—and our wondering about the future, about "what is coming." The first thing to be revealed, though, is not the forest but Shay itself, so that the primary result of the human ability to envision the future is the introduction of tools that modify the environment for human purposes. Like the human tendency to manipulate the land, Shay is ancient and natural—"petrified / rhinoceros"—but it is also strange and ominous, seemingly out of place, and it points towards a possibly deranged "manic" future. "Shay," of course, can be a name for a person as well as a locomotive, and this pun is a pointed one, as the arrival of the colonizing human "we" appears synonymous with the creation of the engine; indeed, individual people are strangely absent from the poem altogether, except in the form of "we" and "Shay." The machine's potentially dark purpose is our own, and to create the engine is to create—and parody—ourselves.

Meanwhile, nature's dreamtime existence, its perpetual making of moss and its cycles of rain and life, are at risk, thrown into doubt by Shay's arrival. The engine serves a similar, if more sinister purpose, to the jar in Wallace Stevens's poem "Anecdote of the Jar." Once placed on a hill, the jar "made the slovenly wilderness / Surround that hill. //

The wilderness rose up to it, / And sprawled around, no longer wild" (Stevens 1923, 60–61). If, as McKay has proposed in his essays on poetry and wilderness, "place" can be understood as "wilderness to which history has happened," or "land to which we have occurred" (McKay 2005, 17), then when we *happen* to the wilderness, the place and time become shaped around our presence, or that of our mechanical ambassadors, and this structure can carry the implications of both domestication and violation.

In its suggestion that the coastal ecosystem, prior to the arrival of the locomotive, exists in a perpetual state of unchanging harmony, the poem seems to run up against two limitations. The first is its apparent erasure of the Indigenous societies whose use of the land far predates the arrival of the apparatuses of Western colonization. A generous reading of this absence would suggest that the poem offers a self-conscious comment on how Western colonization *acts* as though the land is uninhabited. The work of the Shay locomotive, and the colonizing function it represents, will proceed without any recognition of existing cultural worlds, and the poem enters this perspective in order to heighten its critique. McKay's use of the word "dreamtime" is key here; the term ostensibly refers to the eternal time, or the timeless "everywhen," of Australian Aboriginal cosmology, yet the word itself is the coinage of English-speaking anthropologists who misunderstood the concept they were trying to translate (Swain 1993, 14–22). The use of such problematic ethnographic vocabulary in McKay's poem—here made doubly problematic because of its transfer from the Australian Aboriginal context to the coastal North American context—suggests that McKay is parodying the blindness of Western colonization towards Indigenous societies. Still, this gesture towards parody is so subtle, and the absence of Indigenous lives so disquieting, that this aspect of the poem remains difficult to assess. Is the poem commenting on the blindness of Western colonization, or succumbing to it?

The second problem with the poem's representation of a perfectly harmonious pre-Shay coastal ecosystem is its apparent reliance on the idea of the "balance" of nature. Pointing to studies in ecology and evolution which reveal that the view of nature as inherently stable "is basically misleading," John Kricher laments the fact that the "balance of nature paradigm" remains prevalent (Kricher 2009, 23). Greg Garrard notes that "it is more than merely an attractive idea: the balance of nature seems to vindicate the morality of environmentalism" (Garrard 2012, 495). If nature is actually a volatile place, then the case for conservation may begin to erode. McKay's vision of the forest's endless cyclical conversion of "rain / into leaf into berry into bear" does not necessarily account for the volatility of ecosystems and perhaps risks reinforcing the myth of the balance of nature; however, the poem does note that different beetles, "in a string of permutations," have arrived and made their home in the forest over time, lending their "pattern / to the weave" (McKay 2006, 19). The poem's forest, in this sense, is both shifting and stable. The point being made is that even if there is no such thing as total stability in nature, the forest will survive in a much more coherent way, probably for an immense period of time by human standards, in the absence of intensive logging.

In the poem's second section, Shay continues to take shape as both a work of machinery and a living creature:

> Engine, ingenuity: how could we not love it?
> Four-fifths animal, eats wood and water, breathes,
> whistles, relieves itself of pressure with a sigh,
> [...].
>   Asthmatic,
> cheerful, clockwork-clever,
> tough as a troll, tough as a suit of armour that has long
> outlived its knight, Shay is for sure
> the brand-new neolithic monster for the job. (20)

The animalization of the engine lends it a degree of sympathy and suggests that the tool, along with its makers, is not entirely malicious. But at the same time, the animality introduces a degree of wilderness into the machine—understood here in the sense of McKay's definition of wilderness as "the capacity of all things to elude the mind's appropriations" (McKay 2001, 21). Just as a domesticated animal always retains a degree of wildness, a degree of existence outside the human use to which we put it, the engine's "mass of gears and rods" (McKay 2006, 20) is never entirely within human control. Like the workings of a clock, the machine is "clever" in the sense that it is the product of ingenuity, but also in the sense that its consequences for the human relationship to the world lie partly outside the grasp of the minds that created it. Like the suit of armour, Shay will do what it was built to do, but it will also go beyond this intended purpose, inhabiting the future in a way that we cannot fully anticipate. The temporal contradiction in the phrase "brand-new neolithic" speaks to the antiquity of the creation process by which people have brought, and continue to bring, hopeful monsters into existence. Shay—the clockwork machine that is partly animal, or the human that is part clockwork machine—is the latest product of a process as old as the domestication of animals.

In the final section, the poem ends:

When the barge arrives the sleeping shape
will wake and start to breathe and
build a head of steam, accumulating wrath
like a hell-fire preacher. Then,
in a series of sharp
expostulations—work work work—
crank itself ashore. And then
the clock's wound up. (21)

The temporal condition of the poem thus completes its transition from unformed dreamtime to measured modern progressive clock time. The wound-up clock is both the "clockwork" of the Shay engine and the countdown to the destruction of the forest through logging; the measured pace of time, and thus the condition of mortality, has begun to tick for the forest along with the "work work work" of the machine. Shay becomes the engine of death, and we, along with the forest, await the enactment of its (and our) inevitable task. A degree of wilderness exists in us just as it exists in the machine, and we anticipate the horrifying consequences of our own activity as though watching ourselves from the outside even as we act from within. From the moment we scan the first imperative words of the poem—"Summon it from sea mist"—we become implicated in the creation, the conceptualization, of the machine, and just as it is impossible to read the poem without participating in the creative process by which the deadly clock winds up, human existence inevitably involves susceptibility to, and the infliction of, mortality, here embodied within the engine we love as well as dread. In this way the poem serves as a critical comment on, and a lament over, the hazards of the imposition of cultural models of time onto the natural world. With its intensive focus on supposedly linear accelerative development, Western progressive time not only domesticates elements of the natural world, but, at least in the sense of the destruction of particular ecosystems, actually kills certain forms of natural time.

The poem is an example of how thoughtful forms of representation can foster self-conscious reflections on the relationships between ecological temporalities and cultural ideologies of time. It illuminates the fact that while the imposition of cultural time is inevitable, the practices of critical temporal literacy can allow for reflexive re-evaluations of temporal assumptions and can help us identify the limitations of both our own conceptual frameworks and those of the representations around us. More broadly, the process of examining very different forms of cultural texts alongside one another (in this case an advertisement and a poem) reveals both the insights and the mistakes tied to our ways of thinking, just as it illuminates the promises and the threats of the complex interactions between cultural and ecological times. Critical analyses that move freely back and forth between one form and another are well positioned to reflect the difficult everyday work of encountering diverse texts that variously teach us to rethink our understanding of cultural and ecological times, or require us to become the teachers. In enabling us to read, evaluate, and reshape the ideologies of time that are embedded in all forms of cultural and literary texts and practices, ecocritical time studies plays an important role in the projects (now inevitably intertwined) of cultural analysis and environmental sustainability.

ACKNOWLEDGEMENTS

I shared some of the ideas developed in this chapter at two conferences in 2014: Under Western Skies and ALECC (Association for Literature, Environment, and Culture in Canada). I am grateful to those who organized and participated in both events. In particular, I thank Rob Boschman and Mario Trono for their commitment to the UWS conference series and to the resulting books. I am grateful to Brent Bellamy for pointing out to me that, in the Go RVing commercial, the RV trip adds yet another item to the family's to-do list. I also thank the Social Sciences and Humanities Research Council of Canada and Athabasca University for funding portions of the research for this chapter.

Notes

1 One of the limitations to the notion of ubiquitous acceleration as advanced by Rosa and other scholars is that it does not easily account for social inequities and other diverse forms of experience. Sarah Sharma, for instance, notes that adherence to the notion that society is speeding up can obscure "the necessity of tracing how *differential* relationships to time organize and perpetuate inequalities" (2014, 137). While Sharma's critique focuses on diversity in terms of labour and socio-economic class, my focus here is on the diversity of ecological relationships.

2 Estimates of sea level rise continue to evolve over time. Some estimates now anticipate a two-metre rise by 2100 (Gillis 2016). Residents of Lennox Island, a community off the coast of Prince Edward Island that is rapidly disappearing into the sea, fear an even greater rise (CBC News 2016). Such examples add to the temporal complexity of the relationship between ecological disruption and human action.

3 In juxtaposing this reading against my earlier discussion of the Go RVing advertisement, I do not wish to create the impression that literary forms have a monopoly on thoughtful analysis. Daniel Coleman argues in his study of the cultural politics of reading that a simplistic distinction between thoughtful forms of representation such as literary novels or poems, and compromised forms of representation in audiovisual media, conceals the complex possibilities of many texts and forms. He suggests that "we need to read texts that are smarter or wiser than we are on our own" and that doing so requires attention to complexity and variety rather than the assertion of rigid literary boundaries (Coleman 2009, 38). Indeed, the distinction between representations that reinforce problematic temporalities and those that question dominant modes of time is often not clear-cut. My reading of McKay's poem is intended to illustrate these complexities.

References

Bastian, Michelle. 2009. "Inventing Nature: Rewriting Time and Agency in a More-Than-Human World." *Australian Humanities Review* 47: 99–116. ProQuest.

BC Ministry of Environment. 2011. *Climate Change Adaption Guidelines for Sea Dikes and Coastal Flood Hazard Land Use: Guidelines for Management of Coastal Flood Hazard Land Use*. Province of British Columbia. Modified 27 January 2011. http://www.llbc.leg.bc.ca/public/pubdocs/bcdocs2014/541656/guidelines_for_mgr_coastal_flood_land_use-2012.pdf.

CBC News. 2016. "Facing the Change: 50% of Lennox Island, PEI, Could be Underwater in 50 Years." Modified 22 November 2016. http://www.cbc.ca/news/canada/lennox-island-pei-water-ocean-sea-levels-1.3756916.

Coleman, Daniel. 2009. *In Bed with the Word: Reading, Spirituality, and Cultural Politics*. Edmonton: University of Alberta Press.

Cronon, William. 1996. "The Trouble with Wilderness; or, Getting Back to the Wrong Nature." In *Uncommon Ground: Rethinking the Human Place in Nature*, edited by William Cronon, 69–90. New York: W.W. Norton. http://www.williamcronon.net/writing/Trouble_with_Wilderness_Main.html. Web version of chapter accessed January 16, 2015.

Garrard, Greg. 2012. "Nature Cures? or How to Police Analogies of Personal and Ecological Health." *ISLE* 19(3): 494–514.

Gillis, Justin. 2016. "Flooding of Coast, Caused by Global Warming, Has Already Begun." *New York Times*, 3 September 2016. https://www.nytimes.com/2016/09/04/science/flooding-of-coast-caused-by-global-warming-has-already-begun.html.

Girvan, Anita. 2014. "Cultivating Longitudinal Knowledge: Alternative Stories for an Alternative Chronopolitics of Climate Change." In *Found in Alberta: Environmental Themes for the Anthropocene*, edited by Mario Trono and Robert Boschman, 347–69. Waterloo: Wilfrid Laurier University Press.

Go RVing Canada. 2013. "Go RVing Unscheduled English." https://www.youtube.com/watch?v=BEuUOfO-GZE. Modified 12 February 2013.

———. n.d. "Schedule Some Unscheduled Time." Accessed 16 January 2015. https://web.archive.org/web/20150107110621/http://gorving.ca/schedule-some-unscheduled-time/.

Jameson, Fredric. 2005. *Archaeologies of the Future: The Desire Called Utopia and Other Science Fictions*. New York: Verso.

Kricher, John C. 2009. *The Balance of Nature: Ecology's Enduring Myth*. Princeton: Princeton University Press.

Landes, David S. 1983. *Revolution in Time: Clocks and the Making of the Modern World*. Cambridge, MA: Belknap Press.

Levine, Robert. 1997. *A Geography of Time: The Temporal Misadventures of a Social Psychologist, or How Every Culture Keeps Time Just a Little Bit Differently*. New York: Basic Books.

McKay, Don. 2001. *Vis à Vis: Field Notes on Poetry and Wilderness*. Wolfville: Gaspereau Press.

———. 2005. *Deactivated West 100*. Kentville: Gaspereau Press.

———. 2006. "Waiting for Shay." In *Strike/Slip*, 19–21. Toronto: McClelland and Stewart.

Newfoundland and Labrador Tourism. 2011. "Two Brand New Tourism TV Ads." Web. Modified January 17, 2011. https://web.archive.org/web/20151024172459/http://www.newfoundlandlabrador.com/TheLatest/NewsArticle/47.

Nixon, Rob. 2011. *Slow Violence and the Environmentalism of the Poor*. Cambridge, MA: Harvard University Press.

Rosa, Hartmut. 2013. *Social Acceleration: A New Theory of Modernity*, translated by Jonathan Trejo-Mathys. New York: Columbia University Press.

Sharma, Sarah. 2014. *In the Meantime: Temporality and Cultural Politics*. Durham: Duke University Press.

Stevens, Wallace. [1923]1997. "Anecdote of the Jar." In *Wallace Stevens: Collected Poetry and Prose*, edited by Frank Kermode and Joan Richardson, 60–61. New York: Literary Classics of the United States.

Swain, Tony. 1993. *A Place for Strangers: Towards a History of Australian Aboriginal Being*. Cambridge: Cambridge University Press.

Trono, Mario, and Robert Boschman. 2014. "Introduction: Alberta and the Anthropocene." In *Found in Alberta: Environmental Themes for the Anthropocene*, edited by Mario Trono and Robert Boschman, 1–26. Waterloo: Wilfrid Laurier University Press.

CHAPTER 2

A Better Distribution Deal: Ecocinematic Viewing and Montagist Reply

Mario Trono

> It's easy to speak abstractly about the ecology of the image.
> —*Code Unknown*

THE IMAGES ASSEMBLE

The comment above comes from Georges, a disillusioned war photographer in Michael Haneke's *Code Unknown* (2000), a film about the inadequacy of human communication in the face of social complexity. The character speaks to a small group of friends in a café, including his actor girlfriend. They are earnest, educated people worried about war abroad and social injustice at home in their native France. Georges has seen his grim, published photographs of extra-national civilian casualties fail to disrupt the Western political rationalizations and cultures of distraction that enable conditions for war. He refers testily to Susan Sontag's argument in "On Photography" that an ecology of images would preserve the integrity of meanings found where images are first captured. But Georges sees that photojournalism has hit its limits because, as film critic A.O. Scott explains, "[s]ince the 1970s, when the essays in Sontag's book were written, the global glut of images has grown almost beyond measure. In the age of Instagram and Google Earth it is easy to believe—it is sometimes hard not to believe—that every inch of the planet, every human face and patch of wilderness, has been snapped up and uploaded. We have seen it all" (2012). But seen, as Scott understands, from afar and from inside media streams that remove all vestiges of the risk faced by actual travellers who must attune their bodies and their agencies to the bio-political ecologies they visit. That lack of experiential investment—combined with the flattening or eclipsing effects of image ubiquity—can rob images like Georges' of their political potential.

Have environmental images been similarly robbed? Is image-making in general a "polluter of distance" that, by "exposing us to the most distant places of the planet earth, shall lead us to abandon all notions of distance and proximity" in terms of perception (von Amelunxen in Virilio 2009, 11) and in terms of our overweening sense of the medium's powers of demystification? And what of the fact that even deep, experiential immersion in the space-times of any place does not necessarily decode the processual depth of existence? Being there cannot entirely "crack" a place into knowability because, as one reviewer of *Code Unknown* puts it, there is the "incomprehensibility of experience, how it resists encirclement and extends beyond the perimeters of perception" (Bradshaw). And to further complicate matters, images used to communicate an impression of the life world present second-order coding that arises from human image-making practices, thus calling interpreters into vocational being. Critics of ecocinema fall into this category.

Ecocritical film scholars, like others, must "speak abstractly" about image ecologies at times to classify and explain what they observe and experience. And abstraction can become too "easy" as well as dangerous, as a compressive strategy of simplification, when it excludes data, factors, and phenomena that do not fit a preferred, conceptual scheme. As Charles E. Scott says in *The Lives of Things*, the word nature itself takes "us to an abstracting process rather than to the lives of things in their non-discursive, dynamic interactions" (23) and away from "an affirmation of what we find in the world" (183).[1] Do the many abstractions of scholarly image critique affirm the world and help change it for the better? Can ecocinema itself, or ecocinematic forms of interpretation, draw forth changes in human behaviour? Should critics produce more images themselves to accompany their words, and if so, do we risk disappointment like Georges?

In both his situation and our own, it is a question of agency. Georges loses hope when his initial assumption proves false: that reader-viewers back home and away from the war zone can surely be compelled (in one direction and not another) by the right information. He fails to see that humans, along with any given, narrow band of informational flow, are only discrete factors in a vast array of emerging and shifting assemblages of material entities or *actants*, each with their own agency, either in terms of mind endowed with will or matter exhibiting kinds of propensity. In short, he is unaware of Jane Bennett's notion of distributive agency, which holds that complex human–non-human networks are usually beyond human control and that non-conscious things carry unpredictable effects. As Bennett argues in an interview,

> [w]e tend to default to the assumption that the most potent actant in a group is a human being. In some assemblages, that's the case, but in other productions of powerful effects, human strivings are not the key operators. We should recognize that more explicitly, so that we might more accurately discern what was the actual mechanism of a particular effect—such as, for example, a blackout or an oil spill or climate change or a gun culture. (Bennett in Watson 2013, 148)

Or a culture of war, like the one Georges hoped to challenge before being stymied by the sheer density of image flows and the reception point dynamics strung along global networks.

I follow in this chapter Bennett's acceptance of the agential capacity of things, and accept images *as* things that land materially and cognitively before sometimes moving on, often in altered form. Bennett's notion of vital materialism arises from the Latourian construal of materials as *actants* that combine to create assemblages the main force of which, according to Bennett, is distributive agency. Her approach "does not posit a subject as the root cause of an effect. There are instead always a swarm of vitalities at play. The task becomes to identify the contours of the swarm and the kind of relations that obtain between its bits. To figure the generative source of effects as a swarm is to see human intentions as always in competition and confederation with many other strivings" (31). This task, as Bennett explains it, can well direct ecocritics in their appraisals of cinema and in their subsequent efforts to propel their own images along networks in what I hope can become a better distribution deal that carries affects (in the psychological sense) and generates effects.

So in what follows, I begin by separating out for analysis both popular and academic film audiences. I do so amidst a contemporary state of bio-cultural affairs where positive environmental themes in popular films sometimes function industrially against their own messaging and where ecocinema (a genre comprised of formally and philosophically adventurous features, documentaries, and media art installations) is easily attenuated or eclipsed by the techno-capitalist consumerist ethos it most often critiques. And I do so at a time when critics in film studies generally are increasingly generating their own visual texts. Ecocritics of cinema can generate more of their own critically inflected image streams as part of agentive forays into a world of material propensities and embodied intentions. I go on then to advocate a two-part ecocritical approach to commercial cinema:

> *Ecocinematic Viewing* is an experiential viewing habit directed at mainstream, North American films that contain little, compromised, or no environmental content. Popular films that ignore— or inadequately address or thematically co-opt—the interplay between earthly environments and human beings make up the majority of commercial releases. These provide both cover and advocacy for a social order largely consumerist, corporate, and neoliberal in orientation. The construction of space and time in such cinema carries with it narratively embedded fetishes and market-driven assemblages presented through various aesthetics of anthropocentric self-evidence. Such cinema plays into social-material practices now threatening the biosphere, some of which can be seen and understood through the ecocinematic viewing practice I describe. "Ecocinematic" does not refer here to the features of ecocinema, a genre of film, but to an embodied and contextualizing way of seeing moving images, a process-based operation that can take place at almost any point of cultural reception. Once undertaken, this approach becomes generative as viewers can then move towards circulating their own forms of audiovisual reply.

> *Montagist Reply*: This is a viewer-initiated visual production activity that ecologizes a popular film text before spreading online as an assemblage of images, sounds, criticism, activism, art, and entertainment. User-friendly video editing technology is in widespread use, and the adoption of visual essay practices in film studies continues to grow. An environmentally oriented and habitual audio-visual mode of response can help shrink response

times between a film's release date and ecocritical reaction. It can help us quickly join in the flow of late capitalism's media presentations, alongside its speeds, consumptions, and accidents. Such reply would insist on popular, visual imaginaries becoming latticed with ecological understanding. It would move to fill the structuring absences of commercial film texts—considered as part of wider assemblages—with ecological referents. This activity could—barring the considerable challenges presented by copyright legislation—involve remixes, whole movie recuts, GIF sequences, and other remediation techniques practised by online creators, though I primarily advocate here the visual essay. The aesthetic features of such audiovisualized ecocritique would capitalize on the fact that "[p]oliticality is always, on its leading edge, affective. The right has radically understood this, parlaying its strongly performative thinking-feeling of affective politicality into a thirty-year march that has produced many a pragmatic truth" (Massumi 2011, 174). The mode need not be entirely didactic and informational; it could follow Jane Bennett, who has started us thinking about "how to encourage an ecological *sensibility* rather than just an increase in knowledge" (Bennett in Watson 2013, 151). Montagist reply would harvest elements of mainstream, visual regimes through interpretive-creative acts in order to weave popular cinema's audio-fusion effects into broader environmental expression and awareness.

FILM AUDIENCE, ACADEMIC REACH, AND THE PROBLEM OF ECOCONTENT

One of the major challenges for the ecocritical study of cinema involves the question of audience. Can we detect, know, and explain the constellations of affect and effect accompanying film viewing adequately—in even ourselves, not to mention in vast viewing populations? Is pure speculation regarding processes of identification in implied audiences legitimate? Can we point accurately to bio-political consequences that have followed from film's interpellations? Do aspects of story arcs (such as turns in middles and endings) or of film's expressivity (involving lighting and mise-en-scène) demonstrably shape attitudes towards the environment? Especially difficult is the ascertaining of clear connections between audience reception of popularized eco-themes and any resulting cultural behaviours (or lack thereof) that might influence levels of personal consumption, resource/land use, and the kinds of political affiliations that in turn might determine public and private sector actions for or against the environment. Film studies generally must face the possibility that etiologies involving screens, subjectivity, and agency may finally prove too complex to ascertain, track, and predict.

David Ingram's overview of such reception issues well captures the anxieties and pitfalls involved when critics attempt to extrapolate from a film's observable features to speculate on likely forms of ideation. He worries about "ecocritics [who] sometimes make over-deterministic and speculative assumptions about textual effects" (2014, 464) based on nominal or idealized spectators; about claims that film techniques always suture viewers to capitalist subjectivity; and about how easily ideology, as construed by the psycho-semiotic paradigm of film critique, can take us down the rabbit hole of arguments pitting Continental and analytic philosophy against each other (2014, 462–67). Ingram privileges two disciplinary perspectives: (1) in cognitive film studies that of Mike Hulme, who sees risk perception as "socially constructed, with different groups prone to take notice of, fear, and amplify some risks while ignoring, discounting, or attenuating others" (qtd. in Ingram 2014, 470), and (2) work in environmental psychology, a field that "may be beginning to account for ... popular attitudes to the environment in a more empirical way" (Bonnes et al. paraphrased in Ingram 2014, 470–71). As regards such accounts, however, we should concede that "empirical or ethnographic audience studies ... [are not] necessarily neutral or value-free" (Hall qtd. in Ingram 2014, 464), to which I will add that viewers likely process survey questions self-consciously and through idiosyncratic subjectivities largely invisible to circumscribed studies. And in any case, such studies rarely connect evinced perceptions to confirmable actions or inactions.

The fact that film studies finds questions of audience reception bedevilling does not mean, of course, that agit-prop in cinema has never swayed an electorate, that blockbusters never influence public opinion, or that dramatic and inspiring depictions of social issues elicit no affect. But where exactly does eco-affect *go* after environmental themes become mainstream? Does it sufficiently shape movements? Or does it largely evaporate after felt, emotional expenditure, cathartic or otherwise? As my discussion of *Avatar* (2009) below hopefully demonstrates, a popular film's primary power is evident in the consumptions it calls forth and relies on to succeed commercially. And consumption itself, in unsteady-state economies, is what strains environments. While some ecocritics worry about how subjectivity is filmically constituted, a production nexus like Hollywood simply guesses how the process might work and proceeds quickly to formulate its entertainments and disseminate a program for cultural/material consumption. Uncertainty does not preclude action. Since this commercial agency is relentless and quickly reactive to unexpected successes and failures—as if informed by Bennett's notion of distributive agency—it eventually succeeds in terms of profit and influence, even when films are poorly reviewed and said to have lost money (such loss is relative because reports refer to domestic box office only, loss that is negated once global distribution of titles is factored in). This success should cause ecocritics to ask what social effects we presume are called forth when we present our critiques in academic teaching and publication venues as judicially and carefully as possible. What do we really gain in bio-political terms by engaging in such activity when commercial entertainment, comparatively, is far more widely seen and celebrated on the cultural field? Does discussion with colleagues and students of, say, a new materialist approach, which insists on a complex and interdisciplinary emphasis on object-oriented ontology,[2] carry any influence over agentive vectors in space and time beyond usually campus-based contexts? Ecocritical film studies, in light of its implied or announced hopes to contribute to the transformation of human relationships to natural environments, needs to account for *its* economies of distribution and spectrum of hoped-for effects. Since the 2016 election to the American presidency of Donald Trump, a climate change denier who ran in part on a promise to scale back many kinds of environmental initiatives, progressive critique "needs a new populism fast. It's clear what happens if we fail" (Jones 2016). But political models operating on older conceptions of agency and reception require instead, I argue, different ways of distributing the products of our agency.

But before moving in that direction, I want to explain for interdisciplinary readers just how mainstream cinema conceals environmental issues and its own connections to them. Consider what mainstream films present in terms of natural environments. I use the term "natural" here only in the sense that the films themselves seem to—"nature" is usually conceived of and presented as distinct from human agency. Story arcs possess umbilical relation to natural vistas primarily in the form of variations on the pathetic fallacy. With exceptional anthropocentric rigour, human concerns find elemental expression when profilmic locations and digitalized renderings of the same reflect, echo, confirm, and celebrate human agency and desire, even when tragedies and disasters figure prominently in story worlds. We see oceans, rivers, and ponds, tundra, deserts, and forests, canyons, hills, and mountains, currents, tides, and winds, plants and non-human species, the effects of sun and moon all subject to narratives that conceive of natural environments as existing in key—but secondary—relation to human being. Often, natural forces and systems are depicted as antagonists (think of plots involving hostile landscapes, disease epidemics, and predation by non-human animals). Fear, resentment, or excessive technological controls are presented as reasonable responses to environmental exigencies. Often, the audacious privileging of human existence is camouflaged by what appear to be ecofriendly themes. Protagonists may champion environmental causes, human wisdom may follow from communion with nature, spiritual insights may derive from nature's designs, and anthropomorphized creatures may be returned to the wilds after winning our hearts. But these kinds of manifest, environmental film contents do not necessarily carry impetus to environmental right action, or show the way to means, even when environmental messaging is bleak and blunt, as in films depicting ecologically degraded futures or outright environmental apocalypse. They may sometimes carry just the opposite of impetus to action beyond the moment of consumption, as I will explain.

Popular films embed and resonate in socio-textual relation to their promotional attachments and intertexts. They most often suggest, enact, and affirm consumption of themselves by offering quasi-emotional expenditure as a product-experience best enjoyed through an algebra of fleeting attention. Popular cinema's directorial power propels voracious gazes and linked consumptions over legitimating socio-economic networks that treat film as product, product stage, event, experience, window, mirror, site of professional furtherance, auteur vision, cultural authority, and communal touchstone—with all of this subject to market laws regarding planned obsolescence and the forward momentum of new product flows. Film, like literature, "cannot operate outside the systems of mediatized exchange that present it—give it visibility and signifying currency" (Wilson 2006, 562). Discrete moments of popular, aestheticized ecosentiment may be largely negated or contained by the cultural-industrial systems that carry them.

The hallmarks of screen contents and their related promotional networks work together to make a film popular. They ensure that a film's codes of intelligibility mostly accord with those of existing genres. Once a set of cinematic features and conventions are ubiquitous, they (and the production/distribution companies behind them) have the power to eclipse, on the wider cultural field, other kinds of cinematic presentations, including those seen in experimental ecocinema or ecodocumentaries designed for general audiences. Structuring and occluding features of popular films and their promotional contexts include excessive star prominence in framing; related subjugation of place to person; thematic accord with prior commercially successful films; adherence to genre and theme conventions (or minor violations of same presented as pseudo-transgressive kink); product placements and tie-ins; ritualized depictions of dominant culture consumptions; adoption of the most recent film technologies and related special effects (to present older types of interpellation in new specular and spectacular forms); vertically integrated soundtracks and promotion of same; plots vetted by test audiences (who famously and reliably worry over unfamiliar screen contents and routinely dismiss challenges to genre conventions, especially as regards ambiguous or challenging endings); restless editing paces that thrill from frame to frame but that stymie extended reflection; and formulae ranging from catch-phrasing to unearned forms of ideological-narrative closure. The production/distribution span of any upcoming film is now itself a promotional story marked by studios and directors announcing projects, unidentified persons generating subsequent leaks regarding casting, partial trailers appearing years before theatrical releases, red carpet openings, cross-platform media buzz, critical appraisal and reception, and later, sequels, reboots, and a globalized crop dusting of DVDs and digital releases, the promotion of which is outsourced to firms in target countries before global sales figures return home to declare a film as, ideally, the highest grossing movie of all time. All these factors are part of an industrial and cultural ecology for the production and consumption of popular cinema.

Avatar is indelibly marked by every formal, production, and promotional aspect I just described. It is not, of course, impossible that the film's environmental and postcolonial themes might prompt or inspire human agents to act out of concern for ecological awareness and social justice regarding displaced peoples.[3] But it is more likely that *Avatar* offered to audiences a spectacular but finally very light entertainment that functioned primarily as a *simulacrum* of expiation. To what extent did the film actually factor into environmental activism and cultural/political reform during or after its extensive global run? Possibly very little, and there is precedent in North American film and cultural history supporting that possibility. During the period from about 1966 to 1979, a significant number of popular films produced in the United States took up environmental themes, expressively and didactically. *Logan's Run* (1976), *Silent Running* (1972), *Soylent Green* (1973), *Planet of the Apes* (1968), *Omega Man* (1971), *Prophecy* (1979), *Chinatown* (1974), and others connected the excesses of American capitalism and the military-industrial complex to a lack of ecological understanding and to consequences ranging from compromised natural beauty to the disappearance of human and non-human species. Yet once these films had played out their runs in theatres and on late night television—their acting and cinematographic styles soon out of date and their earnestness culturally assigned a bygone or niche status aligned with hippie, green, and other movements—their writers, stars, and audiences found themselves living in a Western ethos epitomized by the conservative governments of Ronald Reagan and Margaret Thatcher (who stood for parties indifferent to environmental issues) within a neoliberal ethos that "expresses inordinate confidence in the unique, self-regulating power of markets as it links the freedom of the individual to markets ... [and that] solicits modes of state, corporate, church and media discipline to organize nature, state policy, workers, consumers, families, schools, investors, and international organizations to maintain conditions for unfettered markets and to obscure or clean up financial collapses, eco-messes and regional conflicts created by that collusion" (Connolly 2012).

Ecothematizing in the 1960s and 1970s appears to have been no match for the corporate lobbyists and investors practising scattershot and quickly adaptive forms of agency focused not on creative expressions, widespread consultations, disinterested debates, scientific considerations but specifically on legislatures, lobbying, and the courts to further strengthen and regularize the legal standing of corporations and the various protections afforded them. The result was industrial and political economies of scale increasingly harmful to natural environments. Ronald Reagan, the one-time actor, exited Hollywood to enter a presidency that Philip Shabecoff, author of *A Fierce Green Fire: The American Environmental Movement*, sees as having constituted, in environmental terms, "eight lost years—years of lost time that cannot be made up and where a lot of damage was done that may not be reparable" (2003, 223). The earnest face of Charlton Heston in the scene from *Soylent Green* (1973) where his character learns that a food source for an overpopulated planet is "made out of people" is now a joke meme online,[4] the metaphor for a First World living at tremendous cost to a Third now mostly forgotten.

Paul Virilio speaks of "cinematic energy" when referring to the vitalities of perception accorded the self. The metaphor is apt; cinema's kinetic, spatial, and temporal flows are in many ways analogous to perception and thought, and its ideo-anamorphic constrictions and dilations present possible worlds and run simulations of moral and ethical dilemmas, as do minds. Perception, though, is compromised in Virilio's view by grey (as opposed to green) ecology, which is the pollution of the self-created world that results when the globalized culture of late-capitalist speed, deadlines, and accidents constricts our cultural experience of space/time through material and virtual compressions. The ellipses of cinematic editing in mainstream films are highly compressive of lived experience. They serve legitimate, narrative purposes, but scenes are becoming increasingly truncated to provide almost immediately the gratifications of dramatic and spectacular reaction and consequence shots. The style is precisely the opposite of Michael Haneke's in *Code Unknown*, where a single tracking shot of events on a Paris street scene lasts more than ten minutes and a static take of a tractor moving across a farmer's field more than three. When a lived process in commercial cinema—such as a fitness regimen or an extended period of book study or journalistic inquiry—needs to be represented, montage and music are deployed to suggest the elation of the *completed* process that viewers will see only a few, brief moments away, *not* the immersion in task and time (or filmic equivalent) that the non-filmic world so often demands. What is rewarded by commercial film text is fast-fleeting attention. Patient engagement is merely an ostensible theme in such montages—or, more precisely, not a theme at all but a process involving viewer experience and quick reward.

Consider how *Avatar* fetishizes speed and consumption, the former at the level of its accelerated editing pace but also in terms of its visual design. When seeking deep communion with their habitat, the Na'vi plug into their resplendent tree of life and gain an *immediate* download of spiritual and material substance. A deep ecology scene in a pre-industrial society takes on the dynamic of accelerated Internet download management through appendages similar to well-designed computer cordage. In contrast, scenes from Ron Fricke's *Baraka* (1992), which also depicts low-tech cultures, use shots of long duration in order to slow and deepen reflection and to allow for more interpretive agency on the part of viewers. The shots contain few kinetic dynamics in order to present visions of environmentally embedded human *in*action, as well as contemplation, as alternatives to heedless enterprise. A handful of fast-motion sequences appear only for didactic purposes regarding the pace of urban life and its unthinking consumptions. The temporal frame of reference is that of geologic deep time. Its slow cinema aesthetic metonymically accords with biospheric cycles and tempos and with related forms of embodied thought. But even such a film as this was promoted with a hyperactive trailer that sped up the "action" to elicit excitement entirely unrelated to the environmental messaging and overall aesthetic. This was apparently considered necessary to gain attention in visual cultures at large.

Popular films, along with the corporate entities that produce them, seek (or seek to create) audiences friendly to their operations. Since "[c]apital is not an abstract category but rather a 'semiotic operator'" (Guattari 2009, 202), a certain kind of subject is called forth. So it is not surprising that "[t]oday, in the regime of neoliberal capitalism, we see ourselves as subjects precisely to the extent that we are anonymous economic units" (Shaviro 2010, 3). A popular film may appear to dramatize and embrace an ethics of environmental care, but, as Lazzarato explains, semiotic motors inside capitalism's visual effusions circumvent consciousness to produce social subjections (Lazzarato 2014). Well-meaning themes combine with spectacle to provide cover for how popular cinema attenuates and directs political and consumer energies. Consider this highly typical perspective (nested at a source, as so many online film reviews now are, inside a culture blog designed to bring viewers, via click bait, to advertising):

> Say what you will about how this movie [*Avatar*] is a cheesy retelling of *Dances with Wolves*. It nevertheless became a global sensation for its ability to tell a simple story of environmental exploitation, violent colonialism, and the quest for social justice. Using cutting-edge 3D and motion-capture software, director James Cameron created an alien world so absorbing that it transformed the way people watch movies—and how they see themselves. (Newitz 2012)

There is here (as in much advertising) a rhetorical bait and switch. Despite the admitted, dated, generic similarities to other films, the "simple story" is given pride of place as the reason for the film's success, even though it was the next-level 3D and CGI effects (not earnest story references to Indigenous displacement, a theme now culturally contained and many decades old in North American cinema) that changed movie watching and how viewers "see *themselves*" (my emphasis). No mention is made of a changed political perspective regarding the seeing of *others*, which is unsurprising given that there is no perceivable connection between an upgrade in screen capture technology and cultural responsiveness to a postcolonial theme. But this connection has been widely assumed, as if *Avatar*, had it been made through traditional cel animation, would have achieved "world box office domination"—a cultural conquest that *ought not to be wished for anyway* by a filmmaker relying on a story about cultural displacement. This sort of confusion over popular social issue films is the ideal state for viewer subjectivity inside what we may reasonably call a neoliberal aesthetics of misdirection. It is only inside the cultural schizophrenia of late capitalism that a film like *Avatar* could get called an *eco* film—a blockbuster production from an industry that is one of California's leading polluters (Rust, Monani, and Cubitt 2013, 10); a film initially consumed throughout North America in multiplexes situated near mall parking lots, in venues seamed with chemically debased confections and featuring tightly controlled viewing environments blasted by product advertising; and a film designed to elicit rapid brain dopamine release to the extent that one critic has described "Post-Pandoran Depression, a condition that results from an unachievable desire for the hyper-real" (Holtmeier 2010, 414). The film represented so little a threat to the prevailing consumerist order that it was considered an excellent investment by Dune Capital, a hedge fund that helped bankroll production and that is run by Steven Mnuchin, a former Goldman Sachs banker and Donald Trump's 2016 choice for treasury secretary (Izadi 2016). How can these negatives be largely invisible to a culture at large? As Žižek well explains, "the very logic of legitimizing the relation of domination [by capital] must remain concealed if it is to be effective" (1994, 8). *Avatar*'s industrial origins and status as neoliberal forcing agent is colourfully veiled by its visual effects and derivative use of story elements from Disney's *Pochahontas* (1995) and *Fern Gully: The Last Rainforest* (1992).[5]

This is the state of affairs in which North American ecocritics of film work when they take commercial cinema as their focus. Sean Cubitt has called on such critics to "earn the formal properties which eco-criticism especially latches onto, as feminism latched onto the gaze" (Cubitt 2013, 279). It is tempting to draw on the formidable feminist, (neo-)Marxist, and postcolonial interrogations of the optics of power that drive dominant visual cultures. Such critical work might, in ways, be worth it. The environmental humanities could become a *sine qua non* in these other fields given how important environmental conditions (and human lives within these) are to questions of gender equality, social justice, and the self-determination of peoples. Yet it is reasonable if perhaps painful to ask what social gains have accrued after decades of what we may loosely think of as politically opposi-tional film theories. Despite the contributions of feminist critics like Kaja Silverman, Theresa de Lauretis, and Laura Mulvey,[6] the world's two largest production centres of popular cinema—Hollywood and Bollywood—still seem, for the most part, structurally and ideologically unaffected in terms of their myriad male gazes, the gender inequities rife in the staffing of studios, the paucity of roles for middle-aged and older women, the lesser amounts of dialogue apportioned to female characters, and the (still) low numbers of female directors and corpo-rate executives. This, even in an environment where the notion of the male gaze (Mulvey's term, born of her landmark analysis of classical

Hollywood cinema) is so well known that it was part of a punchline for a joke made by television writer Jill Soloway at the 2015 Emmy's (Holmes 2015). The #MeToo movement, which seems poised to trigger major changes in Hollywood, is not inflected by high theory. Despite close to a century of (neo-)Marxist film theory and related strands in cultural and globalization studies, the Western world has endured a neoliberal turn characterized by "laws to restrain labor organization and restrict consumer movements ... favorable tax laws for investors; corporate ownership and control of the media ... huge military, police and prison assemblages to pursue imperial policies abroad and discipline the excluded and disaffected at home ... state clean up of disasters created by under-regulated financial and corporate activity; and state/bureaucratic delays to hold off action on global climate change" (Connolly 2012). And despite such fine work in postcolonial media criticism on topics like imperial imaginaries, as in work by Ella Shohat and Robert Stam,[7] neo-colonialism—which forgoes the militaristic occupation of lands in favour of transnational cultural and economic controls—has become a widespread global force.

It may be unfair to attach expectations for social transformation to the efforts of film theorists, researchers, and teachers. But is that not the very hope now informing the rise of the environmental humanities in North American universities, ecocritical film studies included—the hope that these fields, along with their presentations in teaching venues, might intellectually counter grey ecology and trigger collective action?[8] Lorraine Code is right to be worried that "despite the profusion of ecological discourses and despite contestations in the politics of ecology, the *creative* [my emphasis] restructuring possibilities of ecological thinking have yet to be realized" (Code quoted in 2009, 125). I attempt in what follows to present a model for viewing and video authorship that responds to Code's implied call.

ECOCINEMATIC VIEWING AND MONTAGIST REPLY

Ecocinematic viewing lays the groundwork for how ecomontagist reply (specialist and lay) could work. It takes place at the very loci where viewing habits dwell—places under unprecedented scrutiny by companies using mobile telephony to track people's ambulatory and online movements in order to deliver targeted text and image streams to guide consumer behaviour. I recommend an active and omnidirectional viewing response born of habit as Aristotle used the term, as "*hexis* ... an active condition, a state in which something must actively hold itself," a state of "receptivity to what is outside us [and that] depends on an active effort to hold ourselves ready" (Sachs 2016). My idea for a routinized, audiovisual ecoresponse to commercial film accords with contemporary trends in environmental ethics where discriminatory abilities and perceptual faculties get considered in relation to the idea of habits.[9] There is also the biological sense of the word "habit" that refers to the growth patterns through which plants come into being. Just as biological tendency and subsequent event are one in plant life (what Jane Bennett describes as propensity, a kind of agency), so could the viewing of films and audiovisual responses to them function as conjoined phenomena. The action value of any ecocritical engagement lies in the willingness to undertake it, and habit follows from willingness. So, I promote a habit, an eco-interpretive inclination that could become perennially operative inside the retained memories or even rituals of people over time, to the same extent that other ways of seeing and responding have become habitual in human cultures, such as those imagining and envisioning community, bodily well-being, cosmic significance, and love.

At present, entertainment conglomerates have both legal and social licence to saturate our visual fields with discursive, directive, and occluding sonic image streams. Because of copyright provisions, it is difficult for viewing populations to culturally talk back to such omnipresent messaging *using at base the same audiovisual language—including any questionable imagery and framing of same—itself*. Academic film criticism most often relies solely on screen captures (if any images at all), and manifests itself in relatively inaccessible publishing venues (and is alienating to publics when assuming too much prior learning). It can subsequently suffer severe containment on the cultural field, a containment now routinely broken when theorists or their explicators propound via image and text assemblages on platforms such as YouTube, a most welcome development. Remix practitioners must either confine their use of commercial cinematic images to highly specific, non-commercial parameters or risk prosecution for violation of copyright. Academic video essayists/montagists are very similarly restricted despite some leeway when fair dealing (Canada) or fair use (USA) provisions allow for some research or educational permissions to be granted, though this is far from common. Commercial production companies enjoy something that viewers most often do not. The now fully commercialized extension of our exosomatic organs has meant "above all extending the range of our vision, compensating for its imperfections, or finding substitutes for its limited powers. These expansions have themselves been linked in complicated ways to the practices of surveillance and spectacle, which they often abet" (Martin 1994, 3). Montagist reply, were it easily able to flourish, could attune commercial images to environmental views and thought. We need only consider the difference between writing a monograph about genes and actually performing an act of gene splicing to see how much more structurally and impressionistically engaged critique can become when staring down visual and neoliberal aesthetics of misdirection and containment.

Montagist reply could refuse the separation between critical-political discourses and aesthetic-creative ones to be part of the creative restructuring that Code calls for. As Brian Massumi explains in his book *Semblance and Event: Activist Philosophy and the Occurrent Arts*, "[t]he relational/participatory aspect of process could fairly be called *political*, and the qualitative/creatively-self-enjoying aspect *aesthetic*. These aspects ... [need not be] treated as in contradiction or opposition, but as co-occurring dimensions of every event's relaying of formative potential" (2011, 12).

Running critical audiovisual interference in mediatized product flows (which carry few or compromised environmental referents) is part of a reclaiming, through our own creativity, of the same energies of imagination that both capitalism generally and mainstream cinema specifically take for themselves. For we know that "without a *poietic* principle, techno-capitalism cannot generate the 'new,' cannot generate the desire necessary for its current growth" (Wilson 2006, 562). Ecocinematic viewing and montagist response, conjoined, is a technique of existence that resists temporal separation between experience, reflection, and aesthetic-political agency. In my own habitualized, ecocinematic viewing moments, and when essaying in audiovisual formats, I try to let my ideational, affective, and agentive impulses combine, in play, to see what might develop. But distribution of these efforts remains contained.

My current sense of what ecocinematic viewing could entail involves seven features. Each of these is predicated on a kind of open attention to bodily and other contexts beyond the film frame during the moment of viewing. These contexts could easily include the presence of other active screens (like those of laptops or phones) that provide further layers of audiovisual flows to bring even more frames of reference into site-specific, experiential view.

1   *Resisting the dominant.* The point of this practice is to imagine rewriting a movie as you watch, to draw attention to (theoretically at first, in video practice soon after) whatever environmental zones you find germane. Resisting the dominant—the point in a frame to which a given shot is designed to draw your eye—brings viewers into the literal and figurative grounds of a film but outside of authorized viewing points. *Avatar* paradoxically provides the optical illusion of greater access to an imagined reality because of its sophisticated 3D effects. Vision seems liberated from 2D and from narrower kinds of framing. But "the makers of *Avatar* did not want their viewers to be able to choose what to behold, preferring instead to make them view what the filmmakers deemed was the 'right' part of the frame" (Brown 2012, 267). Instead of allowing gazes to wander through the sumptuously rendered alien wilds, they were pushed through the requisite and usual hallways of anthropomorphized, narratively sequenced story elements and authorized foci. While watching any film, we can condition our eye to resist the dominant, to move our attention quickly away from where gaze first "lands" to other parts of the frame and into less authorized looks at profilmically real (or imagined, in the case of animation) vistas. A concern with such an approach might be anxiety over missing the key, visual anchors of a storyline. But it is surprising how this viewing practice allows the narrative to remain entirely apparent while nonetheless placing it in a view *peripheral* to environmental concerns. Gazing for entire films at only backgrounds and settings is a liberating experience. It may seem like one looks where nothing is happening until the ecophilosophical insight arrives that *everything* is happening

there if we think ecologically. Once we suspend belief and focus on, say, soil in a given shot's field depths, we see precisely where the biotic and abiotic factors that provide for all life on earth sit and on which a dramatic planetary stage hosts all the mental phenomena that take place, inside human bodies. One begins to see spaces as dynamic all on their own, outside of centralized human agency, recalling Whitehead's process philosophy and the democracy of trees (Whitehead 1967, 206).

2   *The profilmic.* We could view the profilmic (the zone in front of the camera where actors and locations are set and framed) while remaining mindful of the proposition that all emplacement of human habitat and presence involves some sort of displacement or environmental impact. With a few keystrokes, one can find out during viewing where a given film was shot, and then, while continuing to keep narrative considerations in peripheral view, dwell on specific aspects of screen geographies through our own isolated, viewer-created "shots" within the frame. We could watch for shooting locations with environmental histories that the narrative may representationally colour over. I knew before starting a second viewing of Michael Bay's *Transformers* (2007) that the film had been partly shot in New Mexico, a state that is running out of water and experiencing strained relations between industry and Native American *pueblos*. As well, some scenes were shot near a military base that had to scan for unexploded ordinances in the soil before scenes could be shot of a fictionalized military base under attack. Seen through these interpositions, *Transformers*'s CGI sequences of spectacular, kinetic, machinic compactions and extensions gave the film a tragic, balletic quality. The scene recounting the attack on the base is seen to suggest diegetic war, ongoing ballistic practice for America's foreign wars, and past militaristic displacements of Indigenous cultures in New Mexico over two centuries.

As another example of ecocritical, profilmic connection-making, consider Leonardo DiCaprio, star of *The Revenant* (2015), which was shot near Canmore, Alberta. He was criticized for declaring that the normal chinook winds he experienced during winter filming were dramatic evidence of climate change. DiCaprio's ignorance about Chinooks does not mean that global warming is not taking place, but his error was linked by many to a celebrity environmentalism assumed to be oblivious to environmental science and seen by many as a prime example of wrong-headed opposition to oil pipeline expansion.[10] This debate resonates in a profilmic look at the scene in *The Revenant* where DiCaprio's character Hugh Glass has a riding accident and requires warmth to survive while injured at the base of a cliff. He guts his dead horse to crawl inside the carcass to avoid freezing and emerges the following morning amidst spring-like conditions, gazing up at trees dripping with melt water and shot through with sunlight. The profilmic ecohistory of the shoot becomes interwoven with the film text; Hugh Glass becomes only DiCaprio, simply another interpreter of environment who, like us, gazes at conditions.

3   *Ecolocation.* In front of a film, I practise whole sensorium viewing, remaining aware of my senses during a film's running time. This activity has different intentions behind it than does Laura Marks's idea of treating film as a locus of collective, sensory memories we must excavate in order to re-enrich our bodies to better appreciate "intercultural cinema" (Marks 2000, 243). My aim here is to *de*preciate the suggestive valuations of a unilaterally delivered commercial film product to allow enhanced agency at the point of viewing and to set up subsequent ecomontagist reply. While recently rewatching *Avatar* on Blu-ray in my home, in Calgary, Alberta, I peripherally glimpsed the downtown core a mile or so outside my window (a part of town where many Oil Sands operations are corporately headquartered) and was aware of the dryness of my skin at the same time. I let the film run, leaving its frame contents in view above my laptop screen, in order to read online and discover that Calgary is located in the dry humid continental climate zone. It is an area where hydrological feedback issues, as climate change proceeds, may transform Alberta's expansive northern wetlands into regions of bush and shrub, altering the freshwater cycle and releasing greenhouse gases. There will be "increases in the severity, areal extent and frequency of climate-mediated (e.g. wildfire and drought) and land-use change (e.g. drainage, flooding and

mining) disturbances that are placing the future security of these critical ecosystem services in doubt" (Waddington et al. 2015, 113). My gaze toggled back and forth between TV monitor and laptop screens that rendered each other peripheral in turn. Cooking smells lingering from a barbecue and the sounds made by children playing outside reminded me of Robert Boschman's argument (2014) that a nature/culture dichotomy has negatively affected Alberta's beef production, just as it almost fatally affected Boschman's own daughters when they ate meat infected with *E. coli*. During viewing, I had not simply undertaken an "ecoreading" of *Avatar*; I had cognitively latticed the film with ecological referents while imagining an audiovisual encoding of same. I experienced *Avatar* amid sets of socio-spatial, notional/sensual concerns of an environmental nature. The term "socio-spatial" comes from urbanism research,[11] a field that studies how a culture interacts with its built infrastructure. But I extend the term here to advocate whole-sensorium viewing *on* site of whatever is *in* sight. In this way, one can resist film's suturing of attentions to the spectacular ephemera of globalized, corporate semiosis and remain aware of local environmental conditions and their connections to other ecologies, near and far. Disjunctions between product and place can then be expressed in ecomontagist reply.

4   *Running, body, story, and deep time.* One of the linchpins of the hyper-real is the extraordinary power of commercial image streams to draw on our fascination with time: "[t]he Russian film director Andrei Tarkovsky thought that time itself ... was what people desired and fed upon" (Solnit 2004, 4). The birth of cinema came amidst a process already long under way in which human temporal experience, once linked closely to daily and seasonal cycles, was becoming synchronized instead with the dictates of capitalist space-time. By insisting on the temporal fractures of shift work, industrial deadlines, and the temporalities of commercial media, capitalism co-produced an accelerated film aesthetic in which restless editing paces and cultural aversion to long duration shots meant that commercial, North American cinema soon moved "at the speed of business," as an IBM promotional campaign puts it while claiming that "winning" requires

being "hyperaware" as "markets shift more quickly" (IBM 2014). Hyperawareness of the hyper-real now routinely arrives courtesy of special effects that are themselves news as well as the whole point of entire franchises that connect rootless emotion to enhanced stimuli. This is especially so in the case of 3D and virtual-reality effects, "breakthroughs ... [that are] globally successful when they make a spectacle of emotion" (Baird qtd. in Brereton 2005, 67). While attention spans and shot durations shrink, Hollywood spectacles carry longer running times and extensive DVD extras, which is no surprise given that "[c]apitalism is characterized not by thrift and careful reinvestment, but by excess debt and excessive profit, [and] superabundance" (Wilson 2006, 561).

As part of ecocinematic viewing, we could choose to install in our experience of such films an awareness of our *ecotemporal* locations, wherein a body at rest need not experience any sense of diegetic urgency in an unthinking way. *Avatar* shares with advertising a signal trait: the kind of shot sequence where a stimulating, pleasing visual flashes before being quickly withdrawn. This creates a sense of lack and a related desire to banish that sense by looking to the next frame, then the next, and the next. Such open-endedness and manufactured yearning gets directed at products: "At least since the 1920s, a 'consumer capitalism' has emerged that is interested less in supplying demands based on individual needs and interests, as in creating desire" (Wilson 2006, 561). And, as Slavoj Žižek explains in Sophie Fiennes's 2006 documentary *The Pervert's Guide to Cinema*, "Cinema is the ultimate pervert art. It doesn't give you what you desire, it tells you how to desire." A blockbuster's attenuating temporalities "cannot but occupy both our bodies and minds for the duration of its running time. 'Immersed' as we supposedly are, the 3D (and IMAX) ... are all that we can see and hear" (Brown 2012, 262). But temporal urgency decreases during viewing if one remains aware not only of bodily and diurnal rhythms but also of geologic deep time, conceptually, observing story time as existing in the context of planetary ages. This last might seem an odd thought-viewing experiment, but it is surprising how exhaustingly manipulative films suddenly seem in this frame of mind when they concentrate their affective stressors within shorter and shorter shot durations inside longer and longer running times. In a book chapter titled "The Annihilation of Space and Time," Rebecca Solnit (2004) advances her thesis that the career of early film pioneer Eadweard Muybridge in many ways began a massive shift in how humans perceive time, as his motion studies and resulting photographic experiments factored into the birth of cinema. It began what Virilio calls "the temporal contraction of the inhabited communal geosphere. What one calls in computer science, a temporal convection" (2009, 26). Ecocinematic viewing proceeds instead in its own good time.

5 *Scanning for corporate personhood.* We can "see" films from the outset as semblances that camouflage the modern corporation, as semiotic extensions of the legally delineated corporate veil. These obscure our views of capital, speed, accident, malfeasance, and environmental toll. This occurs even as screen contents mimic, depict, condemn, or affirm these very things in full cultural view, in their themes and aesthetics. Whatever the theme, something initiates a concealment of the film as an assemblage of actants that enact commercial agency. While viewing, we can choose to delineate cinematic spaces of corporate personhood. Ecocinematic viewing and montagist reply could draw attention to corporate personhood and agency per se, to abstract laws that create both corporate entities and their extraordinary rights and powers. Cinema may be used by us to observe and critique abstract legal concepts by linking them to film images. Consider a film's establishing shot of a single, corporately owned office tower. In the same moment, viewers see and "see" (1) the film itself, in the form of light but also as a piece of digital, corporate property on functional display, studio-branded and protected by copyright, and (2) an actual, photographed building (almost certainly corporately owned) that exists in both profilmic reality and the story world, giving the corporate entity a double manifestation in the frame. There is a powerful corporate thereness and aura of inevitability on view. A film overall unfolds as a kind of second-level dramatization of an abstractly empowered but finally material agentive entity connected to corporate power, complicit governments, heavily consumerist economies, neoliberal social relations, and environmental strains.

The modern corporation is an extraordinary assemblage, and its media presentations, including film, are deeply part of it and may be seen as such.

6   *Legal intercoding of film text.* Human activity plays out upon a grid of laws. Persons and other legal entities routinely act in accordance with or contravention of regulations, ordinances, bans, bylaws, embargoes, trade deals, and treaties. Force fields of law, invisible but felt in terms of their material effects, are interlaced throughout most zones of the biosphere. We can choose to see film as a medium through which to reflect on laws affecting lands. Cinema's images of natural environments and industrial/commercial activities are implicitly intercoded with laws and pieces of legislation that allow or disallow various human activities across ecological niches. Cinema has, since its inception, run fantasy simulations of social orders that accord with or defy laws, while philosophers such as Richard Rorty have argued that creative expressions are potentially better at teaching comprehension of legal issues than philosophical critique (Rorty 1989). Creative remixes of an environmental kind could bring into view the laws and regulatory regimes that affect environmental outcomes (both in and outside the film). Montagist reply could connect screen contents to legislative or regulatory processes and efforts to amend them. A reply to Clint Eastwood's *Unforgiven* (1992) could consider, for example, that the film was shot near Turner Valley, Alberta, as was the documentary *Story of Oil* (1947), a black-and-white narrative that rhetorically performs era-defining attitudes towards consumption and energy issues prior to the rise of the modern environmental movement.[12] Ecocritic and law professor Shaun Fluker advised residents of Turner Valley in their 2015 appeal to the Alberta Environmental Appeal Board regarding the question of pollutants from oil and gas infrastructure potentially entering the Sheep River. Media outlets in Alberta covered the story[13] using text and images that could be paired with diagrammed meshes of relevant laws laid over key scenes in the emanation that is *Unforgiven* to create a revelatory picture of profilmic lands in an ecocontext.

7   *Partitioning film narrative, picturing supervenience.* In *The Pervert's Guide to Cinema*, a documentary film (really a kind of extended visual essay), Slavoj Žižek appears on sets re-created from the films he critiques. During his discussion of Hitchcock's *The Birds* (1963), he suggests an interpretive manoeuvre that ecocritics and montagists could usefully adapt. He explains that if we remove the main dramatic element from Hitchcock's film (which would be the birds), whatever is left will comprise the film's latent content. Consider how well that method works with other films: remove the horror element from *The Exorcist* (1973) and it becomes a story of a religious-philosophical examination of a priest who suspects his dead mother has gone to hell; *Alien* (1979), absent the alien, becomes an expressionistic montage of exploited workers in deep space haunted by their working conditions; *The Grudge* (2004), minus demon, morphs into a study in alienation from urban environments; and *It Follows* (2014), without the supernatural stalker, is revealed as a compelling look at one woman's fear of imminent sexual assault and transmission of disease. Partitioning commercial film in this way—in both ecocinematic viewing and ecomontagist reply—would ignore the central conceits of a film to see what else might emerge from the background. Such ecological supervenience would foreground the ontological relationship between anthropomorphic, filmic story and the environmental grounds that make up cinema's lower-level properties.

In *Understanding Machinima: Essays on Filmmaking in Virtual Worlds* (2013), editor Jenna Ng interviews Isabelle Arvers, a curator, exhibitor, and teacher of machinima. In addition to the print version of the book, a video of the interview is available as it took place live inside the online virtual world of *Second Life*. The academic interview took the form of a machinima art piece, creating a *coinstantaneous melding of art and critique* that bests even Žižek's re-created film sets on which he offers critique. Ng brought her critique into the very object of her study, a tremendous illustration of the intra-active processes between mentality and materiality. The "creative restructuring" (Lorraine Code's phrase; see above) of cinematic energy inside eco-remediated montagist reply could manifest in the same way. There is tremendous potential in "putting art and philosophy, theory and practice, on the same creative plane, in the same ripple pool. Art and philosophy, theory and practice, can themselves resonate and effectively fuse" (Massumi 2011, 83). Models for ecocritics to follow are widely available. There are fans, academics, filmmakers, and other artists who are cinematic *bricoleurs* already, as contemporary video practices online increasingly involve restyling, repurposing, and remixing. Easily found, for example, are the Organization for Transformative Works, the essay film *Rohmer in Paris* (2013) by film theorist and montagist Richard Misek, the YouTube channel *Every Frame A Painting* maintained by film editor Tony Zhou, and the Vimeo channel by film studies lecturer Catherine Grant. This does not mean that creators do not struggle with copyright matters. But it is clear that ecomontagist reply has the potential to weave together all manner of scholarly, educative, artistic, activist, and legal/legislative modes if ever the violation of copyright ceases to be the only way for educators and laypersons to significantly qualify and remediate the global glut and saturation of commercial images.

Notes

1   Stacy Alaimo draws attention to this aspect of Scott's work in her 2010 book *Bodily Natures: Science, Environment, and the Material Self.*
2   Object-oriented ontology is a philosophical position that refuses the valuing of humans over non-human objects.
3   See Adamson (2012) and Sideris (2010).
4   See "Heston MegaMix-Soylent Green is PEOPLE!" YouTube. https://www.you tube.com/watch?v=MiHbUaXL_gw
5   I restrict my remarks on *Avatar* to its industrial origins, promotion contexts, and points of reception. In terms of upcoming sequels for the film, Cameron has initiated a number of environmentally friendly shooting and production practices and entities. See Miller (2014) and *EcoCop: Green Film Shooting: COP21 Special Edition* (2016).
6   See in particular Silverman's *World Spectators* (2000), de Lauretis's *Freud's Drive: Psychoanalysis, Literature, and Film* (2008), and Mulvey's *Visual and Other Pleasures* (2009).
7   See, for example, Shohat and Stam, *Unthinking Eurocentrism: Multiculturalism and the Media* (1994).
8   For a fine overview of EH ambitions, see the "Humanities for the Environment 2018 Report—Ways to Here, Ways Forward" by Holm and Brennan (2018), http://www.mdpi.com/2076-0787/7/1/3/htm.
9   For example, Charles J. List (2005), in discussing education and the virtues of wild leisure, cites Aristotle's frequent aligning of nature, habit, and rational principle (363).

10 For example, see Fletcher (2016) and Libin (2016).
11 See Jessop (2008).
12 Available at https://www.youtube.com/watch?v=kENGJhODq0o.
13 For example, see Stephenson (2015).

References

Adamson, Joni. 2012. "Indigenous Literatures, Multinaturalism, and *Avatar*: The Emergence of Indigenous Cosmopolitics." *American Literary History* 24(1): 143–62.

Alaimo, Stacy. 2010. *Bodily Natures: Science, Environment, and the Material Self.* Bloomington: Indiana University Press.

Bennett, Jane. 2010. *Vibrant Matter: A Political Ecology of Things.* Durham: Duke University Press.

Boschman, Robert. 2014. "Bum Steer: Adulterant *E-Coli* and the Nature-Culture Dichotomy." In *Found in Alberta: Environmental Themes for the Anthropocene*, ed. Robert Boschman and Mario Trono, 47–66. Waterloo: Wilfrid Laurier University Press.

Bozak, Nadia. 2011. *The Cinematic Footprint: Lights. Camera, Natural Resources.* New Brunswick: Rutgers University Press.

Bradshaw, Peter. 2001. "Code Unknown." *The Guardian*, 25 May 2001. https://www.theguardian.com/film/2001/may/25/1. Accessed 2 November 2016.

Brereton, Pat. 2005. *Hollywood Utopia: Ecology in Contemporary American Cinema.* Portland: Intellect.

Brown, William. 2012. "*Avatar*: Stereoscopic Cinema, Gaseous Perception and Darkness." *Animation: An Interdisciplinary Journal* 7(3): 259–71.

Code, Lorraine, in Deron Boyles. 2009. "Considering Lorraine Code's Ecological Thinking and Standpoint Epistemology: A Theory of Knowledge for Agentive Knowing in Schools?" *Philosophical Studies in Education* 40: 125–37.

Connolly, William E. 2012. "Steps Toward an Ecology of Late Capitalism." *Theory and Event* 15(1).

Crusto, Mitchell F. 2003. "Green Business: Should We Revoke Corporate Charters for Environmental Violations? *Louisiana Law Review* 63(176): 175–241

Cubitt, Sean. 2013. "Everyone Knows This Is Nowhere: Data Visualization and Ecocriticism." In *Ecocinema: Theory and Practice*, ed. Stephen Rust, Salma Monani, and Sean Cubitt. New York: Routledge.

de Lauretis, Theresa. 2010. *Freud's Drive: Psychoanalysis, Literature, and Film.* New York: Palgrave.

EcoCop: Green Film Shooting: COP21 Special Edition—in Cooperation With Film4Climate. http://regions20.org/wp-content/uploads/2016/08/ECOCOP English.pdf. Accessed January 2017.

Fletcher, Robson. 2016. "DiCaprio's Oscar speech cringe-worthy for some advocates of climate-change action." *CBC News.* http://www.cbc.ca/news/canada/calgary/leonardo-dicaprio-climate-change-oscar-speech-alberta-reaction-1.346 9010. Accessed May 2016.

Gilens, Martin, and Benjamin I. Page. 2014. "Testing Theories of American Politics: Elites, Interest Groups, and Average Citizens." *Perspectives on Politics* 12(3): 564–81.

Guattari, Félix. 2009. "Capital as the Integral of Power Formations." In *Soft Subversions*, ed. Sylvère Lotringer, 244–64. New York and Los Angeles: Semiotext(e).

Holm, Poul, and Ruth Brennan. 2018. "Humanities for the Environment 2018 Report—Ways to Here, Ways Forward." *Humanities* 7(1). http://www.mdpi.com/2076-0787/7/1/3/htm. Accessed 8 August 2018.

Holmes, Sally. 2015. "The Best Feminist Jokes from the 2015 Emmys." *Elle*. http://www.elle.com/culture/news/a30604/the-best-feminist-jokes-from-the-2015-emmys. Accessed 2 September 2015.

Holtmeier, Matthew. 2010. "Post-Pandoran Depression or Na'vi Sympathy: *Avatar*, Affect, and Audience Reception." *Journal for the Study of Religion, Nature, and Culture* 4(4): 414–24.

IBM. 2014. "Innovate at the Speed of Business: Fast IT." YouTube. Posted by Cisco, 4 March 2014.

Ingram, David. 2014. "Rethinking Eco-Film Studies." In *The Oxford Handbook of Ecocriticism*, ed. Greg Garrard, 459–74. Toronto: Oxford University Press.

Izadi, Elahe. 2016. "Trump's Treasury pick Steven Mnuchin is behind some of Hollywood's biggest movies." *Washington Post*, 30 November. https://www.washingtonpost.com/news/arts. Accessed 30 November 2016.

Jessop, Robert. 2008. "Theorizing Sociospatial Relations." *Environment and Planning Society and Space* 26(3): 389–401.

Jones, Owen. "The left needs a new populism fast. It's clear what happens if we fail." *The Guardian*. https://www.theguardian.com/commentisfree/2016/nov/10/the-left-needs-a-new-populism-fast?CMP=share_btn_fb. Accessed November 2016.

Lazzarato, Maurice. 2014. *Signs and Machines: Capitalism and the Production of Subjectivity*. Cambridge, MA: MIT Press.

Libin, Felix. 2016. "Kevin Libin: Leonardo DiCaprio's Oscar for best climate-change drama." http://business.financialpost.com/fp-comment/kevin-libin-leonardo-dicaprios-oscar-for-best-climate-change-drama.

List, Charles J. 2005. "The Virtues of Wild Leisure." *Environmental Ethics* 27(4): 355–73.

Marks, Laura U. 2000. *The Skin of Film: Intercultural Cinema, Embodiment, and the Senses*. Durham: Duke University Press.

Martin, Jay. 1994. *Downcast Eyes: The Denigration of Vision in Twentieth Century French Thought*. Berkeley: University of California Press.

Massumi, Brian. 2011. *Semblance and Event: Activist Philosophy and the Occurrent Arts*. Cambridge, MA: MIT Press.

Miller, Gerri. 2014. "When James Cameron says 'green,' Hollywood studio says 'how green?'" *Mother Nature Network*. http://www.mnn.com/money/greenworkplace/stories/when-james-cameron-says-green-hollywood-studio-says-how-green.

Mulvey, Laura. 2009. *Visual and Other Pleasures*, 2nd ed. New York: Palgrave.

Newitz, Annalee. 2012. "19 science fiction movies that can change your life." http://io9.com/5963166/19-science-fiction-movies-that-could-change-your-life. Accessed 10 August 2015.

Ng, Jenna. 2013. "Agency, Simulation, Gamification, Machinima: An Interview with Isabelle Arvers." In *Understanding Machinima: Essays on Filmmaking in Virtual Worlds*, ed. Jenna Ng, 245–51. London: Bloomsbury.

Rorty, Richard. 1989. *Contingency, Irony, and Solidarity.* Cambridge: Cambridge University Press.

Rust, Stephen, Salma Monani, and Sean Cubitt, eds. 2013. *Ecocinema: Theory and Practice.* New York: Routledge.

Sachs, Joe. "Aristotle: Ethics." In *The Internet Encyclopedia of Philosophy.* http://www.iep.utm.edu/aris-eth.

Scott, A.O. 2012. "Around the world in 99 minutes and zero words." *New York* Times, 24 August 2012. http://www.nytimes.com/2012/08/24/movies/samsara-a-documentary-directed-by-ron-fricke.html. Accessed 2 November 2016.

Scott, Charles E. 2002. *The Lives of Things.* Bloomington: Indiana University Press.

Seymour, Nicole. 2012. "Toward an Irreverent Ecocriticism." *Journal of Ecocriticism* 4(2): 56–71.

Shabecoff, Philip. 2003. *A Fierce Green Fire: The American Environmental Movement*, 2nd ed. Washington, DC: Island Press.

Shaviro, Stephen. 2010. *Post-Cinematic Affect.* Winchester: Zero Books.

Shohat, Ella, and Robert Stam. 1994. *Unthinking Eurocentrism: Multiculturalism and the Media.* New York: Routledge.

Sideris, Lisa H. 2010. "I See You: Interspecies Empathy and *Avatar.*" *Journal for the Study of Religion, Nature, and Culture* 4(4): 457–77.

Silverman, Kaja. 2000. *World Spectators.* Stanford: Stanford University Press.

Solnit, Rebecca. 2004. *River of Shadows: Eadweard Muybridge and the Technological Wild West.* New York: Penguin.

Sontag, Susan. 1979. *On Photography.* London: Penguin.

Stephenson, Amanda. 2015. "Turner Valley residents launch crowd-funding campaign to determine if town's drinking water is safe." *Calgary Herald*, 15 March. http://calgaryherald.com/news/local-news/turner-valley-residents-launch-crowd-funding-campaign-to-determine-if-towns-drinking-water-is-safe. Accessed 8 August 2018.

Virilio, Paul. 2009. *Grey Ecology.* New York: Atropos Press.

Von Amelunxen, Hubertus. 2009. In Paul Virilio, *Grey Ecology.* New York: Atropos Press.

Waddington, J.M., et al. 2015. "Hydrological Feedbacks in Northern Peatlands." *Ecohydrology* 8: 113–27.

Watson, Janell. 2013. Interview: "Eco-sensibilities: An Interview with Jane Bennett." *Minnesota Review* 81: 147–58.

Whitehead, Alfred North. 1967. *Adventures of Ideas.* New York: Free Press.

Wilson, Scott. 2006. "Writing Excess: The Poetic Principle of Post-Literary Culture." In *Literary Theory and Criticism*, ed. Patricia Waugh. Oxford: Oxford University Press.

Žižek, Slavoj. 1994. "The Spectre of Ideology." In *Mapping Ideology*, ed. Žižek. London: Verso.

———. 2006. *The Pervert's Guide to Cinema.* DVD. London: Verso.

CHAPTER 3

Allô, ici la terre: Agency in Ecological Music Composition, Performance, and Listening

Sabine Feisst

Scientists and musicians often pursue separate and seemingly incompatible paths. A closer look, however, shows common concerns and intersections. In illuminating examples of ecologically inspired music, as a musicologist, I hope to pave the way for more fruitful interactions between environmental scientists and scholars, eco-artists, and eco-musicologists who study music in its global, historical, and cultural contexts and investigate relationships between music, culture, nature, and ecology.[1] As skilful and environmentally attuned listeners, eco-musicologists have drawn attention to the slow and fast, subtle and obvious changes in our sonic environment, the fragile acoustic ecologies around the globe; they have illuminated the agency of environmentalist musicians who have provided influential soundtracks to environmental issues and debates; they have discussed sustainability problems in musical practices such as the carbon footprint of music production and the shortage of wood used for guitars, clarinets, and other instruments. They have entered public arenas, playing active roles in community engagement endeavours and environmental activism. In these capacities, eco=musicologists have served as agents of change in both academic and non-academic communities.[2]

Non-human nature and its rich sonic ecologies have long been a fertile source of musical inspiration. Yet since the last century environmental transformation and unprecedented disasters—from the radioactive contamination of Japanese fishermen by the 1954 nuclear fallout on Bikini Atoll to the 2010 devastating British Petroleum oil spill in the Gulf of Mexico—have changed the ways our environments sound and propelled musicians worldwide to develop art that seeks to engender ecological awareness. In this chapter I will focus on four artists from Europe, Australia, the United States, and Canada who as musical agents of change have explored ecologically inspired composition, performance, and listening since the 1970s. I will show how they have used the idea of ecology, related to their non-human environments, and advanced ecological consciousness through their art. I will examine Luc Ferrari's *Allô, ici la terre* (1972), an "ecological" multimedia spectacle; Leah Barclay's *Sound Mirrors*, an environmentally conscious sound installation (2010); David Rothenberg's ecologically motivated interspecies performances; and Hildegard Westerkamp's soundwalks focusing on the acoustic ecology of a chosen environment.

BACKGROUND

Given that all art has political implications, the music under consideration here builds on the tradition of overtly political art, which has a long history, but has gained visibility and significance in the twentieth century in the face of totalitarianism, genocide, war, and social injustice. After the Second World War, politically committed art emerged as a valid alternative to escapist art for art's sake. French existentialist philosopher Jean-Paul Sartre, for instance, promoted *littérature engagée* as a conscious and morally motivated commitment to willed action and dismissed the bourgeois idea of *l'art pour l'art*.[3] Aside from musicians in the folk, pop, and jazz arenas, such classical composers as Hanns Eisler, Dmitri Shostakovich, Luigi Nono, and Christian Wolff created works addressing the Holocaust, capitalism, and racial and gender inequality. With the rise of the environmental movement in the 1960s, composers of classical music, along with many popular music artists such as Pete Seeger and Paul Winter, began to write music that demanded more human responsibility for the planet. As early as 1963, Lou Harrison, a composer based in California, protested nuclear weapons testing in the Pacific in his *Pacifika Rondo*.[4] In the 1970s, such composers as John Cage, George Crumb, Alan Hovhaness, and Toru Takemitsu conceived pieces celebrating the songs of the humpback whale and thus helped save this species from extinction. At the same time, Henry Brant, Cage, Lukas Foss, and many others made musical settings of such texts as *Walden, Journal*, and "Essay on the Duty of

Civil Disobedience" by Henry David Thoreau, whose popularity as a naturalist and critic of development had started to rise exponentially in the 1960s.[5] Also since the 1970s, R. Murray Schafer and Hildegard Westerkamp, pioneers in acoustic ecology, have critically examined the quality of sonic environments and developed soundscape stud-ies at Simon Fraser University in Vancouver. Important environmen-tally engaged musicians in the 1980s and 1990s include John Luther Adams, Annea Lockwood, Pauline Oliveros, and David Rothenberg, who applied ecological and ecocritical thinking to composition, per-formance, and listening. They paved the way for such younger musi-cians as Leah Barclay, Matthew Burtner, and Emily Doolittle, who are currently taking ecological approaches to music-making to new levels.

In music, as in all the humanities, the term *ecology* is used in a broad sense, reflecting, on the one hand, the idea of interconnections of organisms and their inorganic environments, and on the other hand, a critical attitude towards the environment.[6] While only some of the examples discussed below point to analogies between scientific ecolog-ical principles and musical activities, all reveal ecology's connotation of environmental activism, be it through the use of ecocritical texts and subtexts or provocative forms of music-making and deep listening.

Music draws attention to temporal ecologies, perhaps more pow-erfully than any other art form. A time-based art, music is sounded time and makes us aware of the manifold temporal modes that exist: linear and cyclical perceptions of time, clock versus experienced time, slow versus fast passage of time, human versus non-human tempo-ral awareness. Music's regular and irregular, steady, accelerating and decelerating sound patterns influence the rhythms of the musicians' and audience's bodies and their human and non-human environments. As Henri Lefebvre stated in *Rhythmanalysis*: "Everywhere where there is interaction between place, a time and expenditure of energy, there is rhythm."[7] Music is but one layer in a complex polyrhythmic fabric that encompasses the rhythms of human and non-human lives, including lightning-speed digital communication, the accelerating rhythms of species extinction, and the slowly evolving effects of climate change and environmental degradation.

ECOLOGICAL COMPOSITION I—LUC FERRARI'S *ALLÔ, ICI LA TERRE*

Luc Ferrari (1929–2005), a French proponent of tape music, was among the first classical composers to respond musically to the "new environmentalism" in the 1960s. He began his career in the early 1950s, as a pianist and composer, creating traditionally notated modernist works for conventional instruments. But soon after, he began creating so-called *musique concrete*, whose foundation is field recordings, found sounds captured on magnetic tape.[8]

In the 1960s, deeply concerned about the wars in Algeria and Vietnam, De Gaulle's conservative politics, and environmental degradation, Ferrari sympathized with the left and began to merge "the social and the political with musical intentions" in his works.[9] He called these years of his career the "red period": "a period in which I observe society, listen to the landscape and inquire about the voices of others."[10] Skeptical of elitist and abstract music, he strove for "cultural democratization" and accessibility in such tape pieces as *Hétérozygote* (1964) and *Presque rien No. 1* (1969), combining minimally edited acoustic snapshots of sounds from natural and built environments so that listeners could recognize their original identity and use them as a basis for imagining their own stories.[11] He also explored improvisation as a more democratic approach to music-making in such works as his *Spontané* series and *Cellule 75*.

Allô, ici la terre, a so-called ecological multimedia spectacle in two chapters, each lasting about two hours, also falls within this period. Chapter 1 is for slides, tape, and ensemble (1971–2); Chapter 2 is a "spectacle écologique pour l'oreille" for tape, amplified instrumental ensemble, and voices (1973–4). Derived from Argentinian-born composer Roberto Détrée's German text "Also sprach die Erde (Thus Spoke the Earth)," *Allô*'s title is a pun on Friedrich Nietzsche's *Also sprach Zarathustra* and resonates with a line by Victor Hugo: "C'est une triste chose de songer que la nature parle et que le genre humain n'écoute pas (How sad to think nature is speaking and mankind does not listen)."[12] *Allô* originated had its origins in Ferrari's desire to create an anti-war and anti-violence multimedia piece for the 1972 Olympic Games in Munich, tentatively titled *Visage VI ou le visage auquel on a ôté le masque*.[13] But the focus of the project became the earth's environmental health and the prospect that it might run out of time: "le risque de la mort la plus fondamentale qui soit: celle de notre planète" (the risk of the most fundamental death: that of our planet).[14]

Allô's first chapter was conceived as a love poem for the earth, emphasizing all of its beauty but also its fragility. Ferrari suggests that time passes slowly. Nine themes are presented through sounds and images in an intuitive sequence of gradually changing impressions and sensations: light, space, landscape, eroticism, view, hope, movement, diversity, and dance.[15] A large symphony orchestra focuses on very few tones, organized in simple intervals, which are sustained before they change, but only in nuance and very gradually. At the first chapter's premiere in Champigny near Paris in January 1975, Ferrari immersed his audience in these orchestral sounds along with sounds from twelve tape recorders and numerous loudspeakers. He also surrounded his listeners with large screens displaying via forty slide projectors 3,000 nature photos by Jean-Serge Breton in a slow sequence.

The second chapter builds on the first but can also stand on its own. *Allô*'s first chapter appeals first and foremost to the audience's emotions; the second targets their minds and is marked by often fast-paced and heterogeneous rhythms, including sound patterns of nature, human speech, song, and music, as well as rhythms of war and environmental destruction. Conceived as an evening-length Brechtian *Lehrstück*, *Allô*'s second chapter consists of two large units: the first comprises the subsections "Prologue," "The City," "The Countryside," "Production," and "Consumption," and the second, "Water," "Air," "War," and "Utopia."[16] Each subsection features texts on our ecosystems' fragility, health problems, urban waste, industrial farming, water and air pollution, our obsession with production and growth, consumerism, and the Vietnam War's destructiveness. *Allô* concludes with utopian visions pointing to a new type of socialism and to women as the most promising future political leaders.[17]

Ferrari uses significant interrelated "objective" and "subjective" texts.[18] The "objective" texts include environmental news and facts from newspapers, journals, books, and presentations at the 1972 Stockholm Conference on the Human Environment. These are authored by scientists, conservationists, philosophers, politicians, and journalists, including the ecologists Barry Commoner, Garrett Hardin, and Arthur Westing, the marine conservationist Jacques-Yves Cousteau, and philosopher-journalist André Gorz.[19] The "subjective" texts are the lyrics of the *chansons* that interlace the work. Ferrari employs "realistic" sounds, found environmental sounds, and "abstract" sounds, as well as improvised, electronic, and traditionally composed sounds.

The scientific and political texts are narrated live. Captured on tape are electronic, natural, industrial, and war sounds and prerecorded Western and non-Western music, which accompany the narration. A live quintet (flute, piano, organ, electric guitar, and percussion), solo voices, and a chorus perform *chansons* in combination with or alternation with the tape. The musical language of the *chansons* is accessible and unpretentious, and the performers have a certain degree of improvisatory freedom. Coaching the ensemble "Between" for a performance of *Allô* in Cologne in 1974, Ferrari instructed the musicians to sing and play as if "they were under water or without water" in the "Water" section and to sing and play as if "the air were polluted" and as if "they could barely breathe" in the "Air" section.[20] In the penultimate section on "War," a female voice reads excerpts from the 1972 book *Harvest of Death: Chemical Warfare in Vietnam and Cambodia* over a background of taped traditional Vietnamese music and field recordings of war and nuclear explosions, and the live musicians insert such *chansons* as "The Arms Traders."

Allô's second chapter was performed in Germany, Switzerland, and Spain in the 1970s but received mixed reviews and soon lapsed into obscurity. In the 1990s, Ferrari, once a passionate environmental activist, described that chapter, not without resignation, as the "beginning of a reflection on 'green' issues, which, after 20 years, are still trampled on."[21] *Allô*'s fate in France illuminates that country's environmental politics. Despite its pride in its natural environments and in such prominent thinkers as Jean-Jacques Rousseau, France has a long history of troubling environmental policy-making influenced by René Descartes, according to whom non-human nature is insensate and exploitable by humans. Committed to the objectification and commodification of nature, from deforestation in the twelfth century to the emphatic embrace of nuclear power in the twentieth, the French government largely ignored the protests of the 1970s.[22] Although written in an accessible style and for a broad public, *Allô* was presented mostly in venues that catered to an elite audience unwilling to embrace the piece's high-minded message. Ferrari's status as a musical outsider in France may have hindered the work's reception as well. Ferrari's eco-musical legacy, backed by such ecocritical *chansonniers* as Georges Brassens, Jean Ferrat, and Maxime Le Forestier at the time, has been carried forward—albeit on a smaller scale and mostly absent explicit activist tendencies—by other French composers such as Bernard Fort, François-Bernard Mâche, Cécile le Prado, and Eliane Radigue as well as by numerous popular music artists. Yet outside of France, as we shall see, musicians have recently been developing large-scale eco-musical projects that match or even surpass *Allô* in scope and ambition.

ECOLOGICAL COMPOSITION II—LEAH BARCLAY'S *SOUND MIRRORS*

An Australian composer-percussionist creating music for dance, theater, film, and installations, Leah Barclay (b. 1985) builds on Ferrari's work, including his use of technology, field recordings, images, and environmental activism. Believing in the catalytic impact of sound on the listener, Barclay has emphasized environmental issues in many of her works since 2004. She casts her role as that of an "agent of change" and seeks "new processes for a sustainable future."[23] Yet her artistic approach goes beyond the popular notion of ecology as a reflection of one's attitude towards the environment. It seeks to echo the interrelationships of organisms and their environment. Having grown up in Queensland near several rivers and the environmentally threatened Great Barrier Reef and having lived in India, Barclay has been attuned to environmental issues and the acceleration of environmental degradation since childhood.

Barclay's first environmentally engaged work, *River of Mirrors* (2004), for chamber orchestra, paid tribute to sounds of the Noosa Everglades. It was followed by *Confluence* (2005), a multimedia work for cello, live electronics, and two dancers, which also features, similar to Ferrari's *Allô*, field recordings, taped voices, and images offering news and facts on climate change.[24] Her recent works have grown out of on-location environmental research, community engagement, and collaborations with scientists, anthropologists, environmentalists, and policy-makers in Australia, New Zealand, South America, and Asia. They showcase rarely heard recorded sounds of fragile biospheres and live performances, and involve audience participation through interactive labs, workshops, and live and virtual dissemination plat-forms. Catering to technologically inclined audiences, the platforms are often interactive websites in the form of sound maps, featuring geo-tagged environmental field recordings, place-related composed music, and sound-related place stories on an online map. Barclay's soundmaps are often participatory, inviting other people to join her in the various creative place-making efforts.

Members of the global community can help grow her soundmaps by uploading recorded sounds of places as well as stories.[25] Barclay thus provides participatory aesthetic and educational experiences to guide audiences towards sustainable behaviours. Soundmaps can also function as time capsules, storing acoustic snapshots of places. Thus they inspire listeners to engage in temporal travel and to experience how environmental sound is changing over time.

Barclay's *Sound Mirrors* (2010), a sound installation featuring riv-ers and their communities in Australia, India, Korea, and China, is a case in point. *Sound Mirrors*, which received the Australia Council's HELM-Arts Award for innovative environmentally inspired projects, offers a glimpse into the sound of river biospheres around the world and sonifies changing human and non-human environments. Deter-mined to "find a voice for the rivers at a time when it is increasingly important to listen to the environment" and to make communities value their rivers' water, she researched the Australian Noosa, South Indian Pamba, Korean Han, and Chinese Huangpu and Pearl Rivers and their communities.[26] She built relationships with scientist Nata-sha Odges, environmentalist Sandra Conte, and water engineer Ryan Dillon. Barclay conducted workshops with the river communities and took them on soundwalks, a concept that will be discussed in detail below. After that she recorded the rivers' non-human biospheres and the verbal responses of Indigenous people to their river environments using a variety of commercially available and custom-built shotgun and contact microphones and hydrophones as well as digital record-ers. She performed with local musicians, whom she provided with oral and written instructions, some of which allow for improvisation, and recorded this river-inspired music on location.[27] Barclay collected around forty hours of field recordings, which she processed into three hours for use in the compositional process, but included only one and a half hours for use in the multiple versions of *Sound Mirrors*.[28]

Sound Mirrors exists as installations with no specific length or set beginning and end. Its eleven to eighteen tracks run on an Apple computer with customized Max/MSP software and a multi-channel sound system. The software freely chooses, combines, and spatializes the tracks until the system is shut down. There is also a live-performance version of *Sound Mirrors*, as well as a CD version, both of which are titled "Transient Landscapes." Live performances involve real-time mixing of recorded materials and are different each time. The CD version, which offers a linear rather than immersive experience of the work, includes eleven tracks: "River of Mirrors" and "Everglades" (Noosa), "Nakshatra: Divisions of the Sky," "Ritual Bells," "Triloka: Monsoon," and "Backwaters" (Pamba), "Han River" and "Red Cliffs" (Han), "Shimmer" (Huangpu), "Liquid Borders" (Pearl), and "Confluence" (Noosa).

The CD version's first section, "River of Mirrors," pays tribute to the 750-mile-long Noosa River in southeastern Queensland, which is situated in a large, UNESCO-identified Australian biosphere and extends from the Wahpoonga Range at Mount Elliot south to the Pacific Ocean (about 80 miles north of Brisbane). The movement's title, "River of Mirrors," refers to "mirror reflections in the [river's] tannin-stained upper waters."[29] The music itself features recorded sounds from the Noosa River and its flora and fauna, including bird song; the didjeridoo playing and narration of local Indigenous artist Lyndon Davis, a descendant of the Queensland-based Gubbi Gubbi people; and electronics. The music unfolds slowly and organically. First one hears water and bird song, then Davis adds in low didjeridoo tones, evoking a virtual call and response situation between humans and non-human nature. Thereafter Davis verbally comments on humans' relationship with nature and the importance of "belongingness." Yet two-thirds into the piece, Davis's words are electronically multiplied and all the sonic layers are in conflict with one another, which may suggest that humans "just miss the connection" (at 2:14) and overpower nature. The movement, however, concludes on a hopeful note: Davis urges humans to "look out for the earth," and the interactions of didjeridoo, bird, and water sounds evoke a sonic ecology that is not dominated by humans.

Sound Mirrors was premiered as an installation in October 2010 at the public Noosa Regional Gallery in Tewantin, Queensland, and shown for six weeks. The darkened space with stools, pieces of trees, mulch, and leaves suggested a riverbank environment. Visitors could influence the sonic environment through sensors embedded in the floor. Barclay offered guided tours and artist talks to enhance the visitors' experience of the installation. The installation was also presented in Sydney, Melbourne, Korea, and India. The live version has been performed in Australia, New Zealand, India, and Canada.

Soon afterwards, Barclay and Ros Bandt created *Rivers Talk* (2012), which is based on *Sound Mirrors* and Bandt's composition *Voicing the Murray* (1996). Other recent and similarly ambitious projects by Barclay include the *DAM(N) Project* (2011–13) and *River Listening* (2014–), interdisciplinary endeavours exploring aquatic bioacoustics; these have contributed to the expanding series of musical river projects by such composers as Ros Bandt, Eve Beglarian, Bernard Fort, Annea Lockwood, and Garth Paine. What distinguishes Barclay's recent works is their embeddedness in "Sonic Ecologies Frameworks," which has had a considerable impact on communities in precarious regions and on global audiences.

ECOLOGICAL PERFORMANCE—DAVID ROTHENBERG'S INTERSPECIES MUSIC

Besides composers, performers in music have offered examples of ecologically inspired creativity. David Rothenberg (b. 1962), an American philosopher and performer on clarinets, other reed instruments, and electronics with a background in jazz, is among a growing number of performers who are engaging in interspecies communication, sustainable musical practices (carbon footprint reduction and use of sustainably built instruments, etc.), disaster relief benefit activities, and green pedagogy.[30] Rothenberg explores the former, performing live with whales, birds, and insects as well as other non-human species that he considers serious musical partners.[31] As an early influence he credits environmental bandleader Paul Winter and his 1978 album *Common Ground* with its songs using the sounds of humpback whales, timber wolves, and African fish eagles. In 1975, Winter famously joined the Greenpeace V expedition protesting Soviet whalers. On a ship with two other musicians, Melville Gregory and Will Jackson, he played his saxophone to grey whales near Vancouver Island. Flautist Paul Horn and multi-instrumentalist Jim Nollman, and musicians from many traditional cultures (for instance, herding cultures), have preceded Rothenberg in such efforts.[32]

Grounding his interspecies projects in extensive historical and scientific research, collaborations with scientists, and musical experiments, Rothenberg challenges the perception that animals sonically engage only with members of their own species and that animal sounds are purely functional. He even posits that musical communication is more important in the non-human than in the human world and that animals have an aesthetic sense, which may have played a role in evolution. He also suggests that "natural selection is not really survival of the fittest, but survival of the interesting."[33]

The question whether animals have music and use it just like humans in functional and aesthetic ways, and as a pastime, has been asked often and in numerous contexts. Musicologists have considered it seriously since at least the 1980s and coined the term "zoomusicology."[34] French scholar-composer François-Bernard Mâche, among others, has proposed methodologies to analyze and assess animal songs and compare them with each other. The idea that music is not solely a human project has powerful ramifications. It may, for instance, expose the fact that human music often underscores human superiority and that human music tends to frequently dominate acoustic ecosystems. It also raises the question whether animals' musical abilities help make the case for extending personhood status to animals.

In early 2007, equipped with amplification and recording technology and a clarinet, Rothenberg set out to spend musical time with a male humpback whale off the coast of Maui, Hawaii.[35] Rothenberg was in a boat, the whale around 10 metres below the water surface. Humpback whales had been known for their extraordinarily sophisticated songs, which may last for twenty minutes, for interacting with one another musically for more than twenty-four hours at a time, and for changing their songs over the course of several years, but not for showing interest in sounds from other sources. Rothenberg now tried to show that his chosen cetacean musical partner would respond to his sounds and rapidly adjust his song to previously unheard musical materials.[36] He was cautious and persistent in his approach to the whales for weeks and was finally rewarded. Rothenberg and others felt that the result (about twelve minutes) of his sonic encounter with humpback whales could be called music.

The whale's melody has a large frequency scope from 100 to 600Hz, with a few very high sounds at around 4800Hz that are often articulated directly after a very low growl.[37] The high cetacean sounds are somewhat reminiscent of clarinet tones in the third octave, but their pitch often fluctuates. The clarinet has a smaller frequency scope, and Rothenberg normally prefers to sustain high pitches and play in a linear fashion. But he did not try to impose his performance style on the whale, following his fellow interspecies musician Jim Nollman's advice:

> Feel what it means to get on whale time. Don't try to communicate; remain humble to the fact that music—especially "beautiful music"—is a judgment call. That rare bird known as the interspecies musician learns to meet the animal halfway, two species willing to play in the same band, if but for a moment. It frolics with our basic conception of what it means to be both human and animal.[38]

Rothenberg played tones approximating the high warbling whale sonorities and "new whale wails."[39] It appears that the whale may have tried to assimilate Rothenberg's stable high pitch and sustained tones. The two musicians seemed to imitate each other and produce sonic dialogues, overlaps, and continuity.

The experience left a permanent mark on Rothenberg's playing, as he desired to "inhabit the rhythm and shape of the [whale] song" and to save the "music in nature … if only to be able to understand it before it is gone."[40] His elevation of animals as serious musical partners and teachers, from whom he wants to learn, is unusual in the context of Western music and reveals him as a deep ecologist, one who emphasizes the intrinsic value of all species and who questions human superiority. Rothenberg worked with Arne Naess, the father of deep ecology, and translated his writings.[41] Rothenberg's interspecies project, however, should not be considered perfect. His cetacean musical partner, whose ancestors emerged more than 50 million years ago and can look back at a much longer music tradition than humans, has remained nameless, anonymous. Rothenberg controlled the performance situation, the time, and the use of technology. He used the whale's intellectual and aesthetic property without the whale's permission, and his sonic encounter with the whale went against the US Marine Mammal Protection Act of 1972 (which classifies the generation of amplified underwater sounds as harassment of marine wildlife). Furthermore, the human–non-human musical interactions may have just been his but not the whale's perception. Nonetheless, Rothenberg is building on the musical practices of shepherds, the Kulali tribe and Tuva people among others. He also recalls the success of scientists and artists who through the popularization of cetacean songs helped save the humpback whale from extinction in the 1970s, and he continues to raise awareness of the musical worlds of this and other non-human species.

ECOLOGICAL LISTENING—HILDEGARD WESTERKAMP'S SOUNDWALKS

Since the 1970s, Western musicians have explored new engaged and open-minded forms of auditory perception to establish a close relationship between the listeners' inner selves and their environment. Such ecologically inspired listening grew out of the Deep Listening practice developed by American composer and accordionist Pauline Oliveros, as well as the studies in acoustic ecology (or eco-acoustics) pioneered by such Canadian composers as R. Murray Schafer, Barry Truax, and Hildegard Westerkamp.[42] Deep Listening practitioners dismiss passive hearing and experience the passage of time through focal and global attention to sound in order to develop greater body and environmental awareness. Acoustic ecologists critically engage with the sounds of natural and built places, noise pollution, change in our sonic environment, and endangered sounds, and make environmental recordings for study or for use in compositional frameworks, which are often called soundscape compositions.[43] Even though environmental sound captured on tape and presented in different contexts and spaces suggests an ecological anomaly (because the sounds are divorced from their place of origin), it can help raise awareness of our complex and often sonically disengaged relationship to environmental sound and place.

Based in Vancouver, Hildegard Westerkamp (b. 1946) is a composer, performer, and acoustic ecologist who has extensively explored environmental sound and soundscape composition since the 1970s.

Some of her works are sonic portraits of Vancouver locales (*Harbour Symphony*, 1988; *Kits Beach Soundwalk*); others feature the sounds of natural places such as old-growth forests in British Columbia (*Beneath the Forest Floor*, 1992). Westerkamp was one of acoustic ecologist R. Murray Schafer's research associates working for his World Soundscape Project at Simon Fraser University and at the same time involved in the Noise Abatement Project of the Society Promoting Environmental Conservation in Vancouver (1974–75). But above all, she pioneered soundwalking, which she defined as "any excursion whose main purpose is listening to the environment. It is exposing our ears to every sound around us no matter where we are. We may be at home, we may be walking across a downtown street, through a park, along the beach ... Wherever we go we will give our ears priority. They have been neglected by us for a long time."[44] Her practice of soundwalking came out of the World Soundscape Project.[45] Facilitating a deep engagement with one's sonic environment, soundwalks can be performed by both musicians and non-musicians, alone or in groups, barefoot or with shoes, with or without guides, blindfolds, and recording devices. Soundwalks can be based on Westerkamp's verbal instructions or function as orientation in unknown surroundings, as dialogue with a sonic environment, as preparation for or part of a composition, or as composition. Westerkamp urges soundwalkers to pay attention to their own bodily sounds, other human-made and non-human sounds, the sounds' individual characteristics (provenance, volume, frequency, timbre), and their rhythms, situation (foreground and background), and interrelationships within a soundscape. She also asks soundwalkers to critically reflect on the quality of their sonic environments, quality that is often affected by intrusive traffic and machine and mass-communicated sounds.

Westerkamp's *Kits Beach Soundwalk* (1989), an approximately ten-minute work for spoken voice and two-channel tape, which developed from the Vancouver Co-operative radio program "Soundwalking," integrates soundwalking and composition. It is also a time capsule offering a glimpse into the sounds of a place almost two decades ago. *Kits Beach* features Westerkamp's listening to and commenting on the sounds of Vancouver's Kitsilano Beach and reflecting on her role as a soundscape composer and acoustic ecologist. The piece has three layers: subtle nature sounds, urban noise, and Westerkamp's voice-over style commentary. *Kits Beach* opens with a combination of nature and city sounds and Westerkamp's description of her locale: "It's a calm morning. I'm on Kits Beach in Vancouver. It's slightly overcast and very mild for January. It's absolutely wind-still. The ocean is flat, just a bit rippled in places" (00:00–00:42'). Next she focuses on the barely audible sounds of barnacles: "I am standing among some large rocks full of barnacles and seaweed ... The barnacles put out their fingers to feed on the water. The tiny clicking sounds that you hear are the meeting of the water and the barnacles. It trickles and clicks and sucks" (00:53–1:22').[46] Westerkamp then critically exposes how the delicate barnacle sonorities compete with downtown Vancouver's drones, blares, and squawks: "The city is roaring around these tiny sounds" (1:28–1:32'). Later she suggests that the urban roar largely prevents fragile voices from being heard—including a human listener's own breathing, footsteps, and speaking voice. While she as a composer could correct this perceived sonic imbalance captured on tape through such electronic processing means as bandpass filters and equalizers, for many human real-time listeners it may take "too much effort to filter the city out" (2:51–2:55') so that they can perceive the tiniest voices of their surroundings. Yet exactly these sonorities might lead human listeners into their inner worlds and enrich their imagination. About halfway into the piece, the city roar gives way to natural sounds and music by Mozart and Greek experimental composer Iannis Xenakis. The barnacles' sounds seem to have guided Westerkamp into a sonic dream world. The piece concludes with Westerkamp suggesting that rather than entirely blocking out the complex sounds of the environment, listeners should "play with the monster" in order to "face the monster" (9:08–9:14').[47]

In *Kits Beach*, Westerkamp performs and composes ecologically engaged listening in the form of a journey of sonic (self) discovery not only to stimulate the imagination of other listeners but also to change their listening habits and to nudge them towards a critical and activist attitude towards their acoustic environments. In recent years, soundwalks have grown into a significant genre in experimental music and sound art. Outside the musical arenas, they have become an ever more popular touristic activity.

CONCLUSION

All four examples discussed reflect how composers, performers, and listeners have acted as agents of change in ecologically inspired music practices. Ferrari's *Allô, ici la terre* addresses ecological problems through the use of ecocritical texts and sounds and the juxtaposition of different musical temporalities. Despite the work's limited impact on audiences, it is one of the most ambitious early environmentally activist works in classical music. Barclay's *Sound Mirrors* draws attention to ecologically pristine but threatened rivers by making field recordings—time capsules of river sounds—the basis of the work. In its various forms—sound installation, live performance, and audio recording—this work advances environmental awareness on local and global scales. With his interspecies performances, Rothenberg explores whale time and displays a deep ecological outlook, suggesting that human and non-human animals can share the same musical stage, and effectively challenging conventional musical and scientific beliefs. Finally, through her soundwalks, Westerkamp draws our attention to often human-dominated acoustic environments, critiques noise pollution and passive entertainment, and demonstrates that active listening is agentive. All four artists demonstrate that they are, in Bill McKibben's words, "antibodies of the cultural bloodstream, sensing trouble and rallying to isolate and expose and defeat it to bring to bear the human power for love and beauty and meaning against the worst results of carelessness, greed and stupidity."[48] Ferrari, Barclay, Rothenberg, and Westerkamp have listened to our planet's many voices, sensed that we and other living beings may run out of time, and emphasized through their time-based art the need for greater environmental consciousness and stewardship to sustain both human and non-human life on our planet.

Notes

Thanks to Leah Barclay, Peter Michael Hamel, Brunhild Meyer-Ferrari, and Gisela Gronemeyer for providing invaluable information and materials for this essay.

1 Canadian pianist Malcolm Troup used "ecomusicology" as early as 1972 and may have coined this word. Maria Anna Harley was among the first to contemplate eco-musicology as a field of inquiry that considers in a holistic way the interrelationships of human and non-human sounds and studies a wide variety of music cultures in relation to their environments. Harley (1996).

2 See for example Pedelty (2016).

3   Sartre (1948), 13–44.

4   There are, however, some environmentally conscious pieces predating Harrison's work: Percy Grainger's 1899 setting of "The Beaches of Lukannon" from Rudyard Kipling's *Jungle Book* deplores the slaughter of seals in Alaska's Pribilof Islands; Charles Ives's 1912 song "The New River," based on his own text, condemns noise pollution; Maurice Ravel's 1925 opera *L'Enfant et les sortilèges* (libretto by Colette) suggests that the maltreatment of nature has dire consequences; and Hanns Eisler's setting of a poem by Bertolt Brecht. "Vom Sprengen des Gartens." (ca. 1943), part of the *Hollywood Songbook*, critiques the wasteful watering of backyards in Los Angeles.

5   Harding (1992), 291–315.

6   Brent Keogh (2013) provides a concise survey of the different uses of the term "ecology" in musical contexts preceding a discussion of ecology as a metaphor for addressing the sustainability of music traditions.

7   Lefebvre (2004), xv.

8   The term "musique concrète" was coined in the 1940s in France and refers to music that avoids the use of "abstract" notation for performance; it denotes the use of "concrete" sound found in the sonic environment, captured on magnetic tape and "concretely" processed in a studio. This type of music was pioneered by Schaeffer and Henry in the 1940s.

9   Caux, ed. (2012), 148.

10 Caux, ed. (2012). For more information on Ferrari and politics, see Drott (2009), 145–66.

11 Ferrari called this approach "musique anecdotique" (anecdotal music).

12 Born in 1942 in Buenos Aires, Roberto Détrée is a guitarist and composer who settled in Germany in 1965. With the composers Peter Michael Hamel and Ulrich Strunz and the oboist Robert Iliscu, he formed the group "Between," dedicated to eclectic mixes of classical, contemporary, medieval, pop, folklore, and non-Western music. The group performed Ferrari's *Allô*.

13 The title can be translated as "Visage VI or The Face from Which the Masque Is Removed." Between 1955 and 1959 Ferrari worked on a series of five works for various instruments and for tape titled "Visage."

14 Ferrari (1975), 23.

15 Ferrari (1975), 23.

16 Bertolt Brecht and his associates pioneered the *Lehrstück*, the learning or teaching play, as an experimental type of theatre that centres on learning and discovery.

17 Ferrari wrote: "La femme est ici un symbole. Elle signifie que l'homme serait trop usé par les guerres, par le pouvoir et la spéculation, pourri par trop de compromission et par la jouissance du profit trop facile. Elle signifie, en un mot, que le rôle historique de l'homme serait fini. Tandis que la femme pourrait, par sa nouveauté et sa fraîcheur, représenter le future" (The woman is a symbol here. She indicates that the man is worn through wars, through power and speculation and spoiled by too much compromise and the enjoyment of easy profit. She indicates that, in short, the historical role of the man has ended. Meanwhile through her newness and freshness, the woman could represent the future). Ferrari (1975), 23. It is perhaps no coincidence that Françoise d'Eaubonne (1974) introduced the concept of ecofeminism in her book *Féminisme ou la mort*.

18 Ferrari (1975), 23.

19 Other experts whose texts he used include René Dumont, Ivan Illich, Sicco Mansholt, Robert Lattès, William C. Paddock, Egbert W. Pfeiffer, Helma Sanders-Brahms, and Claude-Marie Vadrot.

20 Peter Michael Hamel to the author, email, 18 July 2012.

21 Brunhild Meyer-Ferrari to the author, email, 23 May 2012.

22 Pincetl (1993).

23 Leah Barclay, "Shifting Paradigms: Artists as Agents of Change," paper given at Balance-Unbalance 2011, Concordia University, Montreal, 4–5 November 2011.

24 Leah Barclay to the author, email, 10 August 2012.

25 See Barclay's soundmap *Sonic Explorers*: http://leahbarclay.com/portfolio_page/sonic-explorers/ (accessed 6 July 2013).

26 Leah Barclay, "Liner Notes" for her CD *Transient Landscapes: Select Works from the Installation Sound Mirrors* (Brisbane: Leah Barclay, 2010).

27 The musicians included Lyndon Davis on didjeridoo, Anthony Garcia on guitar, and Richard Haynes on bass clarinet (Australia); Subhash Kumar on mridangam, Santosh Kumar, vocals, Sreenath T.S. on ghatam, and Sreerja on violin (India); Hyelim Kim on taegum, Yoonsang Choi on janggu, and Jung Bong Park on pansori (Korea); and William Lane on viola (China). Barclay performed on moorsing, bawu, hulusi, and percussion.

28 For the mixing, Barclay used Pro Tools, and for part of the editing and processing, Logic Pro. The main softwares for the sound processing were Metasynth, Audiosculpt, Audiomulch, and Max/MSP. She primarily applied equalization, pitch shift, time compression, and expansion as well as reverberation to her materials.

29 Barclay, "Liner Notes."

30 Other notable performers concerned with environmental issues include pianist Soyeon Lee, flautist Michael Pestel, and oboist Brenda Schuman-Post. Such organizations as the New York City-based Ear to the Earth, founded in 2006, facilitate performances of ecologically conscious music. For more information see Sutherland, "Animals in the Mix."

31 Rothenberg's recent recordings include *Whale* Music, TN 0804 (Newark: Terra Nova, 2008), and *Bug Music*, Gruen 122 and TN 1309 (Frankfurt: Gruenrekorder and Newark: Terra Nova, 2013). Excerpts of his performances can also be found on YouTube.

32 Doolittle (2008), 2.

33 Rothenberg (2007). Rothenberg wrote numerous books and articles about his experiences, findings, and ideas about animal music. See for instance, *Why Birds Sing* (2005), *Thousand Mile Song* (2008), and *Bug Music* (2013).

34 See for instance Mâche (1983), Martinelli (2008), and Keller (2012).

35 Rothenberg used an AudioTechnica AT822 microphone with a Sony TCD-7 DAT recorder and a modified Roland Canister amplifier connected to an EsunPride JH001 underwater speaker, two Cetacean Research SQ26-08 hydrophones connected to a Sony MZ-M10 HI-MD Minidisc Recorder, and headphones. Rothenberg (2008b).

36 Payne and McVay (1971; 1979).

37 Rothenberg (2008b), 50.

38 Rothenberg (2007).

39 Rothenberg (2008b), 52.

40 Rothenberg (2008b), 52.

41 Naess (1989).

42 For a concise introduction to acoustic ecology see Wrightson (1999).

43 Author of the 1977 book *The Tuning of the World*, R. Murray Schafer takes credit for coining the term "soundscape."

44 Westerkamp (2007), 49. The first version of this article was first published in *Sound Heritage* 3, no. 2 (1974). See also McCartney (1998).
45 American composer-percussionist Max Neuhaus has conducted "listening walks" in the United States and Canada since the mid-1960s.
46 Westerkamp, *Kits Beach Soundwalk,* https://www.youtube.com/watch?v=hg96n U6ltLk (accessed 10 July 2013).
47 For a detailed examination of this piece, see Kolber (2002).
48 McKibben (2009).

References

Caux, Jacqueline, ed. 2012. *Almost Nothing with Luc Ferrari: Interviews with Texts and Imaginary Autobiographies by Luc Ferrari.* Los Angeles: Errant Bodies.

Doolittle, Emily. 2008. "Crickets in the Concert Hall: A History of Animals in Western Music." *Transcultural Music Review* 12: n.p.

Drott, Eric. 2009. "The Politics of *Presque rien.*" In *Sound Commitments: Avant-Garde Music and Sixties,* ed. Robert Adlington, 145–66. New York: Oxford University Press.

d'Eaubonne, Françoise. 1974. *Féminisme ou la mort.* Paris: Horay.

Ferrari, Luc. 1975. "Allô, ici la terre." *Art Vivant* 54: 23.

Harding, Walter. 1992. "A Bibliography of Thoreau in Music." In *Studies in the American Renaissance,* edited by Joel A. Myerson, 291–315. Charlottesville: University of Virginia Press.

Harley, Maria Anna. 1996. "Notes on Music Ecology as a New Research Paradigm." http://ecoear.proscenia.net/wfaelibrary/library/articles/harly_paradigm .pdf. Accessed 2 July 2015.

Keller, Marcello Sorce. 2012. "Zoomusicology and Ethnomusicology: A Marriage to Celebrate in Heaven." *Yearbook for Traditional Music* 44: 166–83.

Keogh, Brent. 2013. "On the Limitations of Music Ecology." *Journal of Music Research Online* 20: 1–10.

Kolber, David. 2002. "Hildegard Westerkamp's *Kits Beach Soundwalk*: Shifting Perspectives in Real World Music." *Organised Sound* 7, no. 1: 41–43.

Lefebvre, Henri. 2004. *Rhythmanalysis: Space, Time and Everyday Life,* edited by Stuart Elden, translated by Stuart Elden and Gerald Moore. London: Continuum.

Mâche, François-Bernard. 1983. *Musique, mythe, nature ou les dauphin d'Arion.* Paris: Méridiens Klincksieck.

Martinelli, Dario. 2008. "Introduction to Zoomusicology." *Transcultural Music Review* 12: n.p.

McCartney, Andra. 1998. "Soundwalk in the Park with Hildegard Westerkamp." *Musicworks* 72: 6–15.

McKibben, Bill. 2009. "Art in a Changing Climate: Four Years after My Pleading Essay, Climate Art Is Hot." *Grist Magazine,* 5 August. http://grist.org/ article/2009-08-05-essay-climate-art-update-bill-mckibben. Accessed 10 July 2013.

Naess, Arne. 1989. *Ecology, Community, and Lifestyle: Outline of an Ecosophy*, translated and revised by David Rothenberg. New York: Cambridge University Press.

Payne, Roger, and Scott McVay. 1971. "Songs of Humpback Whales." *Science* 173: 585–97.

———. 1979. "Humpbacks: Their Mysterious Songs." *National Geographic* 155: 18–25.

Pedelty, Mark. 2016. *A Song to Save the Salish Sea: Musical Performance as Environmental Activism*. Bloomington: Indiana University Press.

Pincetl, Stephanie. 1993. "Some Origins of French Environmentalism: An Exploration." *Forest and Conservation History* 37: 80–89.

Rothenberg, David. 2005. *Why Birds Sing: A Journey through the Mystery of Bird Song*. New York: Basic Books.

———. 2007. "Interspecies Music: A Guest Essay." *Interspecies Newsletter*. http://www.interspecies.com/pages/rothenberg%20essay.html. Accessed 7 July 2013.

———. 2008a. *Thousand Mile Song: Whale Music in a Sea of Sound*. New York: Basic Books.

———. 2008b. "Whale Music: Anatomy of an Interspecies Duet." *Leonardo Music Journal* 18: 47–55.

———. 2013. *Bug Music: How Insects Gave Us Rhythm and Noise*. New York: St. Martin's Press.

Sartre, Jean-Paul. 1948. *Qu'est-ce que la littérature?* Paris: Editions Gallimard.

Schafer, R. Murray, 1977. *The Tuning of the World*. New York: A.A. Knopf.

Sutherland, Richard. 2014. "Animals in the Mix: Interspecies Music and Recording." *Social Alternatives* 33: 23–29.

Westerkamp, Hildegard. 2007. "Soundwalking." In *Autumn Leaves: Sound and the Environment in Artistic Practice*, edited by Angus Carlyle, 49–54. Paris: Double Entendre.

Wrightson, Kendall. 1999. "An Introduction to Acoustic Ecology." *Journal of Electroacoustic Music* 12: 11–15.

CHAPTER 4

The Environmental Vampire: Terror, Time, and Territory after 9/11

Robert Boschman

> Maybe the proper response to standing on the side of a planet,
> in the open air of its atmosphere, very near to the local star, is
> always terror. Maybe everything humans ever did or planned
> to do was designed to dodge that terror.
> —*Kim Stanley Robinson (2015)*

> To eat the very world.
> —*Justin Cronin (2010)*

Following Nina Auerbach's 1995 statement that every recent West-ern time period has its own vampire, I propose that after 9/11 a new figure, the Environmental Vampire, emerging from bio-political trends and fears concerning terror and the biosphere in the new century, rapidly begins to establish itself in novels, graphic novels, film, and television. The Environmental Vampire constitutes a new agent in the lineage of vampires in vampire narratives. Since at least 1897,[1] the publication year for Bram Stoker's *Dracula*, mainstream readers and audiences have encountered nature/culture narratives involving a nocturnal hominid who transcends modern human temporal con-straints by seductively feeding on and co-opting the diurnal human body. But after 9/11, the intimacy of and with the vampiric Other becomes depersonalized, deterritorialized, and concerned very much with speciation and the question of the post-human future. It is the new vampiric narrative delineating a radical subject shift on a pan-demic scale, a shift that recent theorists have broadly articulated in

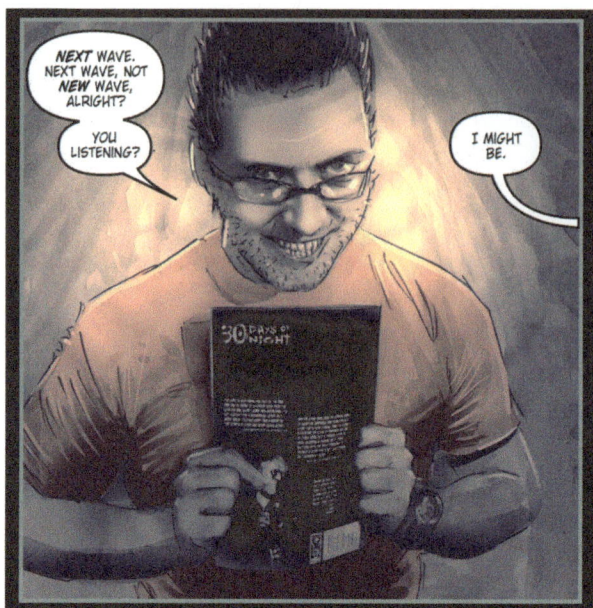

Fig. 4.1

terms of the "ex-Man" (Massumi 1998, 60) and the deconstruction of "species supremacy" (Braidotti 2013, 65).

While it's true that in prose and cinema throughout the twentieth century certain bio-cultural aspects of the vampire figure are present and detectable, the Environmental Vampire grows from and supplants the earlier traditional character. While it's also true that the earlier figure continues to be readily found in popular twenty-first-century texts such as *True Blood* and *Twilight*, the Environmental Vampire represents a noticeable turn towards enhanced agency: what *30 Days of Night* creator Steve Niles calls the "next wave" in vampire story-telling (Niles, DeConnick, and Randall 2007, 21).[2]

Conventionally, the vampire has evoked horror concerning the hominid Other as characterized by its ability to exist in a nocturnal state and to change form continually; but the Environmental Vampire is an even far more disruptive and agentive figure, one that represents pressures on the boundaries of the human in new ways (Niles and Templesmith 2002, 353).[3]

Unlike even the most terrible of twentieth-century vampires, the Environmental Vampire doesn't necessarily care for Western fashions in dress and appearance (in fact, it may not be dressed at all); doesn't

Fig. 4.2

always possess speech (or at least may not communicate in human ways); is not a seducer (indeed, may be asexual, with no recognizable gonads to speak of); and, perhaps most telling, is often depicted with no whites to its eyes (and therefore cannot be engaged by gaze tracking [Flesch 2009, 17]). The Environmental Vampire may indeed not even feature fangs, deploying instead hose-like stinging appendages that emerge from its thorax in order to gulp and appropriate the human subject, with its agency, biology, and energy systems (Del Toro and Hogan 2010, 21). If it does bear the expected teeth, these are over-the-top, exaggerated like the bombastic choppers of some fish found to exist in oceanic trenches three miles deep (Niles and Templesmith 2011, 28; Cronin 2010, 79). Significantly, the Environmental Vampire appears as a collective in the immediate aftermath of 9/11 and/or nests beneath Ground Zero (Del Toro and Hogan 2010, 55–56), or makes reference to that date (Cronin 2010, 42–43, 84; Del Toro and Hogan 2010, 176). Ecologically active and purposeful, the Environmental Vampire kicks over oil pipelines in the Far North and sets them on fire; attacks scientists studying global climate change; and, just for fun and because it can, destroys polar bears with ease (Niles and Temple-smith 2011, 56; Lebbon 2007, 147). In twenty-first-century vampire narratives, the Environmental Vampire evokes the deconstruction of the Westernized human place in nature and reflects the new century and its concerns and anxieties, including its technological advances, its politics, and its terrors.

As this chapter focuses largely, though certainly not exclusively, on titles in the *30 Days of Night* graphic novel series created and overseen by Steve Niles and Ben Templesmith, it is worth noting how the basic narrative features have proliferated through filmic and nov-elization spinoffs since the inaugural title was published in 2002.[4] A 2007 feature film, starring Josh Hartnett as the male protagonist Eben Olemaun, Inuit Sheriff of Barrow, Alaska, was directed by graphic art-ist David Slade from a screenplay by Steve Niles. The film novelization that followed was written by the English horror writer Tim Lebbon. These properties also take the title *30 Days of Night*, and while they follow the storyline of the original 2002 graphic text,[5] they also deviate from and vary certain details and events given in the first rendering, even though Niles and Templesmith, together with IDW Publishing, own and/or control the overall property. The prolific energies at work here delineate, along multiple and multi-dimensional narrative lines, a reinvigorated trope found in the vampire figure regarding terror, time, and territory, as well as—and no less significantly—agency, cli-mate change politics, and energy. Indeed, an important text for my

Fig. 4.3

purposes here is the aforementioned *30 Days of Night: Night, Again* (published almost a decade after the inaugural graphic novel) which focuses directly on agency and climate change (Lansdale and Kieth 2011).

In the *30 Days of Night* universe, the Environmental Vampire rapidly infects textual territory. This pandemic drive is also intra-textual, drilling down into volumes to infect and transform text ecologically. The female protagonist Stella Olemaun is depicted in many texts in the overall series of twelve graphic novels, six novels, and two films, as herself authoring a controversial and dangerous new book titled *30 Days of Night*, as though to stress the new import of terror and territory acting within the territory of text itself.

Stella's Cassandrian message is a source of chronic tension within the series, as she publicly discloses the vampiric invasion of Barrow, Alaska, in a bestselling book that her publisher categorizes as "fiction" without prior consultation (Niles and Templesmith 2011, 118; Niles, DeConnick, and Randall 2007, 25). The new reality is thus constantly contested and denied by the characters themselves. This in turn raises the persistent question of what it is the *30 Days of Night uber*text (not to mention the novels, films, and graphic novels authored by Cronin and Del Toro and Hogan) might be doing in the post-9/11 era. As one *30 Days* character puts it:

The whole world had turned topsy-turvy these last few years, as if the new millennium had been some kind of insanity trigger, and pulling it had changed all the rules. People flew airplanes into buildings and started wars seemingly at random and mailed anthrax and set off bombs in subways and blew themselves up; and skilled politicians self-destructed in new and spectacular ways and everyone worshipped celebrities like gods only to turn on them like jackals at the slightest hint of weakness. (Niles and Mariotte 2007, 239)

Certainly all these texts, through their depictions of the Environmental Vampire, evoke Hannah Arendt's statement that "[w]hat common sense and 'normal people' refuse to believe is that everything is possible" (1951, 138-39). Environmental Vampire texts portray startling shifts in agency that embrace alterity and embody "becoming-animal" (Braidotti 67) even while "normal" characters in these same texts embrace disbelief and experience terror.[6]

Terror, to be clear, is distinct from horror, even though the two exist together in a traditional complex found in a number of genres in literature and film, including horror, gothic, and noir. Terror calls upon and anticipates power deployed by unknown sources. As John Bowen remarks in a discussion of the nineteenth-century Gothic novelist Anne Radcliffe, who is credited with first making the distinction between terror and horror, "[Terror] does not show horrific things explicitly but only suggests them." Terror involves multiple possible focal points in time regarding what may happen and is bound up inextricably with anxiety. It looks with dread felt towards a possible future based on a past horror and is also vulnerable to the Cassandra complex, to the anxieties and turmoil of not being believed, as is the case with the everywoman Stella Olemaun in the *30 Days of Night* universe.[7] Donald Rumsfeld, the Secretary of Defense under President George W. Bush in the aftermath of 9/11, famously evoked temporality in destabilizing terms in relation to terror when he used the term "unknown known" (Morris 2014). Rumsfeld was stoking xenophobia on behalf of the state because "[t]o control a territory is to exercise terror; to challenge territorial extent is to exercise terror" (Elden 2010, xxx). This relation between territory and terror has constituted the chronic global problem of the state since the immediate aftermath of

9/11, when American military interventions such as Operation Global Reach, according to Lawrence Wright, "exposed ... the futility of military power" in its "surgical and proportional response to the terrorist acts" (2006, 5312). If we momentarily follow Rumsfeld, who in 2002–3 was putatively following and circling Al-Qaeda (albeit in Iraq), horror is the known known, one that is already an actuality to be re-experienced in hindsight, replayed perpetually through agentive vehicles like YouTube, and thereby recalled at White House briefings and in the historic present.[8] This aspect of the Environmental Vampire, its relation to the human experience of time, which emerges in a new way in the years immediately following 9/11, is key to understanding this graphic figure's link to global climate change and its erupting temporal urgencies and evocation of anxiety and disbelief.

In literature and cinema, then, terror—in its relation to narrative and temporality—is experienced as a kind of anxious pleasure building towards, and then away from, horror. The latter constitutes what is at last an experienced event, one that is documented and repeated indefinitely. Horror is a fulfillment, the unfolding of pent-up anticipation in the experience of looking that is thereafter acutely remembered.[9] I can think of many people who cannot bring themselves to look at the repugnant images of this new figure, the Environmental Vampire, either in film or in the graphic novel (the latter of which makes for an important place for depicting it after 9/11). Viewers' eyes instinctively turn away or close, especially given the specific bodily features of the vampire encountered in the new century. The Environmental Vampire is a graphically figured bio-consequence of 9/11, with features that largely exclude the eroticized, seductive, conflicted, and well-coiffed bourgeois figures found in texts such as *Twilight* and *True Blood*, these latter having intertextual origins unambiguously located in the nineteenth and twentieth centuries (Weinstock 2012, 278). Written and filmic texts such as *Twilight* only mildly evoke the inarticulate realization that diurnal human life is contingent and vulnerable, and hardly, if at all, incite the emotional and deeply held complex of dread and horror. At its most effective, the twentieth-century filmic vampire evoked the hominid predator rising up through night and within the space of a then-still-powerful religious institution. But by the end of the century, this figure—within its religious space—became hackneyed. Count Yorga certainly couldn't cut it, and Anne Rice's vampires signalled

Fig. 4.4

decadence, especially in their vaunted desire to participate in the Foucauldian author-function.[10] Nina Auerbach notably concluded her study of the vampire in twentieth-century literature and film by asserting that *fin-de-siècle* vampires were tired, all pooped and bled out, and had lost their initiative (1995, 192).

Yet only a few years after this well-timed comment, in 2002, the vampires of Steve Niles and Ben Templesmith, who together created *30 Days of Night,* erupted into graphic novel form on the dark, frozen horizon of Barrow, Alaska, to do something readers had never seen before: seize entire swaths of territory and time, emptying both of their former and expected standards of measurement.

That such a turn occurs at all is, I propose, a specific response to and reflection of an event as catastrophic as 9/11. And with this, we need only think of the many texts published since late 2001 that try to come to terms with the events of 11 September of that year, as well as the events that followed: from Lawrence Wright's Pulitzer Prize–winning *The Looming Tower* (even its title radiates terror and reinvocation) to Laura Poitras's documentary about Edward Snowden, *Citizenfour* (which follows her discovery of Snowden as he reveals his identity to the world before encapsulation in a non-state territory within the Moscow International Airport). The Environmental Vampire, as a narrative trope in the years following 9/11, invokes the sudden dread of the sovereign state and its citizens regarding "non-state actors," from Bin Laden to Snowden to the caravan of Central American migrants on the US southern border in the age of Donald Trump (Elden 2010, xvi). In the works of Niles and Templesmith, Cronin, and Del Toro and Hogan, this figure abruptly appears en masse, a post-anthropocentric cohort bringing terror-horror amidst large-scale environmental catastrophe. The graphic frames of *30 Days of Night* that depict the post-human moment of invasion are occasionally but significantly, as in Figure 4.4, portrayed from an aerial

Fig. 4.5

view that relativizes the heretofore normative human perspective. Catastrophe, to be sure, constitutes a watershed moment involving both an event and the insight regarding that event. It is temporal and linked first to horror (the actual in time) and then terror (anticipatory dread). Catastrophe, as social scientists submit, is distinct from and greater in magnitude than disaster (Quarantelli 2006). As K. Joanne McGlown states, catastrophe "[a]ffects an entire nation and/or parts of the world; requires extensive resource assistance"; and causes "[l]ong-term disruption to the social order, security, or psyche of a nation or its peoples" (2011, 7).[11] It can unfold very quickly (i.e., 9/11); or it can seem to crawl (i.e., global climate change), while it gains inertia and becomes real before our eyes. Catastrophe pans across the local and the individual, and in Environmental Vampire narratives it precludes the appearance of a "monster" who is also a "Count": who is at once chthonic and aristocratic, barnyard and classist. Catastrophe, rather, opens the drawbridge to monster hordes and plebeian anarchy. As the environmental crisis draws out in our imagined and actual worlds in a cascade serially connected to the anticipated next event, it invokes dread and invites further terror as well as terrorism.[12] This is chronic and exhausting. Each moment is a crux that invokes and involves terror and the newly realized hard-to-see Others in linear time, the latter fundamental to Western cultures as seen in terms like *progress* and *development* and *next*.

Operating within this anxious awareness, the Environmental Vampire acts as an agent of a seemingly alternate reality empowered by forces (in part, planetary ones) that are well outside the control of "Man," the Law, the State. The new figure, in fact, deterritorializes sovereign space and expands the nocturnal frame by locating its group activities in the Arctic as the planet tilts away from the sun. It is a darkened and melting landscape. In the post-9/11 period, it isn't the Christian church against which vampires move in our dreams both waking and asleep, arousing audiences with sex-and-death kisses à la Gary Oldman and Winona Ryder (Coppola 1992). Instead, the Environmental Vampire appears in the rapidly shifting cultural nexus built on technology and social media—the latter by definition a group act. We can Google our small blue planet and watch it all unfold. We're predators and victims at the same time, residing on the cusp of the post-human, which Rosie Braidotti calls "becoming-animal," with its "cascade effect that opens up unexpected perspectives." Thus,

> if the crisis of Humanism inaugurates the posthuman by empowering the sexualized and racialized human "others" to emancipate themselves from the dialectics of master-slave relations, the crisis of anthropos relinquishes the demonic forces of the naturalized others. Animals, insects, plants and the environment, in fact the planet and the cosmos as a whole, are called into play. This places a different burden of responsibility on our species, which is the primary cause for the mess. (Braidotti 66)

In this context, the Environmental Vampire appears in its new configuration, precipitated by 9/11 and bursting from its former anthropic boundaries.

In the period since 9/11, Environmental Vampire narratives have proliferated in prose, the graphic novel, film, and television, and many of these specifically mark that date in their texts. Justin Cronin, Tim Lebbon, Jeff Marriotte, and the writing teams of Guillermo Del Toro and Chuck Hogan, as well as Steve Niles and Ben Templesmith, have alluded to 9/11 and/or (bio)terrorism in their stories. In these texts, grossly transformed former humans reveal themselves in aggregate and almost fully to normative public view, like gorgeously enfanged spiders emerging from the parental sac. They turn on and threaten the established order, having previously been hidden from civilized eyes. Political historian Stuart Elden calls the 9/11 date a caesura that emphasizes before and after as well as "American pain" (2010, xvi), and this, I propose, is reflected in the Environmental Vampire.[13] Elden understands the complex term "terror" as one that includes "the terrorism of nonstate actors," which constitutes a small actual proportion of acts of terror but at the same time grabs hold most insistently on the public and state imaginations. The emergence of ISIS in the second decade of the twenty-first century has further complicated analyses of terror, since this organization has employed the word "state" in its name for itself even while attempting to swallow territory in terms of the caliphate. Terror, in its currency as a political noun, is "the use of organized repression or extreme intimidation" (OED). And in this sense terror, we now well know, is profoundly environmental. One recalls the catastrophic ruin of New York City on 11 September 2001, six weeks after which the Patriot Act was passed by the U.S. Congress and, with that, the fact that "three new agencies were given the power to stamp documents 'Secret'—the Environmental Protection Agency, the Department of Agriculture, and the Department of Health and Human Services" (Sassen 2008, 182). We've learned that bioterrorism can be understood as the deployment of organic material through the mail system, which occurred in the immediate aftermath of 9/11; that it can be airborne or borne through blood, human and otherwise; and that our everyday activities in a globalized petro-culture (dependent on planetary "blood") have created, since the eighteenth century, a carbon crisis that involves time and that implicates and undermines established concepts of territory and sovereignty.

꒦

At the core, then, of conventional vampire figuring lies the fact that the monstrously agentive Other is inherently hard to see, a me-not-me, a fabulation of the animal and the human (MacCormack 2012, 304). As readers or viewers encountering the vampire, we both embrace and reject this complicated, entangling, and sometimes dated figure.[14] The conventional vampire is intimate and wants eye contact; we may look and then turn away, gaze but then leave the text for a moment. Though traditionally the vampire is depicted as being like us, a clothed biped, it is awake when we sleep, asleep when we wake. Yet while it sleeps in the earth itself, even the traditional vampire of the twentieth century, with its faded and tired iterations, evokes (to use Braidotti's phrase) "the crisis of anthropos" (66) and our role in that crisis. At the same time, while the vampire occupies the other side of diurnal experience so that its code—perhaps like those of the terrorist and psychopath—is alien, we find ourselves able to identify partly with its ethic, even as we imagine and perhaps partly desire its nocturnal existence. In other words, the vampire narrative traditionally has offered us vicarious experiences (even comic ones)[15] of a fictive creature who represents both what we are (bipedal, big-brained hominids evolved to eat many things, rest at night, and in nature live no more than a few decades) and what we are not (shape-shifting sometimes hominids who drink blood, sleep in dirt during the day, and transcend death). The Environmental Vampire represents erasure to the human subject-position (to our default sense of agency) and comes to consume us, not only in our evolved bodily selves but also in everything erected and maintained within our fragile yet persistent acculturated identities.

Freud, in *Totem and Taboo*, remarks (citing Kleinpaul) that "originally … all of the dead were vampires, all of them had a grudge against the living and sought to injure them and rob them of their lives" (1913, 69). My point in quoting Freud here is to indicate how the figuring of the vampire is produced and contextual, tied first to the human cultures and histories in which it emerges, and how these relate to non-human nature, human mortality, and political constructs. One scholar of the monstrous points out that vampires are "fabulation animal-monsters … [that] cannot be co-opted as they exist only as demands for othering. We can never 'know' that which does not exist, but, like all art and fiction, it does not mean that our ideologies, paradigmatic tendencies, and responses are not affected by experiences of these entities" (MacCormack 2012, 304–5). I suggest that the relationship is more complicated: that historical and bio-cultural shifts do indeed co-opt and appear through the vampire; that major disruptions inform and shape "these entities" (not just the other way around); and that these reflect a fear that biologist Lyall Watson defines in terms of "the risks of [actually] becoming the Other through an outbreak of actions which can best be described as cosmic heresy, natural blasphemy and ecological sacrilege" (1995, 210). Reading Watson from the post-9/11 shifting centre, such sacrilege appears as a "growing catalogue of crimes against nature ... turning some amongst us into something else, something now barely recognizable in human terms. Something that peers balefully back at us with its Nosferatu face" (210).

Environmental Vampire texts, emerging in full force after 9/11, destabilize readers and viewers with the realization that what is being encountered is precisely the hard-to-see pushed to a hitherto unrealized edge, an edge shaped nonetheless by history, agency, and writing. Take Justin Cronin's *The Passage* (2010), for instance. Here vampiric alterity's new-found edge is relayed through the perceptions of a central human character, Peter, who

> had gotten used to the virals' appearance but still found it unnerving to see one close up. The way the facial features seemed to have been buffed away, smoothed into an almost infantile blandness; the curling expansion of the hands and feet, with their grasping digits and razor-sharp claws; the dense muscularity of the limbs and torso and the long, gimballed neck; the slivered teeth crowding the mouth like spikes of steel. (335)

Other post-9/11 texts portraying the Environmental Vampire offer similar descriptions and images. In the *30 Days of Night* movie novelization, Tim Lebbon writes: "They looked like humans with sharks' mouths, though there was nothing human about their behavior" (2007, 123). In another novel from the *30 Days* constellation, a curious human "could see yellowed claws and long teeth and eyes the color of his pee when he drank nothing but Red Bull and Dr. Pepper for three days straight" (Niles and Mariotte 2008, 15). Lebbon again, in yet another *30 Days* novel: "The man's fingers were too long and

Fig. 4.6

tipped with claws. His legs bent unnaturally, like an animal's limbs grafted onto a person. And his face ... it was inhuman. It bore all the normal features, but their combination produced something other than the man it pretended to be" (2010, 23). And later in the same text: "These weren't the fancy vampires from those old Hammer movies" (169). One has "crocodile's teeth crammed into a human jaw" (Niles and Mariotte 2008, 280); another is "doglike" with "sworded teeth, row upon row" (Cronin 2010, 206). Eben Olemaun's first glimpse of them is described thusly: "Just what the hell are they? Eben thought again, and yet again no answer was forthcoming. Thieves, terrorists, wild people, junkies, gangsters ... nothing he thought of could account for what they were doing, and how. Nothing could allow for their strength, their apparent immunity from gunshots, and their teeth and eyes. Long sharp teeth. Deep black eyes" (Niles and Templesmith 2011, 97).[16]

Del Toro and Hogan emphasize internal as well as external post-human change. One protagonist, Eph, is a Center for Disease Control scientist who vivisects a former human, finding not teeth but a stinger that retracts into the throat: "[He] noted the tiny double tip at the end ... The vampire did not suck the blood out of its human victim but instead relied on physics to do the extraction, the second stinger canal forming a vacuum-like connection through which arterial blood was drawn up as easily as water crawls up the stem of a plant" (2011, 78). Such images are particularly visually striking in the inaugural *30 Days of Night* graphic novel published soon after 9/11.

As one Del Toro and Hogan character puts it: "He remembered September 11, 2001, and how the emptiness of the sky had seemed

so surreal back then, and what a
strange relief it was when the planes
returned a week later. Now there
was no relief" (2011, 11). Del Toro
and Hogan title the final installment
in their trilogy *The Night Eternal*,
signalling an endless night at the end
of Western linear time and progress;
they also tie 9/11 to the landing of
the Pilgrims at Plymouth Rock in
1620, as though an unbroken series
of events connects the two distant
happenings (2011, 68). The figure
of the Environmental Vampire, in
the broader historical frame, should

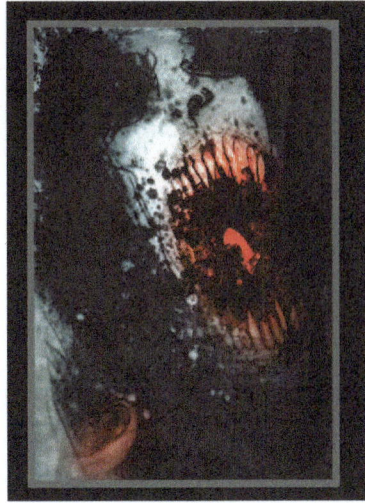

Fig. 4.7

not be unexpected. Such references signal transitions towards post-
human and post-anthropocentric awareness in a culture with a long
and violent history of initially construing the Other as anything outside
the "normative" (e.g., Saints and Strangers circa 1620; white, male,
and middle-class from the seventeenth to the twenty-first centuries).
Furthermore, the Environmental Vampire extends for the new century
Donna Haraway's pre-9/11 categories of race, population, and genome
as well as her statements that the vampire "feeds off of the normalized
human" and that "vampires can be vectors of category transformation" (1996, 322). The additional category, I suggest, is territory.[17]

Stuart Elden charts a careful history of territory as a geopolitical
concept, distinct from and intertwined with both land and terrain,
as "the way used to describe a particular and historically limited set
of practices and ideas about the relation between place and power"
(2013, 7). He quotes the earlier work of Paul Alliès on territory: "territory always seems linked to possible definitions of the state; it gives
it a physical basis which seems to render it inevitable and eternal" (7).
Against this notion, the Environmental Vampire constitutes an invasive
biomass; it metaphorically obliterates territory, which Elden argues is

a bounded space under the control of a group of people, usually
a state, [and] is therefore historically produced ... [T]he notion
of space that emerges in the scientific revolution is defined by

extension. Territory can be understood as the political counterpart to this notion of calculating space, and can therefore be
thought of as the extension of the state's power. Equally the state
in this modern form extends across Europe and from there across
the globe. (2013, 322, Elden's emphasis)

Territory—"the political counterpart to this notion of calculating
space"—is also commensurate with what Elden understands as "a
political technology, or perhaps better as a bundle of political technologies" (322). Such technologies, I submit, are vehicles for agency.

In the inaugural *30 Days of Night* graphic novel of 2002, the
sudden emergence of the Environmental Vampire creates a post-
human event that all but destroys the town of Barrow, Alaska, located
above the Arctic Circle and appearing there as a significant node in
the northern oil pipeline infrastructure: "They came quickly, walking
over the frozen tundra, cutting off communication and escape routes
as they marched" (Niles and Templesmith 2011, 35).[18] The date,
22 November 2001 (just over two months after the events of 9/11), is
embedded in the narrative by way of a hastily written journal left by a
father who tried in vain to defend his family unit against an enemy no
one can identify: "their voices sound like the rasp of a hissing snake"
(Niles and Templesmith 2011, 247).[19] The horror of the predatory
post-human, which nowhere is recognized for what it is until it is too
late, is announced first by a human agent, a white stranger and a drifter
who enters Barrow from the west and on foot to herald the new time
just as a month of darkness is about to begin. This unnamed person,
called indeed the Stranger (Lebbon 2007), a wannabe post-human
and prophet of a new era, walks into the town diner, the Ikos, and
ostentatiously orders raw meat. He in turn arouses the interest of the
law, represented by Eben Olemaun, who has already found a cache of
destroyed mobile phones on the outskirts of Barrow:

Leaving Barrow felt like going out into the wilds. It was a feeling
he experienced every now and then, but never quite this strong.
He knew the dangers of living up here, knew that going beyond
the town, past the drilling sites, and into the desert of snow was
a risk unless you were very well prepared. But tonight it felt as
though there were dangers out there that no one could be prepared for. (Lebbon 2007, 52)[20]

In the invasion that quickly follows, and in the epidemic of texts to come, the vampires created by Niles and Templesmith are referred to as agents of change through various pejorative non-human terms, such as "swarm," "thing," "those things," "rats," "freak" (39, 44, 117, 135, 145).

The Environmental Vampire consumes and occupies the material cultural and industrial reality of the West, represented in Del Toro and Hogan by New York City, where a vampire collective nests beneath Ground Zero; and in *30 Days* by Barrow, Alaska, where blood and oil are removed from their respective conduits. Yet these "next wave" vampires, like Val Plumwood's saltwater crocodile, evoke not only post-human realities but also new possibilities for identity, community, and culture-in-nature.[21] Eben Olemaun, a lawman who is also Inuit and who therefore already knows what it means to be the Other—who is distrusted and resented by whites in Barrow—constitutes the leading edge of change. His last name conjuring both everyman and otherman, Eben radically shifts his position to counter the invasion by injecting himself with the blood of the non-human Other. He thus becomes, to the shock and horror of his wife and comrades, energized and recategorized as one who possesses the physical powers of the vampires but uses that power to defeat them before incinerating himself just as the sun rises. Eben performs such an action as the denouement of the

Fig. 4.8

original graphic text, an action that is both activist and biopolitical in that it involves his own transformation from human to non-human, from diurnal to nocturnal, with the object of redefining the threat that causes terror.

As Eben Olemaun becomes the Environmental Vampire of his own volition, his act invokes homeland security like nothing else. Eventually, however, in the various texts that follow, Eben transitions himself well outside Barrow and what it represents in terms of territory. He does this in order finally to speak for and act on behalf of his developing post-human subject position, rejecting his initial transformative act of physical and epistemic violence on behalf of the state: "I was trying to save Barrow. I was human then, after all—or, to be more precise, at the end there I was more recently post-human, and I still had a human's urges and instincts" (Niles and Mariotte 2008, 276). This statement contrasts with that of the more mainstream vampire, or *humain* (Lebbon 2010, 29), who in the *30 Days uber*text sees the Environmental Vampire as extremely threatening: "They had always known of the existence of the more brutal vampires—like the one she'd just fought, and which she and the others would have to hunt down and destroy—but they viewed themselves as something more controlled, and more natural" (Lebbon 2010, 31). As a bio-political change agent—more than humain but less than outrageously anarchic—Eben Olemaun eventually reshapes the failed Westernized quest for territorial sovereignty and security into a new ecological paradigm. Although in the inaugural graphic novel he is immolated in the dawn that ends the long night in the Arctic, the possibility of his resurrection is left open, and indeed his wife and partner, Stella (also an agent of human law), resurrects Eben and is, in turn, transformed by him. The pair dwell for a while on the outskirts of Barrow, just outside Western space and time.

In their new identities, Eben and Stella configure novel possibilities for environmental equilibrium, a post-human ecology, or what Del Toro and Hogan call "the perfect vampire ecology" (2011, 11). The vampires of *30 Days of Night* seize not only territory but also time, grabbing hold of temporality itself, almost, by bringing their invasive force to bear on the extreme northern hemisphere during its darkest period. They enact the power of climate change in their playful destruction of a polar bear and in their attack on an Arctic research station

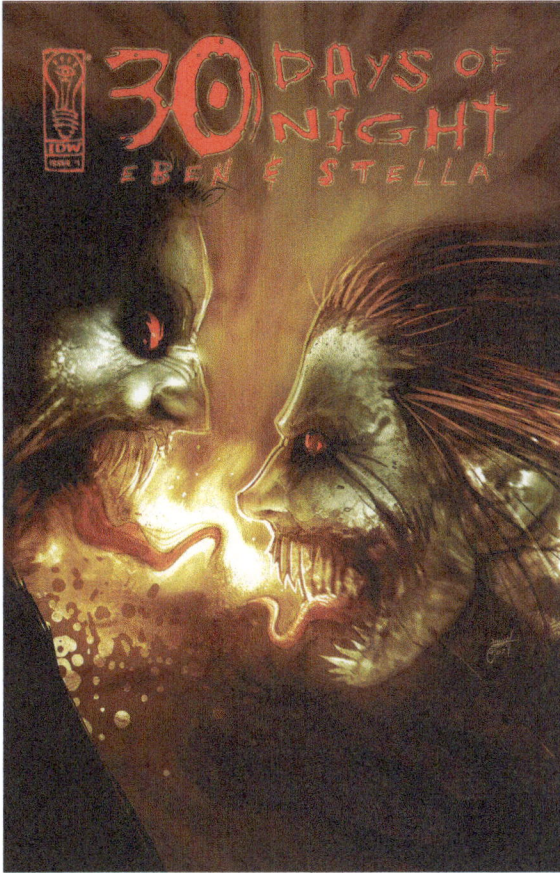

Fig. 4.9

where atmospheric scientists from the south have discovered dramatic evidence of anthropogenic climate change, indeed human history, in the ice itself (Lansdale and Kieth 2011). The conventional countering of the vampire through, for instance, Bram Stoker's Van Helsing, a man of Western Enlightenment science, is no longer a matter of waiting till or counting on the sun's regular return in the cycle of day and night. Humans are adapted to day so that the prospect of extended night evokes terror just as, for vampires, the reverse is true. "Night was man's first necessary evil, our oldest and most haunting terror," writes A. Roger Ekirch in his history of night. "Amid the gathering darkness and cold, our prehistoric forbears must have felt profound fear, not least over the prospect that one morning the sun might fail to return" (3). Ekirch argues convincingly that it's very likely that the nightwatch is actually humankind's oldest profession (75).

By dying as the sun rises, Eben Olemaun re-enacts the traditional Western storyline that sees the human morph and die sacrificially in order to defend and renew the human community. This of course is part of a literary tradition involving catastrophe—and the creators of the *30 Days of Night* series go on to depict the aftermath of the invasion of Barrow, which most notably involves new human–vampire interactions and paradigm shifts culminating in the resurrection of Eben Olemaun and the vampiric becoming and agency of Stella Olemaun.[22] The distinction and interaction between terror and horror again become important here. We live in a time of profound anticipation, partly shaped by 9/11 and—as the new century has begun to build—partly shaped by the growing concern regarding global climate change, evidenced specifically by its impact on Arctic environments. And if the 9/11 date is what *30 Days of Night* begins with, given its 2002 inaugural publication, global climate change is what it points to by 2011. The title itself radiates time limits and duration simultaneously: the limits and duration of the present dilated beyond expectation as humans wait for what comes next. Such moments are felt in the house scenes in the inaugural *30 Days of Night* graphic novel and in its various spinoffs in film (2007) and novelization (2007). Witness the attic scene, where surviving humans find refuge (security, space, territory) and watch vampires kill a polar bear; a similar scene is graphically depicted in *30 Days of Night, Night Again* (2011). The terror/horror emotional complex becomes primary when human sovereignty and territory (domicile, town, municipality) are perceived as imminently threatened; the anticipation of structural extinction drives the character or characters to extremes of agency, performing unexpected actions to pre-empt extinction or demise.

The importance of 9/11 posed by the inaugural graphic novel in the *30 Days of Night* series (along with texts produced by other writers noted here, such as Justin Cronin and the Del Toro/Hogan collaboration) begs questions of related critical issues concerning land, water, and air in their relations to geopolitical space and carbon. In *Arctic Dreams* (1986), Barry Lopez describes how for centuries Europeans regarded the circumpolar region as a place of incipient evil and home of the anti-Christ (17). From Dante to Milton to *Frankenstein* to *Blade*

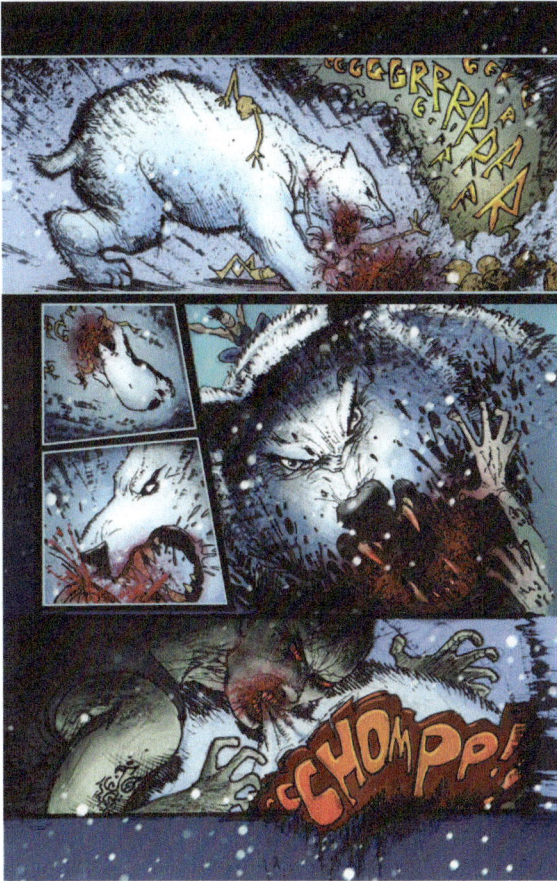

Fig. 4.10

Runner, intense cold has been associated with eternal damnation and/
or lawlessness.[23] The *30 Days of Night* series of texts finds its genesis
in this idea—that is, the idea of the North, which anthropologist Tim
Ingold has argued is becoming everywhere (2011). The Westernized
human and corporate community represented by Barrow, invested in
energy and energy transportation, cedes its territory to a sudden and
violent overthrow by vampires, which are depicted as entirely Other
while also retaining barely recognizable human features to create an
overall effect of terror and horror. The territorial loss portrayed here
is significant, which is why I propose that the categorical bleeding and
mixing first espoused by Donna Haraway as a central feature of the
vampire trope should also include territory. As a concept, territory

emerges from the aftermath of 9/11—and in the new reality constituted by global climate change (its storms, politics, and dislocations)—
as a crucial category that is currently being transfused. Borders and
transnational crises have come into focus as never before. In Canada,
where I live, this is further complicated by Indigenous land settlement
claims, claims that often implicate, in their success or failure before
the courts, latent environmental crises created by corporate projects
such as the Northern Gateway pipeline proposed by Enbridge Pipelines
(and rejected by the federal Canadian Government in 2016). In the
broader context of global climate change and its potential impacts
on both environment and energy, contested lands and their boundaries disrupt state policies and corporate strategies regarding territory
and borders. Atmospheric science also falls under this broad rubric.
Earth's atmosphere and oceans are undergoing transformation because
of carbon output by humans; the circumpolar space is the amplified
signifier of such change, and as it melts, it is undergoing and provoking
territorial challenges and reconfigurations. Meridian lines connect the
Arctic albedo effect to South Pacific coral bleaching.

   30 Days of Night, Night Again (2011) grapples with the Arctic
impact through its depiction of a group of climate scientists who come
under assault at their research station by an invasive force of vampires. Authors Lansdale and Kieth (2011) play with the interactions
between scientists and non-scientists who are under siege together,
probing the slow-moving, self-limiting character of science as a part
of human culture. Human limits, cultural fragility, and the power of
nature are brought together here in the proto-technological figure of
the golem, discovered buried in the Arctic ice, cast over against vampires who have re-enacted their totalizing seizure of time and space
in the Far North. The golem is an ancient Hebrew figure that depicts
human cultural ingenuity and limits. This relic hominid figure, made
of clay and animated at one time, is reawakened but without any real
consciousness or direction other than to seek vengeance indiscrimi-
nately. The golem constitutes a basic depiction of the colonial vision
of the interaction between culture and nature, the tired and out-of-date
nature–culture dichotomy, since without human agency the golem is
an out-of-control techno-cultural act, however old. The golem myth
has links to Prometheus, the Frankenstein creature, and Pinocchio, and
indeed like them is a forerunner of the cyborg and AI found in recent

Fig. 4.11

science fiction and film. *Blade Runner's* Roy Batty and the AI of *Ex Machina* are examples of these.

Stuart Elden implies an evocative link between the terms *terror* and *territory* (Elden 2010, xxviii–xxix). Since, as Neil Smith argues, imperial space is typically perceived as never subjected to deterritorialization (2005, 51), its visual portrayal in *30 Days of Night* is difficult for some readers and viewers to process. Indeed, in teaching the inaugural graphic text during an undergraduate university course on environment and literature, I encountered students who could not bring themselves to process the text and its images. Even Cormac McCarthy's *The Road*, which I also teach and which portrays the almost complete annihilation of culture and nature, has never in my teaching experience produced such a response from students. The inaugural graphic novel's foreshadowing of possible deterritorialization is seen in such things as the stash of stolen cellphones Eben Olemaun discovers on the edge of Barrow, or in the grim sabotage of the town's power station, just as the sun sets on 17 November 2001, for the start of a month-long period of darkness (Niles and Templesmith 2011).

Even in the early stages of the initial narrative, which precipitates a multi-authored textual universe across the next decade, the notion of the American ideal of the individual regulating the state (Hindess 2008, 309) is in jeopardy. That Eben Olemaun will become the Other in order to destroy the Other is, in light of the extensive narratives that follow, a temporary stay against a wider cultural and, with that, territorial transformation. Eben's long nightwatch and initial meta-morphosis involve his coming to terms with the exact, unbelievable nature of the former humans who now constitute an invasive predatory force. In his decision to become Other, Olemaun becomes a radically mutative agent even as the vampiric metaphor itself begins to change into a complex unified concept of nature to be explored throughout the graphic novels and novels that follow. All of this, I suggest, points to a significant reconfiguration of the vampire figure, one that reflects—and is created in response to—the Anthropocene and that also continues to do what vampires seem so readily to perform on our behalf: make cul-tured, alienated westernized humans see themselves in terms of what is conceived of as "natural" and also terrible, what E.O. Wilson calls "the reverse side of nature's green-and-gold ... the black-and-scarlet of disease and death" (2003, 141). The Environmental Vampire invites further readings in terms of bioterror, food security, global epidemiol-ogy, climate change, and post-colonialism, all of these consequences of human agency.

Notes

Graphic novel plates in this chapter courtesy Idea and Design Works LLC.

1   Vampires are a part of folklore predating Stoker's *Dracula*. See Ekirch (2005), 19.
2   Aspects of the Environmental Vampire are also present in *True Blood's* television and book series.
3   This is true, I would argue, even of vampires depicted in comedy (e.g., *Vampire's Kiss*).
4   Especially the first graphic novel, *30 Days of Night* (2002), which lends its name to the ongoing series, as well as *30 Days of Night: Night, Again* (2011), a sequential text by Joe R. Lansdale and artist Sam Kieth.

5 Cited in this chapter from the 2011 omnibus publication of the original trilogy of graphic novels, *Thirty Days of Night* (2002), *Dark Days* (2003), and *Return to Barrow* (2004).

6 Reflected throughout the *30 Days* texts are passages such as this one: "The knowledge of how things really played once the sun went down could make people's heads explode out there in the unsuspecting 'real world'" (Niles and Mariotte 2008, 96).

7 Describing the Cassandra complex in the context of twenty-first-century American racial politics, Frankowski calls it "an identity for the type of social pathology that appears at the critical edge of political discursivity" and is "related to the history of a culture's violence" (2012, iv, 1).

8 Justin Cronin (2010) ties the emergence of the Environmental Vampire directly to the experience of the terror-horror complex in its relationship to time: "Richards remembered the day—that glorious and terrible day—watching the planes slam into the towers, the image repeated in endless loops. The fireballs, the bodies falling, the liquefaction of a billion tons of steel and concrete, the pillowing clouds of dust. The money shot of the new millennium, the ultimate reality show broadcast 24-7" (84).

9 Horror films such as Nicolas Roeg's *Don't Look Now* (1973) are based on this principle. Roeg's film also explores the Cassandra complex.

10 *Count Yorga* (1970) and its sequel are low-budget erotic vampire movies that effectively represent the slew of campy vampire films from this period and typify the decadence of the genre in the last half of the twentieth century. Robert Quarry's Count Yorga is a secondary figure after Christopher Lee's dynamic Dracula found in the Hammer films. Anne Rice's vampires, beginning with Louis in *Interview with the Vampire* (1976), give first-person voice to a series of vampires, most notably Lestat, that arguably become increasingly decadent. In "What Is an Author?" (1966), Michel Foucault pursues the question of the vacuum left by the death of the author. Reading Rice's original first-person narrative, in which the noble, anguished Louis intimates his biography from the perspective of one already dead, provides a fascinating opportunity for analysis in the context of Foucault's argument that, among other things, storytelling forestalls death: "Storytellers continued their narratives late into the night to forestall death and to delay the inevitable moment when everyone must fall silent … This conception of the spoken or written narrative as a protection against death has been transformed by our culture. Writing is now linked to sacrifice and to the sacrifice of life itself" (1467). Louis narrates his life-death story to a reporter late at night in a North American bar and, in doing so, raises the question of decadence itself, for if the figure of the immortal and immutable vampire must tell his story, then vampires participate in time and therefore also in decay (entropy) and death. In other words, Louis tells his story to forestall death as well, which signals vampiric fatigue.

11 McGlown and Robinson (2011) cite the UN definition of catastrophe and that organization's identification of 9/11 as catastrophic (7).

12 It also invites speculative fiction, as Robert R. M. Verchick (2010) points out in a book otherwise devoted to real-world catastrophes: books "help to stretch the imagination and urge the public and policy-makers to conceive of the worst. Then a process of evaluation can take place. Literature, particularly speculative fiction, can help in this process" (2774).

13 As one character in Cronin's (2010) *The Passage* puts it: "His whole life Peter had thought of the world of the Time Before as something gone. It was as if a blade

had fallen onto time itself, cleaving it into halves, that which came before and that which came after. Between these halves there was no bridge" (336).

14 This experience with narrative and character is, of course, not only specific to the vampire. All successful stories offer "an interest in ... what [fictive] others have done and suffered and in the causal relations among the things they have done or suffered" (Flesch 2009, 17).

15 I am thinking especially of Magellan Pictures' *Vampire's Kiss* (1988), starring Nicolas Cage.

16 On eyes, see Watson (1996, 198–200).

17 As this book goes to press, Bruno Latour's *Down to Earth: Politics in the New Climatic Regime* appears in its English translation. Latour addresses the growing significance of territory under Trumpism and its strident denial of climate change: "Have you noticed that the emotions involved are not the same when you're asked to defend nature—you yawn, you're bored—as when you're asked to defend your territory—now you're wide awake, suddenly mobilized" (2018, 8).

18 And additionally, in the same passage: "Eben knew that Barrow was in big, big trouble. Someone had systematically destroyed their means of communicating with, or traveling to, the outside world. That guy back in the holding cell was part of it, but there must be others to have done this much damage. Thieves perhaps, terrorists, some sort of dispute gone bad between oil companies—" (34).

19 Cronin's *The Passage* (2010) ties the Environmental Vampire explicitly to terrorism: "There were things about him that Grey would say were sort of human. Such as, he had two arms and two legs. There was a head where a head should be, and ears and eyes and a mouth. He even had something like a johnson dangling down south, a curled-up little seahorse of a thing. But that's where the similarities stopped" (69). A few pages on, Grey recalls: "Now the whole oil industry was under federal protection, and it seemed like practically everybody he knew from the old days had disappeared. After that Minneapolis thing, the bombing at the gas depot in Secaucus, the subway attack in L.A. and all the rest, and, of course, what happened in Iran or Iraq or whichever it was, the whole economy had locked up like a bad transmission" (84).

20 An additional description of Barrow heightens the theoretical territorial links I am making in this chapter: "Beyond the northernmost limits of Barrow lay the satellite dish tower, bristling with antennae and topped with the dish itself. Not far past the tower ran the Trans-Alaskan oil pipeline, a visible indicator of the riches to be found in the area. The tower boosted all cell phone signals, and was also the relay station for all the radios in Barrow—police, medical, and the several radio hams who liked to keep the outside world apprised of events in this unusual town" (Lebbon 2007, 35).

21 In her well-known written account of being attacked by a saltwater crocodile, Val Plumwood (2002) maintains that "our frameworks of subjectivity" are "structured to sustain the concept of a continuing, narrative self," one that is radically interrupted and often terminated by a sudden encounter with an apex predator like a crocodile. The croc that took Plumwood into a death roll that she somehow managed to survive demonstrated for her, and for her readers, the disparity between the civilized subject position adopted by many humans and the "reality" of the "outside" represented by her would-be killer, who meant to eat her. The prospects raised by terror and terrorism function similarly. Plumwood recounts in her essay how the experience changed her mind: "the story of the crocodile

encounter now has, for me, a significance quite the opposite of that conveyed in the master/monster narrative. It is a humbling and cautionary tale about our relationship with the earth, about the need to acknowledge our own animality and ecological vulnerability."

22 In the *30 Days of Night* universe, representatives of the law are characterized as prime candidates for post-anthropocentric transformation. Del Toro and Hogan in their *Strain* trilogy, and Justin Cronin in *The Passage*, focus on similar transformative characterizations of law enforcement figures.

23 Dante's *Inferno*, Canto XXXIV; Milton's *Paradise Lost*, Book II; Shelley's *Frankenstein*, Chapter 24; Ridley Scott's *Blade Runner: The Final Cut*, Chapters 9 and 10.

References

Alighieri, Dante. 1997; rpt. *The Divine Comedy: Canticle I, Inferno*, translated by Henry Wadsworth Longfellow, edited by Dennis McCarthy. *Project Gutenburg*. http://www.gutenberg.org/files/1001/1001-h/1001-h.htm. Accessed 12 June 2016.

Arendt, Hannah. 1951. *Totalitarianism: Part Three of The Origins of Totalitarianism*. New York: Harcourt, Brace, Jovanovich.

Auerbach, Nina. 1995. *Our Vampires, Ourselves*. Chicago: University of Chicago Press.

Bierman, Robert. 1989. *Vampire's Kiss*. Magellan Pictures. DVD.

Bowen, John. 2015. "Gothic motifs." British Library. https://www.bl.uk/romantics-and-victorians/articles/gothic-motifs. Accessed 7 June 2016.

Braidotti, Rosie. 2013. *The Posthuman*. Cambridge: Polity Press.

Coppola, Francis Ford. 1992. *Bram Stoker's Dracula*. Columbia Pictures. DVD.

Cronin, Justin. 2010. *The Passage: A Novel*. Toronto: Doubleday Canada.

Del Toro, Guillermo, and Chuck Hogan. 2009. *The Strain*. New York: HarperCollins.

———. 2010. *The Fall*. New York: HarperCollins.

———. 2011. *The Night Eternal*. New York: HarperCollins. eBook.

Ekirch, A. Roger. 2005. *At Day's Close: Night in Times Past*. New York: W.W. Norton.

Elden, Stuart. 2010. *Terror and Territory: The Spatial Extent of Sovereignty*. Minneapolis: University of Minnesota Press.

———. 2013. *The Birth of Territory*. Chicago: University of Chicago Press.

Flesch, William. 2009. *Comeuppance: Costly Signaling, Altruistic Punishment, and Other Biological Components of Fiction*. Cambridge, MA: Harvard University Press.

Foucault, Michel. 1966. "What Is an Author?" In *The Norton Anthology of Literary Criticism*, 3rd ed. (2018), edited by Vincent B. Leitch et al., translated by Donald F. Bouchard and Sherry Simon, 1394–1409. New York: W.W. Norton.

Frankowski III, Alfred. 2012. *The Cassandra Complex: On Violence, Racism, and Mourning*. Ann Arbor: ProQuest LLC and UMI Dissertation Publishing. UMI 3523334.

Freud, Sigmund. 1913. *Totem and Taboo*, translated by James Strachey. London: Routledge.

Garland, Alex. 2015. *Ex Machina*. Universal Pictures. DVD.

Haraway, Donna J. 1996. "Universal Donors in a Vampire Culture: It's All in the Family: Biological Kinship Categories in the Twentieth-Century United States." In *Uncommon Ground: Rethinking the Human Place in Nature*, edited by W. Cronon. New York: W.W. Norton.

Hindess, Barry. 2008. "Sovereignty as Indirect Rule." In *Re-envisioning Sovereignty: The End of Westphalia?*, edited by T. Jacobsen, C. Sampford, and R. Thakur. Oxford: Ashgate.

Ingold, Tim. 2011. "The North Is Everywhere." *Niche*. http://niche-canada .org/2011/09/07/the-north-is-everywhere. Accessed 12 June 2016.

Kelljan, Bob. 1970. *Count Yorga*. MGM. DVD.

Lansdale, Joe, and Sam Kieth. *30 Days of Night: Night, Again*. San Diego: IDW.

Latour, Bruno. 2018. *Down to Earth: Politics in the New Climatic Regime*. Trans. Catherine Porter. Cambridge, UK: Polity Press.

Lebbon, Tim. 2007. *30 Days of Night: A Novelization*. New York: Pocket Star Books.

———. 2010. *30 Days of Night: Fear of the Dark*. New York: Pocket Star Books.

Lopez, Barry. 1986. *Arctic Dreams*. New York: Vintage.

MacCormack, Patricia. 2012. "Posthuman Teratology." In *The Ashgate Research Companion to Monsters and the Monstrous*, edited by Asa Simon Mittman with Peter J. Dendle. Surrey: Ashgate.

Marriotte, Jeff. 2009. *30 Days of Night: Light of Day*. New York: Pocket Star Books.

Massumi, Brian. 1998. "Requiem for Our Prospective Dead! Toward a Participatory Critique of Capitalist Power." In *Deleuze and Guattari: New Mappings in Politics, Philosophy, and Culture*, edited by Eleanor Kaufman and Kevin Jon Heller. Minneapolis: University of Minnesota Press.

McCarthy, Cormac. 2006. *The Road*. New York: A.A. Knopf.

McGlown, K. Joanne, and Phillip D. Robinson, eds. 2011. *Anticipate, Respond, Recover: Healthcare Leadership and Catastrophic Events*. Chicago: ACH Management Series.

Milton, John. 1667. *Paradise Lost*. http://www.paradiselost.org. Accessed 12 June 2016.

Morris, Errol. 2014. "The Certainty of Donald Rumsfeld (Part 1)." *New York Times*, 25 March 2014. Web. Accessed 6 June 2016.

Niles, Steve, Kelly Sue DeConnick, and Justin Randall. 2007. *30 Days of Night: Eben and Stella*. San Diego: IDW.

Niles, Steve, and Jeff Mariotte. 2006. *30 Days of Night: Rumors of the Undead*. New York: Pocket Star Books.

———. 2007. *30 Days of Night: Immortal Remains*. New York: Pocket Star Books.

———. 2008. *30 Days of Night: Eternal Damnation*. New York: Pocket Star Books.

Niles, Steve, and Ben Templesmith. 2011. *30 Days of Night Omnibus*, vol. 1. San Diego: IDW.

Plumwood, Val. 2002. "Prey to a Crocodile." *Bealtaine* 30. http://www.aisling magazine.com/aislingmagazine/articles/TAM30/ValPlumwood.html. Accessed 15 June 2016.

Poitras, Laura. 2014. *Citizenfour*. Praxis Films. DVD.

Quarantelli, E.L. 2006. "Catastrophes are Different from Disasters: Some Implications for Crisis Planning and Managing Drawn from Katrina." http://under standingkatrina.ssrc.org/Quarantelli. Accessed 8 June 2016.

Rice, Anne. 1976. *Interview with the Vampire*. New York: Ballantine.

Robinson, Kim Stanley. 2015. *Aurora*. New York: Little, Brown.

Roeg, Nicolas. 1973. *Don't Look Now*. Casey Productions. DVD.

Sassen, Saskia. 2008. *Territory, Authority, Rights: From Medieval to Global Assemblages*. Princeton: Princeton University Press. ProQuest ebrary. Web. Accessed 29 July 2015.

Scott, Ridley. 2007. *Blade Runner: The Final Cut*. Warner Bros. DVD.

Shelley, Mary. 1831. *Frankenstein, or the Modern Prometheus*. Project Gutenburg. http://www.gutenberg.org/files/84/84-h/84-h.htm. Accessed 12 June 2016.

Slade, David. 2007. *30 Days of Night*. Columbia Pictures. DVD.

Smith, Neil. 2005. *The Endgame of Globalization*. London: Routledge.

Verchick, Robert R.M. 2010. *Facing Catastrophe: Environmental Action for a Post-Katrina World*. Cambridge, MA: Harvard University Press.

Watson, Lyall. 1995. *Dark Nature: A Natural History of Evil*. London: Hodder and Stoughton.

Weinstock, Jeffrey A. 2012. "Invisible Monsters: Vision, Horror, and Contemporary Culture." In *The Ashgate Research Companion to Monsters and the Monstrous*, edited by Asa Simon Mittman with Peter J. Dendle. Surrey: Ashgate.

Wilson, E.O. 2003. *The Future of Life*. New York: Vintage.

Wright, Lawrence. 2006. *The Looming Tower: Al-Qaeda and the Road to 9/11*. New York: Knopf.

II.
TIMELINES
AND INDIGENEITY

Ice fishers seen from the St. Mary's Reservoir Dam embankment in Southern Alberta. Here in 2013 the tools of pre-Clovis people dating from the ice-free corridor approximately 13,300 years ago were discovered and dated, along with bones of camels and horses. (Photo courtesy Robert Boschman)

CHAPTER 5

"We are key players ...": Creating Indigenous Engagement and Community Control at Blackfoot Heritage Sites in Time

Geneviève Susemihl

INTRODUCTION

Head-Smashed-In Buffalo Jump, in the Porcupine Hills of southern Alberta, is one of the oldest, largest, and best-preserved buffalo jumps in North America. Because of its extraordinary archaeological, historical, and ethnographical value, combined with its prairie setting and outstanding interpretive potential, it was designated a UNESCO World Heritage Site in 1981. Inscribed under criterion vi,[1] which recognizes its direct association with "the survival of the human race during the pre-historic period" (ICOMOS 1981), this site is a primary illustration of the subsistence hunting techniques of Plains Nations.

Layers of bison bones buried up to 10 metres below the cliff represent nearly 6,000 years[2] of use of the buffalo jump by Indigenous peoples. Covering 1,470 acres, this site has four distinct components—the gathering basin, the V-shaped drive lanes, the cliff kill site, and the campsite and processing area. Each has different archaeological remains associated with communal buffalo hunting, ranging from drive lane cairns and projectile points to butchered bone and fire-broken rock. The site exemplifies the culture and society of the Plains Peoples for many centuries before the European settlement of the region, and thus represents a complex range of Indigenous identities, ideologies, and social relations.

Since 1972, UNESCO[3] has been encouraging the identification, protection, and preservation of the world's cultural and natural heritage.[4] As a World Heritage Site, Head-Smashed-In Buffalo Jump

(HSIBJ) is on a list of more than 1,000 outstanding cultural and natural sites around the world. World Heritage Sites are meant to "belong to all the people of the world, irrespective of the territory on which they are located" (UNESCO, n.d., *World Heritage*). Visitors recognize these sites as profoundly significant for the stories they tell and the information they offer about the past. Thus these sites are highly valued for educational purposes: they inform and educate local, regional, and international communities about the past, present, and future of humanity.

For Indigenous people concerned about their cultural survival, World Heritage Sites support empowerment and capacity building (UNESCO 2005). Indigenous ownership, control, and community involvement at HSIBJ have significant implications for Indigenous empowerment and capacity building as well as for the stories the site tells and the manner of their telling. In this chapter I look closely at the government-owned site and at how local Indigenous people are participating in its administration and operations.[5]

INDIGENOUS EMPOWERMENT AND CAPACITY BUILDING THROUGH HERITAGE

Heritage, tourism, and museum studies must confront many sensitive issues regarding ownership and control, community involvement and management, interpretation, and representation of heritage. These issues have been discussed fairly extensively. Many writers in the field of Indigenous cultural tourism have focused on the political nature of cultural tourism, in that the protection of cultural identity and traditions is central to its development and management. Walsh (1992) contends that superficial portrayals of the past separate people from their own heritage as well as from their understanding of their cultural and political present. Kirshenblatt-Gimblett (1998) explores the "agency of display" (16) and shows how objects and people are made to perform "meaning" (3) through the very fact of being collected and exhibited. She writes that although heritage is marketed as something old, it is actually a new mode of cultural production, one that revitalizes places, economies, and vanishing ways of life.

Smith (2003) discusses Indigenous cultural tourism as part of the expanding heritage tourism industry and shows that cultural tourism has exacerbated the commodification of heritage. She calls for more community-based cultural tourism initiatives. Notzke (2006) examines community involvement in tourism development, arguing that the use of social space and the assignment of certain roles to tourists play an important part in the management strategies of Indigenous hosts. Kramer (2006) charts the fluid character of material culture and analyzes the ambivalent reactions to the ownership, appropriation, and repatriation of both tangible and intangible culture. For her, the loss of cultural objects amounts to proof that culture is valuable (i.e., as a source of external affirmation) and thus helps assert a collective national Indigenous identity, which, among other things, supports the reclamation of traditional territory and self-determination.

Ownership and control are inevitably linked to education and empowerment, since it is the owner who determines what heritage is being protected, and how, and what stories are being told. When one culture decides what is significant and worth protecting in the heritage of another, there is always the possibility that injustice will arise. While the argument has been made that cultural heritage belongs to the public and "should be used for the greater good of contributing to the knowledge of humankind" (Asch 2009, 395), many Indigenous peoples assert that increased protection and control of cultural heritage significant to them is fundamental to the continuity, revival, and survival of their cultural identities.

In Canada, the final authority when it comes to control over cultural heritage is divided between the federal and provincial governments. Since First Nations are not provinces, the law assumes that they do not have ultimate authority over decisions that affect their heritage (395). Consequently, where Canadian authority assigns control outside of First Nations, as "in the case of statutory 'ownership' of material culture by the federal or provincial governments, or heritage sites located off reserve land, First Nations seeking greater control are left to make claims, to negotiate, and ... to make 'demands'" (395).

Yet the division of powers in Canada seems to be evolving, and more and more First Nations are taking control of their cultural heritage. Awareness is growing of how the spiritual, political, social, racial, educational, gendered, and economic strengths of individuals and communities can be nurtured through heritage. Thus, heritage is becoming more important to First Nations, and the value of community-led cultural heritage management is being explored by more and more First Nations and governments (Bell and Paterson 2008; 2009). Empowerment,[6] which can be both a process and an outcome, can be viewed as a process of social action whereby individuals, organizations, and communities gain the expertise to modify their social and political environment and thereby improve their quality of life. It also encompasses participation, education, and opportunities to use the acquired knowledge in ways that contribute to society. In this way, "the acquisition of knowledge and skills [is linked] to social needs and mobilization" (Biancalana 2007, 24).

Heritage, then, can be viewed as an asset of economic, social, cultural, and political significance. In this regard, when heritage sites are associated with Indigenous cultures, attention needs to be paid to the relationship between the Indigenous community and the site itself (Biancalana 2007, 7). The links that bind heritage to Indigenous communities are related to spiritual values, historical significance, and traditional occupations. Heritage thus has a strong relationship with identity and individual and collective memories. Memory, being an essential element of individual and collective identity (LeGoff 1992, 98) as well as continuity with the past, offers "certainties, allowing us to draw a line in which our present can fit" (Biancalana 2007, 6). Our collective memory is formed in part by historic environments, which contain an infinity of ancient and recent stories, written in stone, brick, or wood or inscribed on the landscape, and these become the focus of community identity and pride. Heritage sites thus provide mnemonic features, as Armstrong (2007), Calloway (2003), Eigenbrod (2005), Nelson (1993), and Lutz (2007) express. This communal memory is part of Indigenous culture.

Since Indigenous heritage sites are part of a living culture, the levels of significance applied by the state when assessing non-Indigenous cultural heritage do not align easily with assessments of Indigenous heritage (NSW Heritage Office 2011, 28). So it is important to investigate the community's understanding of the heritage and for the Indigenous community to participate in the preservation of heritage sites. Participation, however, needs to involve more than making communities the beneficiaries of a tourism project. Jobs are an important benefit, but they do not replace empowerment. Consultation is not enough;

communities must participate in the decision-making. Processes must be initiated to ensure that communities manage their own growth and resources; project managers must identify local leaders; local organizations must get involved; the community's key priorities must be identified; the expectations of local people must be met; their concerns and ideas must be listened to (Notzke 2006, 186). Organizations tasked with preserving the heritage site must consider integrating Indigenous communities into this process. UNESCO meets this demand with its mission of encouraging the participation of the local population in the preservation of their cultural and natural heritage. The practical work of meeting this objective, however, falls on national and provincial governments.

A people's ethnic origins, heritage, and culture are expressed through their unique intellectual, scientific, and spiritual achievements as well as through their art. Every nation or ethnic group has its own way of expressing itself, developed out of necessity, inspiration, and geographic realities. Cultural heritage projects have the responsibility to tell stories not only about the past but also about the present, and thus help shape the future. Patil writes that "the apparently ever-increasing importance of heritage in an individual and collective sense relates directly to an on-going series of challenges to, and explorations of, notions of identity and belonging in a highly mobile world" (Patil 2017, 21). For Indigenous people like the Blackfoot in southern Alberta, a close connection with their cultural heritage is a source of empowerment and strength in a highly challenging postcolonial world.

"A PLACE SHARED": OWNERSHIP AND MANAGEMENT OF HEAD-SMASHED-IN

Head-Smashed-In Buffalo Jump was first explored in 1938 by members of the American Museum of Natural History. Since the 1950s it has been a site of systematic excavations that has considerably enriched our knowledge of ancient weapons and tools and transformed our understanding of how game animals were a source of clothing and shelter as well as food. After the government purchased the land from proprietors who had been ranching it for decades,[7] HSIBJ was designated a Canadian National Historic Site in 1968, a Provincial Historic Site in 1979, and a World Heritage Site in 1981.

To tell the story of the Plains hunters, a $9.8 million Interpretive Centre was built in 1987, which depicts and interprets the ecology, mythology, lifestyle, and technology of the Blackfoot peoples through archaeological evidence.[8] Besides the exhibitions, the centre offers a ten-minute documentary, "In Search of the Buffalo," featuring a re-enactment of a buffalo drive and related activities, which is shown throughout the day. There is also a permanent photo exhibition, "Lost Identities—A Journey of Rediscovery," which pictures Blackfoot people of different communities; and there is a gift shop that sells Blackfoot arts and crafts as well as souvenir items. Most visitors explore the exhibitions on their own, but also available is a guided tour that leads them through the centre and the theatre and along the clifftop trail, introducing Blackfoot culture and history and the mechanics of the buffalo jump. Blackfoot interpreters guide these tours. For students, educational programs are provided,[9] designed to "complement the Alberta Learning Curriculum" (Alberta Community Development 2004).

Head-Smashed-In Buffalo Jump is owned and managed by the Government of Alberta with only minimal Indigenous involvement in decision-making. With around 60,000 visitors per year,[10] HSIBJ is the "flagship" of the four Southern Alberta Historic Sites.[11] The provincial government provides the funds to operate the building (including salaries for the government staff, a small annual goods and services budget, and regional marketing); the site itself generates revenues from admissions, the gift shop, educational tours, a tipi camp, and a percentage of the café income.[12]

That said, most of those who work at the centre, including the site interpreters, are Blackfoot people, who play a key role in the site's operations. Currently there are six governmental staff positions. The site manager, the head of finances and visitors' services, and the head of education are non-Indigenous; Piikani hold the positions of marketing/program coordinator, head of interpretation, and lead guide. Also, for the winter season of 2011–12 seven Piikani contract workers were hired as interpreters and gift shop employees, as well as for the front desk and the office. In the summer, as many as thirteen additional people are hired as interpreters, shuttle bus drivers, and extra staff.[13]

Applications from non-Indigenous people are considered, but
Blackfoot staff are preferred, because they are "what people want to
see."[14] For the interpreters, "it is also preferential if they speak Blackfoot, because visitors like to hear the Blackfoot language." Besides
direct employment as staff and contract workers, HSIBJ provides some
business opportunities for local Blackfoot. Drummers and dancers
from all three Blackfoot communities are hired for performances every
Wednesday in summer and for special events on National Aboriginal
Day, and art by local artists and artisans is purchased for the gift
store.[15] As HSIBJ is a government-run facility, government rules apply.
Concepts of Indigenous time, spiritual values, and traditional practices
have made way for "Western" management[16] concepts, and this has
sometimes made working together a challenge for both sides.[17]

When it comes to operations, marketing, and educational programs, Blackfoot knowledge is taken into consideration and Blackfoot
staff influence decisions as they relate to their positions, as program
coordinator Quinton Crow Shoe notes:

> It's a place shared by the First Nations and the Government of
> Alberta, who are stringent in what they do here, because it's tax
> payers' money, and so we have to follow those rules. But at the
> same time, it gives us the opportunity to share stories, knowl-
> edge, and ideas, to have input. Right now, we are talking about
> redesigning the building, and ... I'm very fortunate that as a First
> Nation, as an employee, I am involved in this process ... Our
> input is being valued.[18]

Crow Shoe sees HSIBJ as a cooperative venture "between the Plains
Blackfoot people and the Alberta government": "We are telling the
story, we are sharing the culture, and we are key players, we work
here. It would be a museum otherwise."[19] Others reject this perspective
and oppose the current ownership and management of HSIBJ, for they
view the site as part of their tribal heritage. Some Blackfoot[20] resent the
fact that "the government is draining money out of the region and of
Native culture."[21] However, neither the Piikani nor the Kainai Nation
have the means to operate the site on their own.

"CLAIMING THE SITE AS THEIR OWN": INDIGENOUS
COMMUNITY INVOLVEMENT

HSIBJ is of great spiritual significance to the Blackfoot people, as Blackfoot culture is based on a long and intimate relationship with the land, and the landscape has always been part of Indigenous traditions. In spite of government ownership, the Blackfoot have come to claim the site "as their own" (Brink 2008, 290), and HSIBJ has become a place of weddings, funerals, medicine bundle openings, meetings of elders, and many other ceremonies that reflect the esteem in which the place is held. While the buffalo jump method was abandoned around 1850 (Brink, 257; Verbicky-Todd 1984, 132), the descendants of the Blackfoot have always resided within a few kilometres of HSIBJ. The knowledge of its existence and use, together with the traditions, have been passed on meticulously to successive generations, and the jump has never ceased being a proud piece of the past for the local Blackfoot (Brink, 257). For the government, though, the use of HSIBJ as "a community centre" is rather difficult, as the site had been designated an archaeological site. Regional director Ian Clarke explains:

> One of my problems with the place is that it has become a Blackfoot cultural centre, and my sense is that it is not really that. It is a World Heritage Site because of the buffalo jump and what the buffalo jump means. And the buffalo jump and its content are more important than the Chicken Dance. The Chicken Dance is colourful, and it's an interesting representation of Blackfoot culture, but it doesn't tell us the whole element of what this place means. This place is about how Native people of this country survived thousands of years by their ingenuity and their knowledge, and that is what I think people need to understand.[22]

This conflict between the government and the Blackfoot partly originates in different views of the concepts of archaeology and heritage. The designation of HSIBJ as a World Heritage Site is predominantly based on archaeology, and culture came in "officially" only in 1987, when the centre opened. But archaeology is a Western concept, one that encompasses both a record of the past and the interpretations and values that people today apply to that record. But many Indigenous people do not view archaeological artifacts or sites as things of the past; instead, they see these things as active elements of their contemporary world. These objects and places and the stories attached to them are valued as much for their "heritage" as for "being repositories of beings and powers of importance within their worldview" (Nicholas 2006, 218). This has major implications for understanding the critical reactions that some Indigenous communities have to archaeology; it also identifies the need for alternative heritage management strategies. As the story of HSIBJ predates Blackfoot culture, it represents Plains Peoples in general—so argues the government—and therefore "it is not really a Blackfoot story, but it's really a Plains people story."[23]

Buffalo hunting was the primary subsistence activity of the nomadic Plains peoples, and the buffalo was, for them, the most sacred of all animals. Their nomadic life was a consequence of their need to secure success in hunting. Since they relied entirely on the buffalo, the Plains tribes developed highly efficient hunting techniques. Using their deep knowledge of the topography and of buffalo behaviour, they killed their prey by chasing them over a precipice; the carcasses were later carved up in the camp below. To encourage the buffalo to run to the desired point along the precipice, the hunters built drive lanes of piles of rocks or other materials, which herded the buffalo to the jump-off point. These hunts drove hundreds of animals; it is reported that sometimes more than 1,000 buffalo were killed at a time (Krech 1999, 131). Depending on the quantity of meat, the distance to the camp, and the means of transportation, sometimes all of the meat was used; other times, the hunters butchered the buffalo "lightly," taking tongues and humps only, or "harvesting" only a few and leaving hundreds to rot where they had fallen (132–4).

Communal hunts were of great importance not only for the survival of the Plains peoples but also for their safety and their social and cultural life. Besides providing essential supplies of food, communal hunting served a number of social purposes in Plains culture. It enabled the people to live together in large bands, which rendered them "less vulnerable to their enemies, and also facilitated the maintenance of tribal cultural traditions" (Verbicky-Todd 1984, 11). Many groups came together to work cooperatively to make these kills possible. The hunt was also an exciting event where "families and friends were reunited, marriages arranged, stories and experiences shared, trade good exchanged, business conducted, ceremonies held, songs sung, prayers offered" (Brink 2009, 9). All of this was preceded by pre-hunt rituals and concluded with feasting and celebrations that sometimes lasted several days.

The drives did not always succeed. Often the buffalo broke through the drive lanes. The Plains people, however, were eager not to let any buffalo escape. They believed that bison possessed many of the same attributes as people and that they "were aware of the world around them, perceived the behaviour of humans, and recognized patterns of actions and their consequences" (Brink, 157).[24] Buffalo that escaped the fall recognized that they had been tricked and would help other buffalo avoid the trap (158; Krech, 147; Verbicky-Todd, 120). Plains hunters thus "tried to kill all the animals ... because they had to ensure their own future and that of the generations to come. It was not an option, not a decision of conservation or waste; it was the crux of survival," writes Brink (158). Also, wounded and disoriented animals posed a serious danger to the people at the site.

The Head-Smashed-In site was named after a tragic event, the story of which has been handed down through generations of Piikani. Once while the hunt was being prepared, a young boy, still too young to join as a hunter, wanted to watch and crawled beneath the cliff, just below where the great beasts would soon be plunging over the edge. The mounting carcasses of the successful hunt buried him; only after the animals had been butchered was he discovered with his head smashed against the rock. Although this story might be of value for tourism, it has been a constant source of consternation among the Blackfoot elders, who believe that HSIBJ should actually be assigned to another site nearby and that the current location has been given the wrong name. They are quite aware, though, that it would be economically unwise to change the name at this point (Brink, 26).

Involvement with the site is important for the cultural and spiritual capacity-building of the Blackfoot, who maintain a close relationship with their traditions and beliefs. While many Piikani and Kainai have become closely involved with the centre, other Blackfoot believe that the government is exploiting and utilizing Indigenous knowledge and see Blackfoot contributions to the site as a betrayal of their culture. Piikani site interpreter Trevor Kiitokii says that "many of my people feel I'm selling out our culture. They don't like that I'm working here."[25] Others, however, believe it is important to share their knowledge, as Quinton Crow Shoe explains: "I'm not selling out, I'm sharing. Because when I don't share, the colonialists are going to win."[26]

"LOOKING FOR AUTHENTIC INDIANS": STORYTELLING AND NARRATIVES OF THE SITE

When the government developed the storyline for the galleries, they consulted Piikani elders. The government did not, however, consult the elders of the Siksika and the Kainai, and thus the narrative was developed without their voices. The Kainai elders believe that "their people had built and used the buffalo jump every bit as much as the Piikani" and do not always agree with the stories of "their fellow traditionalists of the Piikani" (Brink, 284). This again raises questions of who has the right to interpret another culture and thereby "gain authenticity for one's case" (Braun 2007, 199). This is also reflected in the guided tours. Although a general guideline exists for teaching the programs, each Blackfoot guide approaches the topic differently and offers unique personal insights into Blackfoot culture.[27] While one guide explained that the Blackfoot "never had wars back then," another mentioned that the Blackfoot were "a very war-type people." And even though they are expected to talk only about traditional Blackfoot culture and ways of life, the interpreters also point to the current and historical challenges faced by the Blackfoot people, such as alcohol abuse and the legacy of residential schools. In doing so, they often put their jobs in jeopardy.

Non-Indigenous tourists visit the site with certain expectations. Most come to see the cliff, watch the film, and walk through the centre. They are looking for "authentic Indians," that is, a stereotypical image of the Plains Indians as mounted warriors and buffalo hunters (Susemihl 2007; 2008), so they are sometimes disappointed with the "modern, non-authentic" guides.[28] A survey on visitors' satisfaction found that most important for tourists at Head-Smashed-In was authenticity (64%), followed by an educational experience about people's current ways of life (45%), Native ownership and operation (38%), ticket prices (17%), and entertainment (12%) (Notzke 2006, 86). The general degree of visitors' satisfaction with their experience at the site was rather high; such criticisms as there were tended to focus on the absence of live bison and the desire for more personal contact with and guidance by Indigenous staff (86). About half the visitors questioned either took a guided tour or spoke to the guides individually.[29] All of those surveyed found the guides friendly and knowledgeable. Visitors to the site, however, do not learn about the location of the Blackfoot reserves or about the ownership or management of the site, and many are unsure whether the site is located on a reserve. When visitors were surveyed after visiting the site, it was found that almost one third thought the Blackfoot owned the site and the building (30%), a slightly smaller percentage thought it was owned by the federal or provincial government (25%), and a roughly equal number were not sure (25%). Ten percent of the respondents thought the Blackfoot and the government shared ownership (10%), and another 10 percent thought it was owned by UNESCO.[30]

The various and somewhat controversial (and competing) narratives of the Blackfoot as ecologically sensitive people, stereotypical "Indians," people of the past, and contemporary custodians of their cultural heritage are communicated through interpretions of the site, including the stories told there. These narratives are strongly affected by and dependent on ownership and control of the heritage, community involvement, and visitors' expectations. The stories told and the histories passed on—not only to non-Indigenous visitors but also to a younger generation of Blackfoot—are important in terms of Indigenous empowerment and capacity building, as children and youth learn through heritage about their tribal history and traditions and build an understanding of their culture and identity. The site offers schools and youth groups many different educational programs, including tipi camp sleepovers,[31] but none of these are directed solely towards Indigenous people. The Piikani and to some extent the Kainai have been involved with the heritage site, and their communities have gained certain cultural, spiritual, and economic benefits; that said, there is still limited socio-political empowerment through the UNESCO site for the Indigenous heirs of the ancient buffalo jump.

COMMUNITY CONTROL: INDIGENOUS ENGAGEMENT AT BLACKFOOT CROSSING

At Blackfoot Crossing Historical Park, the Blackfoot are facing a different situation. BCHP opened in 2007 and is the largest (6,000 acres) First Nations–owned and operated museum in Canada today. It is located on the Siksika Reserve, 110 kilometres east of Calgary. A traditional bison-hunting site for the Blackfoot, this crossing on the Bow River was where, in 1877, Treaty No. 7 was signed between the First Nations of what is now southern Alberta and the Canadian government on behalf of the Crown. It is also here that Crowfoot, chief of the Siksika, died and was buried, as was Poundmaker, chief of the Cree. In 1925 this traditional gathering place and the site of the treaty signing were declared National Historic Sites of Canada. BCHP has recently been recommended for World Heritage designation (Hassall 2006, 35).

Here, the Siksika Nation has built a remarkable facility—a gateway to the history and culture of the Blackfoot Confederacy, with excellent exhibits and performances. Initiated by the Siksika Nation, Blackfoot Crossing was built with community money and is run and staffed by community members. This took decades of negotiating and planning by the Siksika in concert with local people and Canadians at large. Widespread interest in the Treaty No. 7 Commemoration in 1977 intensified the vision of the Siksika Nation to build a unique, world-class tourist attraction that would engage visitors in authentic cultural experiences with the Blackfoot people (BCHP 2017). The Historical Park includes a museum, monuments to Chief Crowfoot, Poundmaker, and Treaty No. 7, a tipi village where guests can camp overnight, hiking trails from the eighteenth century, and the remains of an ancient Mandan earth lodge village. The $25 million museum (62,000 square feet) accommodates a Blackfoot heritage exhibition, archives, a library, a theatre, conference facilities, a gift shop, a restaurant, and sacred keeping places. The park also offers an ecotour through the picturesque valley, which contains one of the few remaining pristine prairie river ecosystems.

The Siksika Nation created Blackfoot Crossing Historical Park to preserve and strengthen their culture, educate people about their culture and history, promote the Siksika territory as a tourist destination, encourage economic development, and strengthen socio-political relations with industry and government in the region (Sisco and Stewart 2009, 29). President and general manager Jack Royal strives to "break down stereotypes surrounding Aboriginal-owned businesses and wants anyone visiting the museum to walk away feeling impressed by the experience." To that end, he encourages his staff to "build their knowledge of the Siksika culture and history in order to be able to educate visiting guests."

The gallery exhibits and storylines were developed by a Siksika team. Staff, elders, specialists, and non-Siksika contractors brought together expertise and knowledge from the wider community. Throughout the process, elders acted as consultants and advisers and held the power of veto (Onciul 2015, 113–14). The exhibits' panels use the Blackfoot language, in addition to English, and stress the devastating results of colonization for Blackfoot cultures and peoples. But while emphasizing the living Blackfoot cultures today, they do not detail contemporary problems. Onciul argues that the "desire to create a positive self-representation" is "understandable and natural" and "important in creating community pride" (132). The power and capacities of the Siksika Nation today are presented in a film that visitors are invited to watch, so as to firmly place the culture in the present, whereas the film offered at HSIBJ frames its interpretation in terms of the past. And while at HSIBJ most visitors do not travel through the Piikani reserve, at BCHP they encounter the Siksika community first-hand.

BCHP is owned and managed by the Siksika Nation and can be seen as "an example of citizens' power" (Onciul, 18), referring to the zenith of Arnstein's model of community engagement (Arnstein 1969). However, not all community members can be equally involved, and not all are interested; thus, as Onciul argues, the "idea of citizens' control is misleading" (114). Representation is always a challenge, for it means combining Blackfoot and Western practices of heritage management as well as engaging with the wider Siksika community, although only some of its members have influence in and over the museum. Besides, in order to be officially recognized as a museum and thus to be considered for collection loans and the repatriation of Siksika materials from other museums, BCHP must meet the requirements of the Alberta Museums Association. As a consequence, it has "hybridized its curatorial practice, blending Blackfoot protocol with Western standards" (Onciul, 155).

As a "grassroots community development," the site is a powerful means of Indigenous representation and empowerment. However, Siksika have not been visiting "their" heritage site as often as anticipated, even though they have free entry, and "there are some critics of the centre who really don't see the centre as representing the Siksika Blackfoot culture," as VP of Marketing and Public Relations Shane Breaker notes (quoted in Onciul, 115). Clearly, formal community control requires more than community empowerment; heritage sites that involve the community still face criticisms from local people.

CONCLUSION

Heritage is our legacy from the past, what we live with today, and what we pass on to future generations; it is an irreplaceable source of life and inspiration. Cultural heritage is an important source of empowerment and capacity building for Indigenous communities. Actions outside the boundaries of a heritage site affect its preservation even while communities are partners and agents in the conservation process.

Head-Smashed-In Buffalo Jump is a popular tourist attraction along the Cowboy Trail of southern Alberta, with thousands of visitors each year, and is owned and managed by the Government of Alberta. The situation of the Blackfoot at HSIBJ constitutes an unresolved political conflict, as well as a legacy of colonialism, given that the land was ceded through Treaty No. 7. Because of that treaty, the Blackfoot have no legal title to the land of their ancestors, including to the site of HSIBJ, since it was negotiated at a time before the Task Force on Aboriginal Languages and Cultures.[32] Since the Blackfoot do not own the land and have no authoritative voice in its management, there is no co-management of HSIBJ. Yet the site's attractiveness depends significantly on Blackfoot presence and involvement in its operations. That involvement is not based on formal agreements with the Piikani or Kainai but consists of the contributions of Blackfoot individuals involved in its operations. Their attitude towards HSIBJ, though, does not represent the position of the Piikani or Kainai in general. Without official cooperation, there is no general employment policy for Blackfoot and no revenues go to the Indigenous community. Moreover, improvements in Indigenous employment, living conditions, and education, as well as agency, all seem to exist only on an individual basis. While Blackfoot working as guides, dancers, or sales personnel at HSIBJ identify with the site, there is little identification with the site among many young Blackfoot. Many visitors believe that the Piikani are the owners and managers of the site and thus cannot understand Blackfoot complaints about their life circumstances. They seem to think that the Indigenous community makes use of the revenues generated by the site.

HSIBJ was developed with community consultation from the Piikani (seen as "tokenism" by Arnstein), using "participatory methods that spanned the engagement spectrum at different points in the process and adapted its practice to accommodate some Blackfoot protocol" (Onciul 138). By contrast, Blackfoot Crossing Historical Park has been developed and is run by the Siksika community, which has been indigenizing its management and storytelling and retains control of funding.

These examples demonstrate how representations of Indigenous heritage have been developing in recent decades. Moreover, as museums are part of a Western philosophical tradition (Cruikshank 1992, 6) and as such constitute an "institutionalized invention of the colonizing culture" (Ross and Crowshoe 1996, 253), heritage sites and museums always have to negotiate with colonial (and stereotypical) constructs of "otherness," which permeate assimilation policies and popular culture and also influence Indigenous identity and historical perceptions. "The ability for Blackfoot voice to be heard within the museum is limited by the audience's ability to understand what is said" (Onciul, 163). For improved cross- and inter-cultural communication to take place, it is necessary to develop a "nuanced understanding of how people who are deeply connected to their land" confront these questions and "how these struggles link to a much larger postmodern narrative," as Lea Rekow argues in her chapter in this volume regarding the mining legacy the Navajo Nation is having to confront. Similarly, Nancy C. Doubleday states in her chapter that a "source of community formation and validation" is "necessary to support the development of agency in individuals and groups," and claims that "we are at a point where we need to advance our conceptual, technological, and communicative approaches, and deepen collaborative abilities." As Onciul makes clear, this is also true for community control, which needs to be considered carefully, for it does not empower all community members equally, but "requires models of inter-community engagement" (Onciul, 116).

In this respect, Indigenous voices have recently been raised throughout the First Nations communities of southern Alberta, and major changes and developments in Indigenous awareness and in building agency and engagement among Blackfoot communities throughout Alberta are happening. Through the REDx Talks presented by the Iiniistsi Arts Society and hosted by Mount Royal University in Calgary since 2015, a platform has been established where Indigenous people can share their accomplishments and shatter myths, and the constructive impact of this on young people is growing. According to Blackfoot curator Cowboy Smithx, the speaker series "act[s] as a tribute to the loss of children in residential schools, language and cultural genocide."[33] In his recent documentary *Elder in the Making* (2015), Smithx travels across Blackfoot territory, exploring and experiencing his traditional homeland. Natural and cultural heritage, including the buffalo jump, is playing a crucial role in the new generation's search for identity and belonging.

For Indigenous empowerment and capacity-building through Cultural World Heritage, Indigenous involvement in decision-making is essential. For heritage preservation to succeed, it is vital that all stakeholders, including communities, be involved in conservation processes. Heritage, together with education, can become a basis for constructing meanings. Visitors, however, must be able to read the context of a site if they are to derive significant meaning from it. This conception of heritage aligns with the basic human needs for orientation and identity, thereby creating vital relations between humans and their environment so as to give sense and order to the world. In this context, it becomes essential that the stories about Indigenous people provided at the sites be told from Indigenous perspectives reached by consensus. At HSIBJ, Indigenous storylines about the contemporary situation of the Blackfoot need to be included, for heritage always educates about the present as well as the past. Thus, heritage needs to be "moving forward, creatively re-defining itself, seeking engagement in critical agendas relating to conflict resolution" and inter-cultural dialogue (Aguda, Tamayo, and Barlan 2013, 91). In this way, heritage tourism "can relate to the emerging generations of tourists who seek not only to passively observe the past, but to learn from it and, where appropriate, challenge and change it" (91).

The challenges posed by storytelling and education also show that tangible and intangible cultural heritage cannot be examined separately, for material culture can be understood and appreciated only in the context of appropriate knowledge of the natural world, with all the narratives that come with it. The ways in which intangible properties are valued and thereby order social behaviour are of vital importance for the relationships among people engaged with such properties. These systems of intangible property shed light on "the basic principles and objectives underlying their protection within the appropriate customary cultural frameworks" (Thom and Bain 2004, 43). Understanding and protecting tangible and intangible cultural heritage, including traditional knowledge, requires a systemic approach, and cultural protocols around the regulation of traditional knowledge are inherent in traditional knowledge systems. Only if these are respected can the expected teaching follow. Only then will Cultural World Heritage—which is fundamental to the continuity, revival, and survival of Indigenous cultural identity—be an asset in the empowerment and capacity-building efforts of First Nations.

Notes

1 Head-Smashed-In Buffalo Jump was inscribed in 1981 under criterion vi "to be directly or tangibly associated with events or living traditions, with ideas, or with beliefs, with artistic and literary works of outstanding universal significance." UNESCO (n.d.), *The Criteria for Selection*.

2 The site was used for the slaughter of bison from 3600 BC to 2600 BC, then intermittently towards 900 BC, and finally, continuously from AD 206 to 1850. ICOMOS (1981).

3 UNESCO (2003, 2005).

4 In 1972, UNESCO adopted the *Convention Concerning the Protection of the World Cultural and Natural Heritage* in order to identify cultural and natural heritage worldwide, and to provide organized international protection for World Heritage Sites. As of March 2012, 189 State Parties had ratified the World Heritage Convention. Currently, the World Heritage List includes 962 properties, including 745 cultural, 188 natural, and 29 mixed properties in 157 States Parties. UNESCO (n.d.), *World Heritage List*.

5 The study is part of a larger research project on world heritage and Indigenous empowerment. In 2011 a faculty research grant of the Canadian government enabled me to travel to HSIBJ to examine the site and conduct interviews with the Blackfoot, representatives of the Canadian government, and scholars of cultural conservation and heritage studies. My journey proved to be a venture into

the historical, scientific, cultural, and spiritual world of the Blackfoot, and I am grateful for the insights and teachings of my guides and interview partners. The interviews I conducted followed a structured questionnaire, dealing with four main areas of interest. The first set of questions concerned UNESCO in general; the second concentrated on ownership, management, and control of the heritage site; the third dealt with community involvement and identity; and the fourth with education and tourism. Each interview was recorded and partly transcribed. Using a structured questionnaire, I also conducted a visitors' survey, inquiring about visitors' expectations and impressions of the site, access to information, and their general understanding of cultural heritage, UNESCO, and the site. In 2012 I was fortunate to visit Blackfoot Crossing Historical Park, undertaking studies and research there.

6   The origins of the theory of empowerment are associated with the Brazilian humanitarian and educator Paolo Freire (1971; 1973), who proposed liberating the oppressed people of the world through education. See also Biancalana (2007) and Hur (2006).

7   As a consequence of the signing of Treaty No. 7 there has been no question concerning ownership of the land. Treaty No. 7, concluded on 22 September 1877, was an agreement between the British Crown and several, mainly Blackfoot, First Nations of southern Alberta that established a delimited area of land for the tribes as reserves, promised annual payments and provisions from the British government to the tribes, and guaranteed continued hunting and trapping rights on the "tract surrendered." In exchange, the tribes ceded their rights to their traditional territory, of which they had been recognized as the owners. Aboriginal Affairs and Northern Development Canada, *Copy of Treaty No. 7*.

8   The exhibition "Napi's World" explores and explains the geography, climate, and vegetation of the Northwestern Plains. "Napi's People" details the ancient culture and way of life of prehistoric Plains peoples. "The Buffalo Hunt" presents how a buffalo jump worked and describes pre-hunt ceremonies as well as how the buffalo were gathered, driven, and killed at the cliff. "Cultures in Contact" charts the decline of traditional buffalo hunting, with the arrival of Europeans and the introduction of the horse and gun. A re-created archaeological dig and slide at the exhibit "Uncovering the Past" explores the science of archaeology and shows how we study and learn about the past.

9   The educational programs at HSIBJ include "Buffalo Tales" (ages 4–6), "Living Long Ago" (grades 1–3), "Living Off the Land" (grades 4–6), "History Underground" (grades 5–9), "Social Organization" (grades 7–9), "Sticks and Stones" (all grades), and a facility tour (all grades). A portable education program, the "'Contrast' Resource Kit," featuring a slide show, photographs, and a teachers' guide and documenting the lifestyle of the Blackfoot before and after the Europeans' arrival in southern Alberta, is available to schools.

10 Visitors' numbers have been decreasing, from 150,000 in 1987 to 90,000 in 2001 to 60,000 in 2011. Interview with Deloralie Brown, Head of Finances and Visitors' Services, 13 October 2011 (unpublished).

11 Historic sites in southern Alberta include, besides HSIBJ, the Frank Slide Interpretive Centre, the Remington Carriage Museum, and the Lougheed House Historic Site.

12 The author has precise information about budgets, expenses, and revenues.

13 Usually Kainai contractors are also hired, but since their reserve is 50 kilometres away and driving takes much of their earnings, only a few apply. The Piikani

Reserve is about 40 kilometres away, and most of the staff are carpooling. There are no employees from the Siksika as their reserve is 200 kilometres north of Fort Macleod.

14 Interview with Deloralie Brown. Since employment at HSIBJ requires a driver's licence and a reliable vehicle, and many Blackfoot have neither, there are sometimes also openings for non-Indigenous applicants.

15 The Peigan Crafts moccasin factory in Brocket, established in 1976, was the first in Canada to manufacture authentic First Nations moccasins, mukluks, and mitts. It closed in 2011 because of falling support within the community. Interview with Ian Clarke, Regional Director, Southern Operations, Historic Sites and Museums Branch, 12 October 2011 (unpublished).

16 Interview with Catherine Bell, University of Alberta, 14 October 2011 (unpublished).

17 According to Deloralie Brown, Indigenous reliability and punctuality are tested now and again: "When an elder dies, everybody goes to the funeral, and everybody of the staff is gone."

18 Interview with Quinton Crow Shoe, Site Marketing and Program Coordinator, HSIBJ, 14 October 2011 (unpublished).

19 Interview with Quinton Crow Shoe.

20 The Piikani Nation, the smallest Blackfoot tribe to sign Treaty No. 7 in 1877, count about 3,900 people today, about 1,900 of them living on reserve at *Peigan 147*, an land parcel of 430.31 km² between Fort Macleod and Pincher Creek in southern Alberta. The Kainai Nation (or Blood) count around 10,000 members; their reserve, *Blood* 148, around Stand Off, is currently the largest in Canada, with 3,850 inhabitants on 1,414.03 km². The Siksika Nation, with a total population of 3,000, has its home at the reserve *Siksika 146*, where around 2,700 people live on 696.56 km². Statistics Canada. (http://www12.statcan.gc.ca).

21 Interviews with Ian Clarke and Edwin Small Legs, guide at HSIBJ, 16 October 2011 (unpublished).

22 Interview with Ian Clarke.

23 Interview with Duncan Daniels, Head of Regional Marketing, Government of Alberta, 12 October 2011 (unpublished).

24 Generally, Indigenous peoples regard animals as non-human relatives and teachers. They are highly respected for their spiritual abilities and must be treated according to certain protocols. Relationships with animals are of great importance to Indigenous peoples. In this vein, Morgan Zedalis and Sean Gould, in their chapter "Reacting to Wolves: The Construction of Identity and Value" in this volume, discuss the importance of wolves as they relate to peoples' identity and the historical and political effects of human-environment and social interactions. See also Bruchac (1992); Deloria Jr. ([1973]2003); and Kuchylski. McCaskill, and Newhouse (2003), among others.

25 Interview with Tevor Kiitokii, guide at HSIBJ, 13 October 2011 (unpublished).

26 Interview with Quinton Crow Shoe.

27 I participated in the "Living Off the Land" program on 13 October 2011 with Edwin Small Legs, and on 18 October 2011 with Trevor Kiitokii.

28 Person interviewed during a survey at HSIBJ (2011).

29 A survey of thirty visitors was conducted at HSIBJ by the author in October 2011.

30 Visitors' statements as to the ownership include: "The tribe itself owns it, which is amazing. I know that they run it, and it's well done"; "I hope that the Natives do, but I think the government does"; "I think it is shared. Is it on the reserve?";

"The First Nations own it, through the government." Survey at HSIBJ, October 2012.
31 See HSIBJ (2012), *Education Programs 2012.*
32 The Task Force on Aboriginal Languages and Cultures was appointed in 2002 by the Minister of Aboriginal Heritage. In June 2005 it published *Towards a New Beginning—A Foundational Report for a Strategy to Revitalize First Nation, Inuit, and Métis Languages and Cultures,* in which it recognized the importance for Indigenous people "of maintaining a close connection to the land in their traditional territories, particularly wilderness areas, heritage and spiritual or sacred sites" and recommended "their meaningful participation in stewardship, management, co-management or co-jurisdiction arrangements."
33 Edwardson (2015).

References

Aboriginal Affairs and Northern Development Canada. *Copy of Treaty No. 7.* http://www.aadnc-aandc.gc.ca/eng/1100100028793. Accessed 17 February 2012.

Aguda, Lesley Allaine E., Ma. Rosario Tamayo, and Leoncio Barlan Jr. 2013. "Effects of Heritage Tourism to the Municipality of Taal, Batangas, Phillipines." *Educational Research International* 2(1): 91–95.

Alberta Community Development. 2004. *Dynamic Diverse Destinations in Southern Alberta: Education Programs 2002–2004.* Calgary.

Armstrong, Jeannette. 2007. "Kwatlakin? What Is Your Place?" In *What Is Your Place? Indigeneity and Immigration in Canada,* edited by Hartmut Lutz with Thomas Rafico Ruiz, 29–33. Augsburg: Wißner Verlag.

Arnstein, Sherry R. 1969. "A Ladder of Citizen Participation." *Journal of the American Planning Association* 35(4): 216–24.

Asch, Michael. 2009. "Concluding Thoughts and Fundamental Questions." In *Protection of First Nations Cultural Heritage: Laws, Policy, and Reform,* edited by Catherine Bell and Robert K. Paterson, 394–411. Vancouver: UBC Press.

BCHP (Blackfoot Crossing Historical Park. *About Us.* http://www.blackfootcrossing.ca/about-us.html. Accessed 28 October 2017.

Bell, Catherine, and Robert K. Paterson, eds. 2008. *First Nations Cultural Heritage and Law: Case Studies, Voices, and Perspectives.* Vancouver: UBC Press.
———. 2009. *Protection of First Nations Cultural Heritage: Laws, Policy, and Reform.* Vancouver: UBC Press.

Biancalana, Renata Neves. 2007. *The Importance of Heritage in Community Education: The Case of Serra da Capivara National Park—Brazil.* MA thesis, Cottbus: Brandenburgische TU, Cottbus.

BCHP (Blackfoot Crossing Historical Park. *About Us.* http://www.blackfootcrossing.ca/about-us.html. Accessed 28 October 2017.

Braun, Sebastian F. 2007. "Ecological and Un-ecological Indians. The (Non)portrayal of Plains Indians in the Buffalo Commons Literature." In *Native Americans and the Environment: Perspectives on the Ecological Indian,* edited by

Michael E. Harkin and David Rich Lewis, 192–208. Lincoln: University of Nebraska Press.

Brink, Jack W. 2008. *Imagining Head-Smashed-In: Aboriginal Buffalo Hunting on the Northern Plains.* Edmonton: Athabasca University Press.

Bruchac, Joseph. 1992. *Native American Animal Stories.* Golden: Fulcrum.

Calloway, Colin G. 2003. *One Vast Winter Count: The Native American West before Lewis and Clark.* Lincoln: University of Nebraska Press.

Cruikshank, Julie. 1992. "Oral Tradition and Material Culture: Multiplying Meanings of 'Words' and 'Things.'" *Anthropology Today* 8(3): 5–9.

Deloria Jr., Vine. [1973]2003. *God Is Red: A Native View of Religion.* Golden: Fulcrum.

Eigenbrod, Renate. 2005. *Travelling Knowledges: Positioning the Im/Migrant Reader of Aboriginal Literature.* Winnipeg: University of Manitoba Press.

Edwardson, Lucie. 2015. "REDx Talks spurs cultural conversation." *Metronews.ca*, 7 October. http://www.metronews.ca/news/calgary/2015/10/06/redx-talks-spurs-cultural-conversation.html. Accessed 7 July 2016.

Freire, Paulo. 1971. *Pedagogy of the Oppressed.* New York: Seabury Press.

———. 1973. *Education for Critical Consciousness.* New York: Continuum.

Gerberich, Victoria L. 2005. "An Evaluation of Sustainable American Indian Tourism." In *Indigenous Tourism: The Commodification and Management of Culture*, edited by Chris Ryan and Michelle Aiken, 75–86. Amsterdam: Elsevier.

Gibson-Pugsley, Monica. 2016. "Documentary: Elder in the Making." http://www.allsaints-agassiz.ca/pages/documentary-elder-in-the-making-march-9-2016. Accessed 24 November 2018.

Hassall, Kate. 2006. *Partnerships to Manage Conservation Areas through Tourism: Some Best Practice Models between Government, Indigenous Communities, and the Private Sector in Canada and South Africa.* The Churchill Trust. http://www.compassodyssey.net/documents/Kate_Hassall_Churchill_Fellowship_May_2006.pdf. Accessed 28 October 2017.

HSIBJ. (Head-Smashed-In Buffalo Jump). n.d. http://history.alberta.ca/headsmashedin/default.aspx. Accessed 7 September 2012.

———. n.d. *Buffalo Tracks: Educational and Scientific Studies from Head-Smashed-In Buffalo Jump.* Alberta. http://www.history.alberta.ca/headsmashedin/docs/buffalo_tracks.pdf. Accessed 31 May 2013.

———. 2012. *Education Programs 2012.* http://history.alberta.ca/headsmashedin/educationprograms/docs/hsibj_edprogrmflyer_12.pdf. Accessed 7 September 2012.

Hsiung, Chris (producer/director/editor). *Elder in the Making.* 2015. 95 min.

Hur, Mann Hyung. 2006. "Empowerment in Terms of Theoretical Perspectives: Exploring a Typology of the Process and Components across Disciplines." *Journal of Community Psychology* 34(5): 523–40.

ICOMOS. 1981. *World Heritage List No. 158—Advisory Body Evaluation.* http://whc.unesco.org/archive/advisory_body_evaluation/158.pdf. Accessed 25 May 2013.

————. 1993. *Cultural Tourism: Tourism at World Heritage Cultural Sites: The Site Manager's Hand Book*. Sri Lanka: National Committee of ICOMOS.

Kirshenblatt-Gimblett, Barbara. 1998. *Destination Culture: Tourism, Museums, and Heritage*. Berkeley: University of California Press.

Kramer, Jennifer. 2006. *Switchbacks: Art, Ownership, and Nuxalk National Identity*. Vancouver: UBC Press.

Krech, Shepard, III. 1999. *The Ecological Indian: Myth and History*. New York: W.W. Norton.

Kulchyski, Peter, Don McCaskill, and David Newhouse, eds. 2002. *In the Word of Elders: Aboriginal Cultures in Transition*. Toronto: University of Toronto Press.

Le Goff, Jacques. 1992. *History and Memory*. New York: Columbia University Press.

Lutz, Hartmut. 2007. "'To Know Where Home Is': An Introduction to Indigeneity and Immigration." In *What Is Your Place? Indigeneity and Immigration in Canada*, edited by Hartmut Lutz with Thomas Rafico Ruiz, 9–28. Augsburg: Wißner.

Mathew, Jose. "Importance of Heritage." http://www.indiastudychannel.com/resources/8742-Importance-heritage.aspx. Accessed 27 June 2012.

Nelson, Robert M. 1993. *Place and Vision: The Function of Landscape in Native American Fiction*. New York: Peter Lang.

Nicholas, George P. 2006. "Decolonizing the Archaeological Landscape: The Practice and Politics of Archaeology in British Columbia." *American Indian Quarterly* 30(3–4): 350–80.

Notzke, Claudia. 2006. *"The Stranger, the Native and the Land": Perspectives on Indigenous Tourism*. Concord: Captus Press.

NSW Heritage Office. 2011. *Assessing Heritage Significance*. Sidney: NSW Heritage Office. http://www.heritage.nsw.gov.au/docs/assessingheritagesignificance.pdf. Accessed 22 January 2012.

Onciul, Bryony. 2015. *Museums, Heritage, and Indigenous Voice: Decolonizing Engagement*. London: Routledge.

Patil, Trupti. 2017. "Historical Monuments and Tourism Development (A Case Study on Shirval)." *Scribd*. https://de.scribd.com/document/32748732/Heritage-tourism. Accessed 28 October 2017.

Piikani Nation. http://www.piikanination.com. Accessed 27 January 2013.

Ross, Michael, and Crowshoe, Reg. 1996. "Shadows and Sacred Geography: First Nations History-Making from an Alberta Perspective." In *Making Histories in Museums*, edited by Gaynor Kavanagh, 240–56. London: Leicester University Press.

Sisco, Ashley, and Nicole Stewart. 2009. *True to Their Visions: An Account of Ten Successful Aboriginal Businesses*. Report November 2009. Ottawa: Conference Board of Canada. http://abdc.bc.ca/uploads/file/09%20Harvest/10-131_TrueToTheirVisions_WEB.pdf. Accessed 28 October 2017.

Statistics Canada. 2006. "Community Profiles: Blood 148." http://www12.statcan
.gc.ca/census-recensement/2006/dp-pd/prof/92-591/details/page.cfm
?Lang=E&Geo1=CSD&Code1=4803802&Geo2=PR&Code2=48&Data
=Count&SearchText=Blood&SearchType=Begins&SearchPR=01&B1=All
&Custom=. Accessed 01 Feb 2013.

———. 2006. Community Profiles: Piikani 147. http://www12.statcan.gc
.ca/census-recensement/2006/dp-pd/prof/92-591/details/page.cfm?Lang
=E&Geo1=CSD&Code1=4803801&Geo2=PR&Code2=48&Data=Count&
SearchText=Piikani&SearchType=Begins&SearchPR=01&B1=All&
Custom=a. Accessed 1 February 2013.

———. 2006. Community Profiles: Siksika 146. http://www12.statcan.gc
.ca/census-recensement/2006/dp-pd/prof/92-591/details/page.cfm?
Lang=E&Geo1=CSD&Code1=4805802&Geo2=PR&Code2=48&Data
=Count&SearchText=Siksika&SearchType=Begins&SearchPR=01&B1=All&
Custom=. Accessed 1 February 2013.

Susemihl, Geneviève. 2007. "The Visual Construction of the North American
Indian in the World of German Children." In *Visual Culture Revisited*, edited
by Nicole Leonhardt et al., 267–91. Köln: von Halem.

———. 2008. "The Imaginary Indian in German Children's Non-Fiction Literature." In *Aboriginal Canada Revisited*, edited by Kerstin Knopf, 122–56.
Ottawa: University of Ottawa Press.

Task Force on Aboriginal Languages and Cultures. 2005. *Towards a New Beginning: A Foundational Report for a Strategy to Revitalize First Nation, Inuit, and
Métis Languages and Cultures: Executive Summery.* Ottawa. http://knet.ca/
documents/ALCC-Task-Force-Executive-Summary.pdf. Accessed 7 September
2012.

Thom, Brian, and Don Bain. 2004. *Aboriginal Intangible Property in Canada:
An Ethnographic Review.* Ottawa: Industry Canada. http://www.ic.gc.ca/eic/
site/ippd-dppi.nsf/vwapj/Thom_Final_Report_e_proofed_28feb05.pdf/$FILE/
Thom_Final_Report_e_proofed_28feb05.pdf. Accessed 26 June 2012.

Thomas, K.W., and B.A. Velthouse. 1990. "Cognitive Elements of Empowerment:
An 'Interpretive' Model of Intrinsic Task Motivation." In *Academy of Management Review* 15(4): 666–81.

UNESCO. 2003. *Convention for the Safeguarding of the Intangible Cultural Heritage.* Paris.

———. 2005. *Convention on the Protection and Promotion of the Diversity of Cultural
Expressions 2005.* http://portal.unesco.org/en/ev.php-URL_ID=31038&URL
_DO=DO_TOPIC&URL_SECTION=201.html. Accessed 31 May 2013.

———. n.d. *Head-Smashed-In Buffalo Jump.* http://whc.unesco.org/en/list/158.
Accessed 19 January 2013.

———. n.d. *The Criteria for Selection.* http://whc.unesco.org/en/criteria. Accessed
12 January 2012.

———. n.d. *World Heritage.* http://whc.unesco.org/en/about/. Accessed 31 May
2013.

———. n.d. *World Heritage List*. http://whc.unesco.org/en/list. Accessed 9 February 2013.

Verbicky-Todd, Eleanor. 1984. *Communal Buffalo Hunting among the Plains Indians: An Ethnographic and Historical Review*. Archaeological Survey of Alberta Occasional Paper no. 24. Edmonton: Alberta Culture Historical Resource Division.

Walsh, Kevin. 1992. *The Representation of the Past: Museums and Heritage in the Post-Modern World*. London: Routledge.

CHAPTER 6

Mapping the Mining Legacy of Navajo Nation

Lea Rekow

This chapter describes how the communities on and around the Navajo (Diné) Nation are integrally connected to the legacies of coal and uranium and how these industries have produced a vast social and physical landscape overwhelmingly defined by agencies of extraction and milling for the purposes of energy and defence. It illustrates how this has resulted in a feeble Indigenous economy dependent on mining royalties and exposes the interrelated corporate/government system that operates with impunity to profit from it. It reveals a people who systematically endure the fallout from the poisoning of their lands and bodies, the subjugation of their rights, and the poverty in which they are trapped. It depicts the policy advocacy and environmental reclamation work tirelessly being undertaken by Indigenous and regional stakeholders themselves to confront these injustices, mitigate the damages, and demand accountability. Finally, this chapter explores how a panoply of active and abandoned industrial sites tied to uranium and coal extraction continue to be agentive through time in this region of the United States.

The impact of coal and uranium on the cultural ecology of the Navajo Nation and its surrounding tribal areas is demarcated by landscapes that have been bulldozed, blasted, ripped open, dug into, contaminated, and at times recontoured to create entirely new forms of post-mining terrain. To witness these scarred wastescapes—the mountainsides of radioactive sands, the shimmering ponds of toxic chemicals, the carcasses of abandoned mines, the murky rivers that cut their way through the desert, the skeletal remains of processing mills, the towering stacks of power plants spewing noxious smog, and the mammoth burial cells that entomb the most lethal elements of this legacy—is spectacularly tragic. These visual markers link the patterns of natural resource consumption to the waste and contamination associated with them. From the air, it is easier to grasp the immense scale of it all and how it has led to the environmental annihilation of the American Southwest. At this scale, it also becomes strikingly evident why the mining, defence, and energy industries require the implementation of a land policy framework to legislate and enforce far more stringent forms of environmental accountability.

However pervasive the impact of resource extraction is on the southwestern United States as a region, it is the Navajo who suffer most. The Navajo Nation is the largest recognized tribe in the United States, in terms of both population and size. The reservation sprawls across the Four Corners region, covering a vast desert area the size of West Virginia. It is a semi-autonomous zone bound to Indigenous, federal, and corporate land title issues, which in turn are tied to the legacies of coal and uranium mining.

The Navajo people govern some of the most sought-after resources in the world and should be among the most affluent of populations, but instead they experience extreme social exclusion and oppression as a people. In this region the average annual income is a little more than $7,000 and half the population is unemployed (Navajo Nation Division of Economic Development n.d.) and living without running water, electricity, indoor plumbing, or refrigeration (Bray 2005). Since the late 1920s, the Navajo have been offered false assurances of prosperity from leasing their lands to outside corporate mining interests; this has steadily weakened their economy and devastated their lands (Singer 2007, 29–31).

The uranium mined to develop the first atomic weapons for the Manhattan Project and the Second World War, and later harvested for the United States' massive nuclear weapons stockpile and nuclear energy facilities, has burdened the southwestern United States, and in particular its Indigenous peoples, with a legacy of radioactive waste and contamination. Private companies leased tribal lands and hired Navajo workers to mine and process uranium, which was delivered to the Atomic Energy Commission (AEC) from the end of the Second World War through to the mid-1970s. All uranium mined during this time was sold to the AEC, the exclusive agent for uranium in the United States from 1946 until 1974, when it closed in the face of harsh criticism, to later be reorganized under the umbrella of the Department of Energy.

Uranium mining on the Navajo Nation continued until 2005, when a moratorium was declared. However, leases for uranium mining continue to threaten hundreds of thousands of acres that surround tribal land. Though uranium mining is currently suspended, coal mining continues to proliferate. This has resulted in a failed Indigenous economy dependent on fossil fuels. More than half the Navajo Nation's annual revenues are derived from coal mining (Navajo Nation Division of Economic Development n.d.), thousands of Navajo rely on the industry for employment, and coal taxes provide the only funds for reclamation. Yet despite the revenues brought in by the coal industry, more than 40 percent of the Navajo Nation's 183,000 inhabitants live below the poverty line (Navajo Nation Division of Economic Development n.d.).

New Mexico has the United States' fourth-largest coal reserves. The massive 5,600-acre Chevron-owned McKinley coal mine near Gallup, New Mexico, bulldozed and blasted its way through 200 to 500 feet of surface material daily between 1962 and 2010, to produce an estimated 175 million tons of coal, until its reserves were largely exhausted and the mine became unprofitable.

The McKinley mine provides a template for understanding some of the health and environmental costs linked to surface removal in the region. Coal mining is associated with chronic heart disease, kidney disease, higher mortality rates, and lung cancer (Hendryx and Ahern 2009, 541–50). Around 2 percent of surface miners are diagnosed with black lung disease after working for no more than a year at the mines (National Institute for Occupational Safety and Health 2011).

Environmental fallout includes pollution from radioactive contaminants and heavy metals, which chronically degrade arroyos and aquifers. The loss of piñon pines and topsoil has destabilized the region further by making it more vulnerable to erosion. Surface removal has also led to a severe decline in wildlife species due to loss of habitat. The industry engages in some reclamation through reseeding, but there is no evidence that replanting mitigates the devastating environmental impact wrought by surface removal (Palmer 2010, 148–49). And as Chevron's PR representative admitted as he drove me around the excavated remnants of hills and valleys, it just isn't possible to rip up

Fig. 6.1. McKinley coal mine.

these grand, hardy natives that have taken hundreds of years to grow, and expect to create ecological stability by spreading grass seeds.

During the mine's long lifespan, the coal from McKinley was transported by rail to fuel half a dozen coal-fired power plants that power the state of Arizona, including the notorious Cholla plant, which ranks eighth on the list of most polluting coal plants in the United States (Institute for Southern Studies 2009) and releases more than three and a half million pounds of fine particle pollution into the Arizona airshed every year.

Producing electricity by burning coal is a simple process. Chunks of coal are crushed and burned to boil water to create steam that spins turbines to generate electricity. In an average year, a typical coal plant generates a lethal array of almost four million tons of airborne toxins. For example, a coal plant can produce up to 750 pounds of mercury per year, this while only one seventieth of a teaspoon of mercury in a midsize lake will render the fish unsafe for human consumption (Union of Concerned Scientists n.d.). At the time of this writing, there were 589 coal-fired power plants operating in the United States, with another 153 new plants scheduled to open by 2030 (Shuster 2007).

A couple of hours' drive north of McKinley, the San Juan and Four Corners coal-fired power plants are located on the rim of the reservation, within fifteen miles of each other between the towns of Farmington and Shiprock, New Mexico. These generating stations have historically been two of the heaviest polluters in the United States. They emit a continuous stream of particulate matter as a haze of thick smog that drifts more than 150 miles over the nation's most prized natural heritage sites, to settle throughout the valleys of the Grand Canyon, Canyonlands, Arches, Bryce, and Zion National Parks.

Fig. 6.2. Stream bordering McKinley mine.

Since the early 1960s, the Four Corners plant has generated electricity for 1.7 million houses across the non-Native Southwest, bypassing the residents of the Navajo Nation, half of whom still have no electricity in their homes, even though the transmission lines run overhead. Until late 2013, when three of its five stacks closed, the nitrogen oxide (NOx) that belched from the plant produced more NOx than any other coal-fired power station in the United States. This pollution has contributed to more than 40 premature deaths, 800 asthma attacks, and other associated health problems that cost taxpayers around $340 million annually (Nichols n.d.). The sting is that the plant brings in only $225 million in economic revenue per year (Arizona Public Service 2013).

When NOx is overproduced, it damages forests and crops and eats through both natural and synthetic materials. It also causes tissue and cell damage, respiratory problems, eye irritation, nausea, and headaches. The San Juan generating station dumps almost 15 million tons of toxins into the Four Corners airshed every year, including 28,000 tonnes of NOx (Sands 2005). But times are changing for this

Fig. 6.3. San Juan power plant.

generating station as well. The plant's owner, the Public Service Company of New Mexico, is now decommissioning two of its coal-burning stacks.

The coal-burning process also severely impacts clean water reserves. Coal plants use billions of gallons of pristine water every year as a coolant, as well as to extract and wash coal, to generate the steam to make electricity, to manage the coal ash pollution produced during the burning process, and to produce coal slurry to use as liquid fuel.

The Peabody Coal Company uses vast quantities of water in its mining operations at Black Mesa. For 50 years, Peabody has siphoned enormous quantities of water from the Navajo Aquifer (N-Aquifer), the sole source of drinking water for the Navajo and Hopi people. *The water has taken tens of thousands of years to accumulate and cannot be replaced on a human time scale.* In addition, it is now laced with arsenic and other pollutants.

Peabody owns two coal mines on Black Mesa—the Black Mesa coal mine, which closed in 2006, and the Kayenta mine, which is still operational. For more than 40 years, the Black Mesa mine piped coal slurry more than 270 miles through Arizona to the now

Fig. 6.4. Four Corners power plant, with coal ash ponds in the foreground.

decommissioned Mojave Generating Station in Nevada, to provide inexpensive electricity to the cities of Las Vegas, Los Angeles, and Phoenix. During the mine's operating life, the N-Aquifer was depleted by approximately 17¼ billion gallons of water (Yurth 2011). That's about enough water to fill 20,000 football-field-size swimming pools, each three metres deep.

Peabody continues to use almost 400 million gallons of water from the N-Aquifer annually in its operations at Kayenta, lowering its levels by more than 100 feet per year (Department of Economic and Social Affairs 2007). The Hopi and Navajo tribes are paid $1.1 million per year for that water, but within a generation their natural springs and watering holes have become little more than dustbowls. Peabody refuses to accept responsibility for the aquifer's radical depletion (Yurth 2011).

In addition, thousands of Navajo and Hopi have been ordered to relocate from Black Mesa without any compensation or assistance. This has forced many families into bankruptcy—families that are also witnessing the destruction of their culture, their burial grounds, and the sacred lands they have occupied for more than 7,000 years (Macmillan

2013). In 2008, Peabody was granted permission to expand its mining operations at Kayenta by an additional 19,000 acres. The federal
government waved environmental regulations in order for them to do
so. A long court battle between the Office for Surface Mining and environmental activist groups ensued to have the life-of-mine expansion
revoked (Berry 2011). Unfortunately, the fight is not over.

Since 2014, Diné families on Black Mesa have been raided by
SWAT teams, drones, and armed helicopters. These assaults are exe-
cuted by Hopi Rangers and backed the federal Bureau of Indian Affairs
and the Department of the Interior. The justification is that herd sizes
are exceeding their permit limits and that overgrazing is harming the
land during a time of extended drought. For their part, the residents
point to Peabody's mining operations as the principal cause of land
degradation and drought and claim that the raids, funded and insti-
gated by the federal government, are part of an ongoing campaign
to access coal resources by forcing Navajo off the land. Peabody's
proposed mining expansion falls on the land targeted by the impound-
ments (Minno Bloom 2014).

In 2016, after a prolonged industry downturn, Peabody filed for
Chapter 11 bankruptcy protection. This has allowed Peabody to seek
protection from its creditors, reorganize its holdings, and receive $800
million in financing so that it can continue conducting business as
usual. Operations at the Kayenta mine have been continued on the
basis of its "cash positive" status.

Kayenta provides fuel to the coal-fired Navajo Generating Station
(NGS) in Page, Arizona, near Lake Powell. The NGS in turn provides
the power to divert the waters of the Colorado River hundreds of
miles through the desert via the Central Arizona Project to artificially
fabricate a livable Arizona. The legacy of the Cold War uranium frenzy
is also channelled in this river water—water that is now contaminated by millions of tons of radioactive tailings that lie beneath Lake
Powell, and that were deposited for decades into the Colorado, La
Plata, and San Juan rivers (Spangler and Spangler 2001). Ironically,
NGS's majority stakeholder is the Federal Bureau of Reclamation,
whose mission is "managing water in the West" (US Department of
the Interior Bureau of Reclamation n.d.). Just as ironically, the motto
for the Central Arizona Project is "your water, your future" (Central
Arizona Project n.d.).

Enormous quantities of water continue to be contaminated throughout the Southwest. The San Juan and Four Corners power plants siphon away the water of the San Juan River as it snakes through the stark desert landscape. Along the way, that river is polluted by mercury and selenium produced by the Navajo Mine (Center for Biological Diversity 2012), the only mine owned by the Navajo Nation itself (Guerin 2014). Ten miles downstream, its waters are further poisoned by a massive plume of radioactive contamination that leaches into it from a monolithic uranium disposal cell the size of 80 football fields that sits perched on the riverbank.

Since the 1940s, the extensive mining and milling of uranium ore for defence and energy purposes has produced a prolific legacy of abandoned uranium mines (AUMs) in the region. The process for extracting usable uranium from the earth is highly toxic. Uranium ore found in sandstone is finely ground and leached with sulfuric acid. The acid removes the sought-after isotopes, and the leftover waste sands ("tailings") are, for the most part, left literally blowing in the wind. These tailings contain 85 percent of the ore's original radioactivity and 99.9 percent of its original volume. There are now at least 200 million tons of tailings scattered around the West (US Environmental Protection Agency n.d.). Meanwhile, the liquid produced by the acid milling process, known as liquor, is dissolving into highly radioactive isotopes. The tailings and liquor need to be isolated in order to reduce health risks, but this has yet to be achieved with any long-term success.

The vast majority of tailings piles, holding hundreds of millions of tons of radioactive waste, are located in the Four Corners/Navajo Nation region. Radioactive waste lies in the open on abandoned mines and mill sites and cascades down the sides of mountains and mesas; it finds its way into water supplies and even building materials. Unreclaimed, highly contaminated sites laden with unsecured radioactive waste are mostly left unmarked, and access to them is often unrestricted. The seductive natural beauty of these sites can be deceptively inviting to both people and wildlife; it is almost impossible to tell that these sites are highly radioactive.

The Indigenous Southwest's vast, breathtaking desert landscape is saturated with radioactive contamination. Few visitors who stand on the observation deck at the Monument Valley Visitor Center, marvelling at this great natural heritage, realize that the mesa to their right,

Fig. 6.5. Monument Valley abandoned uranium mine with radioactive waste in foreground.

Mitchell Butte, is an abandoned uranium mine that has deposited its fair share of alpha dust particles into the fresh wilderness air (Chenoweth 1995).

Less than 20 miles away, on the edge of the highway that leads to the Monument Valley Visitor Center, lies the immense Uranium Mills Tailings Remediation Action (UMTRA) site of Mexican Hat. Over the course of its operating life, the Mexican Hat mill processed about 11.4 million pounds of Uranium 308 for the AEC. Ore operations ceased in 1965, but a sulfuric acid plant continued to operate until 1970. The site initially contaminated 550 acres, but by the time it was cleaned up by the Navajo themselves in the early 1980s, an additional 250 acres had been saturated with radiation. Eventually the area was consolidated into a huge above-ground disposal cell and covered with rock and gravel. The cell is 72 acres in size and contains four and a half million tons of highly radioactive material. Surface reclamation was completed in 1995; however, the groundwater remains highly contaminated, affecting more than half a million cubic yards. As of this writing there is no proposal or funding to mitigate this contamination.

More than 1,300 abandoned mine sites have been inventoried on the land of the Navajo Nation. The Navajo Abandoned Mine Lands Reclamation project (AML) has reclaimed 264 coal mines, 913 uranium mines, and 33 copper mines, mostly through surface remediation (Navajo Abandoned Mines Lands Reclamation Program n.d.), and this has helped stabilize the land and reduce some of the dangers associated with various forms of environmental contamination. The AML has developed radiation level standards for cleanup; however, there are still no federal guidelines in place to regulate this procedure.

Reclaiming an AUM begins with identifying a site, which is then assessed for its hazard levels. Open pits with standing water; sites used for watering livestock, recreational swimming, and illegally dumping trash; and entrance portals used for livestock containment or frequented by campers are all high priorities for reclamation. AML reclamation protocol typically follows a conventional sequence. Sites are inventoried, there is a field evaluation by engineers, technical specifications are designed to comply with National Environmental Policy Act Requirements, and contracts are signed.

Once the AML moves forward with a reclamation job, the area is mapped to measure radiation levels so as to mitigate them wherever possible. Navajo contractors are hired and heavy machinery is brought in. As Elizabeth A. Povinelli observes regarding former mining communities in Australia, "[T]hose persons who were left to inhabit the toxic fields made by others are given the job of removing them" (Povinelli 2016, 88). Through the AML, the Navajo make decisions about a host of environmental challenges such as how to reduce or monitor the migration of radioactive contaminants, how to divert or channel contaminated stormwater, how to block wildlife habitats, and how to control the transference of pollutants in grazing areas. These kinds of problems require ingenious solutions. At times, AML workers have had to climb precariously steep mesas to seal mine portals using polyurethane foam spray. The AML staff physically walk behind their bulldozers and excavators, reading radiation levels while the machinery is at work. Equipment and site access points are restricted and monitored closely. Workers wear radiation uptake readers to monitor their exposure. They change their clothing in a secure area and are scanned for radiation at the end of each shift.

Fig. 6.6. Phytoremediation project, Cane Valley.

All of this takes time. It can take up to two years to reclaim a site. However, reclamation at best usually means burying or covering accessible areas with anywhere between a few inches to a few feet of rock and gravel. Reclaimed or not, rarely are sites marked by signage, unless they are so large that they have to be, as is the case with the region's four colossal UMTRA sites. Smaller abandoned mine and mill sites may sit unidentified and unrestricted for decades.

Reclaimed sites are designed to blend with the surrounding natural environment. They are often replanted with hardy native species, though replanting can be a complicated matter. Plant and tree roots may permeate the radon barrier and compromise the reclamation. On the other hand, vegetation can help stabilize reclaimed areas by reducing the amount of windborne radioactive dust particles.

Any form of mining cleanup involving radiation contamination is very costly. Costs per site range up to $17 million, with an additional $14,000 per year in maintenance costs (US Department of Energy 2014). Cleaning up groundwater contamination is even more costly and almost impossible to achieve. A significant groundwater phytoremediation experiment is under way at the Monument Valley UMTRA

Fig. 6.7. Cows grazing on burial cell, Monument Valley.

site in Cane Valley, Arizona. Around three hectares of native saltbush and greasewood shrubs have been planted on the site of this former uranium mill. The shrubs root down into a large plume of radioactive groundwater and suck up the ammonium and nitrates that run below the surface (Waugh et al. 2010). They are irrigated with contaminated water that has been diluted with clean water that has been pumped to the site through the desert from several miles away. When the plants reach several feet in height they are harvested and burned, and their radioactive ashes are buried in the desert before replanting begins.

There are three AUMs along the boundaries of the Cane Valley UMTRA site. Mine portals at the top of several mesas have been closed, though inaccessible radioactive waste remains in many open areas. Arroyos have been reinforced to help direct runoff, though the runoff itself remains contaminated and there is a problem with the contaminated groundwater seeping into the phytoremediation project. At the boundary of the UMTRA site, a last vestige of mining culture remains as a visible testament to the sinister nature of radioactive contamination: a radioactive heap of corroded tin cans, the remnants of miners' meals. Nearby, a long, snaking disposal cell several metres high, itself resembling a mesa, runs two kilometres along the boundary of a bordering AUM. Beside it is a large burial cell, home to a herd of beef cattle that graze on its meagre grasses.

Burial and disposal cells are large areas where radioactive waste has been consolidated. In the Southwest, livestock graze on burial cells near open piles of tailings and next to AUMs. Contaminated soil and water ingested by livestock create health risks because of the radiation that lodges in the animals' organs. This contamination enters the food supply chain when the livestock are slaughtered. In the United States, trace-back protocols and requirements in the beef and dairy industry are in their infancy and do not extend to screening for radioactive contamination. This makes it nearly impossible to identify radioactively contaminated mutton or beef, let alone trace it back to its source.

Radioactive contamination has infiltrated and impacted the Navajo community in many ways. Navajo graze their livestock on these sites, and children live and play within close proximity to more than 1,000 open tunnels, pits, and radioactive waste piles. Radioactive tailings have been found in school playgrounds, and people still use those tailings to build their homes (Klauk n.d.). Radioactive rocks from the mines have been used to construct walls and foundations, and the tailings are used as the aggregate ingredient in cement. This practice was widespread throughout the 1950s and 1960s, and it continues today: people continue to live in homes or use building materials without knowing they are contaminated. The government and mining industries knew this posed a health risk as early as 1959, yet the information was withheld from Native families and mine workers until well into the 1970s (Brugge and Goble 2002, 1410–19). The AML has tested radiation levels in numerous commercial properties and homes. In one test that I witnessed, a family's home had to be abandoned because of the high levels of radiation being emitted from its rock walls. The family's new home, in construction when tested, was also found to be "hot" because of the contaminated sand that had been mixed with the cement. The family now lives in a small, prefabricated trailer next to both abandoned dwellings. Next door, the foundation of a neighbour's shed has also tested hot. It is a tragic situation, because many families have few resources with which to rebuild or relocate. Many homes, businesses, schools, and agricultural areas are still located directly on or next to hot zones and contaminated sites. In fact, it is difficult to find a Navajo family that has not been touched by the legacy of uranium mining.

Former miners often recount heartbreaking stories of uranium exposure—about the oppressive heat of the mines and how they drank from cold, clear streams of radioactive water to cool down, or about setting off dynamite before lunch and how dust from the blast settled over them while they ate. They also talk about how, as children, their fathers took them to work to play in the mines, and how their mothers swept clouds of radioactive dust from their homes.

Carl Holiday is the AML's health physicist. His work involves recognizing and evaluating the health and safety hazards associated with ionizing radiation. He has 30 years' experience of working with the impacts from the uranium industry. He grew up in Monument Valley, where his grandfather, father, father-in-law, and uncles all worked in the mills and mines—and all died of radiation exposure. Carl also worked in a uranium mine for a year when he was young. As a medical professional, he knows he is now at the age when his own exposure-related health problems will most likely start to appear.

Neither Carl nor any of his family members were ever notified by the government or the mining companies of the hazards of radiation exposure, or of the environmental health impacts associated with mining and milling waste. He is cynical about the federal Radiation Exposure Compensation Act, which theoretically dispenses compassion pay to compensate families of uranium miners suffering from or lost to radiation-related diseases (Uranium Miners Millers and Transporters Benefits Center n.d.). The funds, which award up to $150,000 to persons meeting stringently documented criteria (Harrison n.d.), ran out in May 2000. Since that time, the Department of Justice has been handing out IOUs. So far, 275 have been issued to successful claimants (Wise Uranium Project n.d.). As far as Carl is concerned, no amount of money can make up for the loss of his family.

New Mexico has the second-highest amount of uranium reserves in the United States. Its largest deposits are in the northwestern corner of the state in the Grants Mineral Belt (also known as the Uranium Belt). Hundreds of AUMs exist in and around the Uranium Belt, including the former Jackpile mine, which was once the largest open-pit uranium mine in the world. Scores of families still live close to the mines, grazing livestock in the area. Church Rock, New Mexico, lies just to the

Fig. 6.8. Breached Church Rock tailings pond (spill site) next to State Route 566 (left back).

west of the Uranium Belt. It is beautiful red rock country, home to a scattered community of 350 Navajo families whose income is derived mostly from raising small herds of sheep and cattle.

Church Rock is the site of the third-largest radioactive spill in history, after Chernobyl and Fukishima. It was the biggest single release of radioactive contamination on American soil apart from the atomic bomb tests. The incident occurred on 16 July 1979, 34 years to the day after the first atomic test at Trinity, New Mexico, and only weeks after the Three Mile Island nuclear accident. The Church Rock spill occurred when United Nuclear Corporation's (UNC) uranium mill tailings disposal pond breached its wall and spilled 1,100 tons of radioactive mill waste and approximately 93 million gallons of mine effluent into the Rio Puerco. The pond had been knowingly built on geologically unstable land. Safety violations had been many.

Ultimately, for the UNC, the spill meant minimal loss of revenue. Within five months the same pond was back in use. Some changes were made to the structure, but constant seepage continued—up to 80,000 gallons of highly radioactive liquids per day.

In 1982, UNC announced the temporary closure of its Church Rock operations because of the depressed uranium market and simply walked away from the environmental fallout. The site continues to sit unmarked and unreclaimed. A cleanup was performed in accordance with state and federal criteria, but only about 1 percent of contaminated waste was removed, and almost none of the radioactive liquor was pumped out of the water supply. In 1981 the community stopped receiving trucked water. This left farmers with little choice but to resume using the river water for their small herds.

Teddy Nez and his family have been living in Church Rock for four generations. His family, like many in the area, suffers from chronic health issues. The Nez family homestead is sandwiched between an enormous disposal cell and a massive unreclaimed AUM (abandoned by Kerr-McGee). Just five miles down the road are an abandoned UNC uranium processing mill and the Church Rock pond spill site, currently owned by General Electric. This small community lives with one million cubic yards of mine waste alone, making it possibly the largest and most formidable zone of radioactive contamination on the Navajo reservation (Frosch 2014).

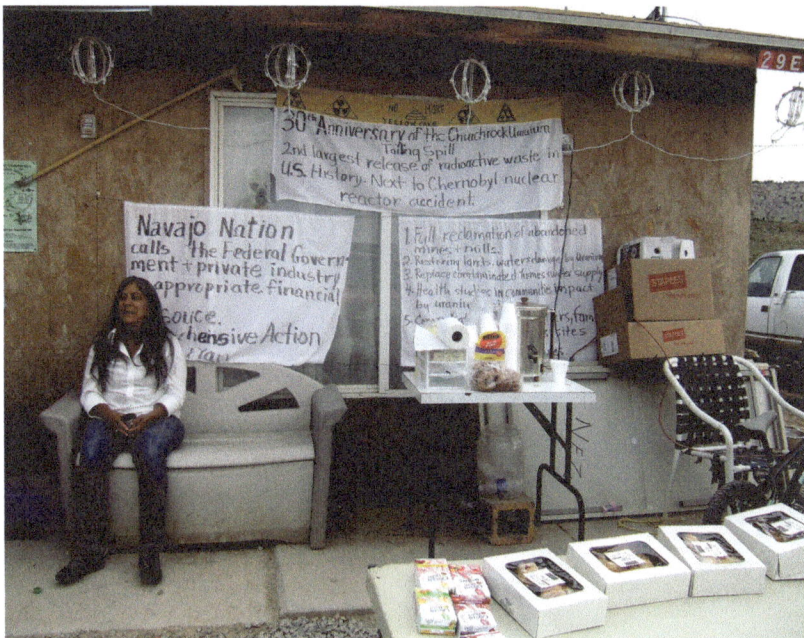

Fig. 6.9. Preparing to commemorate the thirtieth anniversary of the Church Rock spill.

On the thirtieth anniversary of the Church Rock spill, the community gathered at the Nez home to walk in protest to the infamous spill site as part of a Uranium Legacy Remembrance and Action Day. Residents, health specialists, politicians, and activists discussed the critical situation at Church Rock. Joe Shirley, Navajo President at the time, renewed the tribal moratorium on uranium mining. Community members spoke about family members and friends they had lost to kidney disease or various forms of cancer, and about their sick and dying cattle and sheep.

Resident Louise Jim spoke of the "ugly feeling that you could smell" near the pond the morning it breached its walls. He described how his Uncle Bob, who had crossed the Puerco that morning, began suffering from sores that appeared on his feet, and how he later passed away from cancer. He also spoke about his mother-in-law, who had passed from kidney disease, and his father, a uranium miner who was taken by lung disease (Jim 2009).

Chris Shuey, from the Southwest Research and Information Center (SRIC), shared the initial results of data gathered for the first ever health impact report conducted in the region, which links residents' illnesses to uranium exposure. Shuey stressed the danger the Church Rock spill site continues to pose. The area remains critically contaminated and unmarked. With each passing year, it grows less and less identifiable, with brush growing up to cover the waste. Thirty-five years on, there are still no plans to clean up the spill site.

As local resident and anti-uranium activist Larry King points out, the pond didn't only discharge radioactive water on the day of the spill. And it isn't just the Church Rock community that has been affected. For more than 20 years the mine was releasing untreated contaminated waste into the river that runs through Gallup, west to Holbrook and the Little Colorado River, down to Winslow, Arizona (Gault 1989). Another plume of contaminated groundwater has been moving north for more than 30 years. To date, there have been no comprehensive health studies in any of these surrounding areas, and all requests for them have been ignored by both government and industry.

North of Church Rock, less than 20 miles from the Four Corners and San Juan power plants, is Shiprock, New Mexico. The entire town is a hot zone. It has been suggested that the best solution would be to entomb the whole area in cement and relocate the residents. Home

Fig. 6.10. UMTRA disposal cell at Shiprock. Evaporation pond in background, San Juan River in foreground.

to about 20,000 Navajo, Shiprock is scarred by a monolithic 77-acre disposal cell that sits squarely in the centre of the town. Perched on the edge of the San Juan River, adjacent to the town park and fairgrounds, the cell is accessible via Uranium Boulevard, which forks off the "Devil's Highway" (formerly Route 666).

The colossal cell rests on the former site of the Kerr-McGee mill, which processed about one and a half million tons of uranium ore for the AEC between 1954 and 1968. Surface cleanup of the site began in 1983, fifteen years after it had closed. This involved consolidating tailings, soil, debris, the mill structure, and fifteen additional commercial and residential buildings into a mammoth burial chamber holding 2.8 million cubic yards of radioactive materials. For decades the cell has been leaching a plume of radioactive groundwater that covers an area more than a mile and a half long and about three-quarters of a mile wide, which has deposited about 1.2 million cubic yards of contaminated water into the San Juan River and the adjacent floodplain.

Remediation of Shiprock's groundwater began in 2003. This project consists of some drainage channels, a monitoring system, and two wells that pump contaminated water from the town and aquifer into a pond next to the disposal cell. The pond is periodically evaporated; the radioactive residue is then removed and buried in the desert. The process is minimally effective. It is hoped the river will provide some natural flushing to reduce radiation levels over the next hundred years.

Shiprock is also subject to frequent and very intense windstorms that blow radioactive dust particles known as alpha emitters into schools, businesses, and residential districts, as well as onto land used for grazing, agriculture, and recreation. These particles, once ingested or inhaled, become lodged in the lungs. Inhaling alpha emitters is one of the most dangerous ways to absorb radiation.

I finished my fieldwork in Shiprock after being caught between the disposal cell and the evaporation pond during a particularly intense dust storm. It wasn't the first I had been caught in. How much alpha radiation lodged in my lungs from the dust particles that whirled around me, caking me with fine grit? And what about the gamma radiation? It was impossible for me to shield my body from the hot spots and high radiation zones I had been exposed to. And the wellwater I drank—was it laced with radon, as almost all of the wells are? I continue to experience anxiety over the potential health impacts of it all.

Radioactive contamination is a very real threat to us all (Nuclear Information and Resource Service n.d.). It blankets our planet, having plumed out from nuclear tests, waste deposits, mines, and nuclear accidents. Fallout rains down on us, and radiation swirls around the planet in dust particles and ocean currents. It flows down our rivers and migrates around the Earth in the holes drilled by worms, on the wings of birds, and in aquatic wildlife. It is transported across the nation by the complex labyrinth that is the US highway system. It lives in our atmosphere and in our food chains.

Private mining companies dodge accountability through a range of legal loopholes; indeed, they are tacitly encouraged to do so by compliant federal and state laws. They continue to avoid liability for the health and environmental costs of their industry by generating red tape, bankrolling long and expensive legal battles, obfuscating ownership, breaking the chain of title, declaring bankruptcy and setting up shop elsewhere, and finding ways to be grandfathered into SMCRA laws.

The crippling legacy of environmental racism faced by the people of the Indigenous Southwest is relatively unknown to North America's general public. It is difficult for outsiders to grasp the cultural, economic, and environmental destruction these industries bring to Indigenous communities, or to comprehend the scale of the betrayal they have suffered at the hands of the US federal government and scheming corporations. Nor is it easy to understand many of the land use and economic policies that are sanctioned by the tribal governing bodies. The Navajo Nation and the tribal lands of the Southwest are isolated cultural ecologies that are bearing the brunt of US energy policy, economic impoverishment, corporate impunity, and widespread corruption.

Programs like the Navajo AML are under constant threat of losing funding, while managing as best they can to mitigate the environmental costs of mining. The Office of Surface Mining Reclamation and Enforcement is now attempting to eliminate all AML funding entirely. This will affect not only the Navajo but the Hopi and Crow as well. On-the-ground reclamation will stop, as will community infrastructure projects and technical assistance. The Navajo Nation is fighting the issue at the federal level, for congressional action will be required to amend the SMCRA in order to abolish the program. If funding is cut, the Navajo Nation will lose an important natural resource stewardship department that provides critical services to its people.

Sometimes, though, there are victories in the struggle for corporate accountability. Kerr-McGee was recently ordered to pay the highest ever compensation settlement—a record $5.15 billion to cover reclamation costs to address its toxic legacy of 80 years. Of that amount, $4.4 billion will be used to clean up toxic waste (*Russia Today News* 2014), including 50 AUMs and processing sites in and around the Four Corners area. The remaining three-quarters of a million dollars will go to compensate more than 8,000 individuals for health issues related to Kerr-McGee's mining and milling activities. The number of mining-related deaths is staggering. At the Shiprock uranium mine alone, 133 of the 150 Navajo miners employed there were dead from lung cancer or fibrosis by 1980 (Ali 2009, 2).

Many North Americans may not have heard of Kerr-McGee, unless they recall the name from the 1983 Mike Nichols film *Silkwood*.

Karen Silkwood was a Kerr-McGee employee and anti-nuclear activist who had allegedly been exposed to radiation while working at one of its uranium-processing facilities. Silkwood died a mysterious death while en route to deliver documents containing records of Kerr-McGee's health and safety violations to a *New York Times* reporter. And Silkwood was only one of the many activists who have been battling the mining, energy, and defence sectors since the 1970s.

The many environmental justice and advocacy groups active in and around the Navajo Nation have formed a micro-political web that continues to grow, even while it suffers many setbacks. Its many threads may not have altered the dominant macro-political/corporate structure, but each strand constitutes an important struggle to make a positive impact on the health, safety, and well-being of communities in the American Southwest. Its successes are being achieved through small regional alliances that work collectively to reach concrete goals.

Various clean-energy professionals, health workers, and activists are working together in the Four Corners area to develop and implement renewable technologies, promote energy efficiency, generate green jobs, create energy security, and compile health impact reports about their communities. Based in Arizona, the Grand Canyon Trust's Native American program develops community-based renewable energy initiatives in an attempt to help the Navajo diversify their economy. In 2013, this trust, working with the Black Mesa Water Coalition (BMWC), the Just Transition Coalition, the Sierra Club, and others, secured a $3.5 million revolving fund to help establish renewable energy projects on Hopi and Navajo lands. That revenue comes from the sale of acid rain allowances (US EPA Acid Rain Program) from the closed Mohave Generating Station (Umberger 2013). In addition, BMWC and its sister organization, Beautiful Water Speaks, were instrumental in restricting Peabody Energy's access to the N-Aquifer in 2006. BMWC continues to fight for environmental justice against Peabody and to introduce renewable energy and green jobs to the community.

New Energy Economy (NEE) is a New Mexico–based not-for-profit group that works statewide with tribal communities to create energy security and promote education around renewable energy and energy efficiency. NEE pioneered New Mexico's carbon pollution reduction program—the most advanced policy of its kind in the United States. After intense lobbying, the organization pushed through two bills in 2010 that legislated a landmark cap-and-trade agreement and limited New Mexico's greenhouse gas emissions. The sad ending to this story is that early in 2012 the Environmental Improvement Board repealed these two climate regulations (Nikolewski 2012). Even so, these actions point to how grassroots organizations can influence policy. NEE now installs solar electric systems in tribal areas; this has allowed money previously spent on electricity to be redirected towards health, education, and green economy initiatives (New Energy Economy n.d.).

The Multicultural Alliance for a Safe Environment (MASE) is a coalition of organizations and groups that network with uranium-impacted communities such as Church Rock to block new uranium development on tribal lands. MASE also works with the Red Water Pond Road Community Association (RWPRCA), a group of Diné families who are advocating for the reclamation of Church Rock lands and water (Teddy Nez is a member). After more than three decades of relentless community advocacy, the Environmental Protection Agency (EPA) has finally announced plans to clean up the former uranium mine next to the Nez household at an estimated cost of almost

Fig. 6.11. Church Rock Uranium Legacy: Remembrance and Action Day.

$45 million. But the news is not all good. Reclamation does not extend to the infamous Church Rock spill site, where the contamination is so severe that the Nez family and most of the Red Water Pond Road community are facing the prospect of having to leave their land forever (Frosch 2014).

Both MASE and the RWPRCA have worked closely with the SRIC to provide critical information to tribal communities about the environmental and public health impacts of energy development and resource exploitation in the Southwest. These groups are also part of the Diné Network for Environmental Health, which worked with the Church Rock chapter of the Navajo Nation to launch the Church Rock Uranium Monitoring Project, which produced the only report to date assessing the environmental health problems associated with uranium in the area. Preliminary data from the report make it clear that living in close proximity to a uranium mine or mill site increases vulnerability to kidney disease and diabetes (Shuey 2007).

In Crownpoint, New Mexico, the Eastern Navajo Diné Against Uranium Mining coalition works to block new uranium mines on Navajo land. In the Grants Uranium Belt/Ambrosia Lake region, the Laguna-Acoma Coalition for a Safe Environment, a group of *pueblo* residents, is working to protect Mount Taylor, a cultural heritage site under constant threat from uranium mining.

Post '71, a group of former uranium workers (miners, millers, ore haulers, and drillers) based in Grants, are documenting the health effects of uranium on workers since 1971. They are also advocating for amendments to the legislation that has excluded post-1971 uranium workers from compensation under the 1990 federal Radiation Exposure Compensation Act. Also located in Grants is the Bluewater Valley Downstream Alliance, a group of residents and property owners directly affected by groundwater pollution and radiation releases from dozens of uranium mines, tailings piles, and mills in the area. For more than a quarter century this community has been fighting various mining corporations, the New Mexico Environment Department, the Nuclear Regulatory Commission, and the EPA in an ongoing effort to remediate the region's aquifers, though they have yet to achieve any significant results.

These grassroots alliances illustrate how advocating for more environmentally responsible land use practices—being part of the development and stewardship of our land on many scales—can help combat a culture of irresponsible land use and corporate greed. But in order for these alliances to be effective at instituting protocols for better environmental protection and public health and safety, a massive shift in land use development and management paradigms is required, one that brings issues of inequality, accountability, and transparency to the forefront of policy conversations.

Due to the pressure exerted by these micro-movements, some issues are now being given federal attention. Since 2008, the EPA, in partnership with the Navajo EPA, the Department of Justice, and other federal agencies, has been working with $100 million of federal funding to address some of the health risks associated with uranium contamination on the Navajo Nation. More than half the money to date has been spent on cleaning up uranium mines, securing safe drinking water sources, and demolishing and rebuilding contaminated homes. In addition, the EPA is attempting to use the Superfund law to compel mining corporations to clean up some of their abandoned mine sites.

These examples all show that regional restructuring through the perseverance of micro-political initiatives, together with intervention at the state and federal levels, is having some positive impact with regard to land use issues on and around the Navajo Nation. What my years working in the Southwest has starkly revealed is that even the most massive land use issues can remain hidden in plain sight. They have also given me a nuanced understanding of how people who are deeply connected to their land can tackle these problems and how these struggles link to a much larger postmodern narrative that is striving to redress the economic, legal, and environmental frameworks and policies that surround and protect energy and defence industries globally. Ultimately, in order to secure a more environmentally progressive relationship to the land and how it is used, we must press to move forward with ambitions to create a cleaner energy future—one that protects our health and safety—and that makes sense in the twenty-first century.

Note

All photos in this chapter courtesy of Lea Rekow.

References

Ali, Saleem H. 2009. *Mining, the Environment, and Indigenous Development Conflicts*. Tucson: University of Arizona Press.

Arizona Public Service. 2013. "APS Completes Purchase at Four Corners Power Plant." *Market Watch*, 31 December. http://www.marketwatch.com/story/aps-completes-purchase-at-four-corners-power-plant-2013-12-31-815900.

Berry, Carol. 2011. "Coal Mine Records Made Accessible." *Indian Country*, 13 February. http://empowerblackmesa.org/press/2011_02_13_Coal_Mine_Records_Made_Accessible.pdf.

Bray, Scott. 2005. "The Campaign for the Advancement of the Navajo (CAN): The Case for Giving." Navajo Institute for Social Justice. http://www.nisj.org/articles.html.

Brugge, Dough, and Rob Goble. 2002. "The History of Uranium Mining and the Navajo People." *American Journal Public Health* 92(9): 1410–19.

Center for Biological Diversity. 2012. "Suit Filed against Expansion of Navajo Coal Mine Polluting Four Corners Region." *Common Dreams*, 16 May. http://www.commondreams.org/newswire/2012/05/16-13.

Central Arizona Project. n.d. http://www.cap-az.com.

Chenoweth, Wilhain. 1995. "Geology and Production History of the Mitchell Butte Uranium-Vanadium Mine, Navajo County, Arizona." Arizona Geological Survey, Contributed Report 95-B. January.

Department of Economic and Social Affairs. 2007. "United Nations International Expert Group Meeting on Indigenous Peoples and Protection of the Environment Report." Khabarovsk, Russian Federation. 27–29 August.

Frosch, Dan. 2014. "Nestled Amid Toxic Waste, a Navajo Village Faces Losing Its Land Forever." *New York Times*, 20 February, A10. http://www.nytimes.com/2014/02/20/us/nestled-amid-toxic-waste-a-navajo-village-faces-losing-its-land-forever.html?_r=0.

Gault, Ramona. 1989. "Navajos inherit a legacy of radiation." *In These Times*, 13 September. http://www.unz.org/Pub/InTheseTimes-1989sep13-00006.

Guerin, Emily. 2014. "Navajo Nation bets on coal: A tribe digs into a dying industry." *High Country News*, 22 March. https://www.hcn.org/issues/46.5/navajo-nation-bets-on-coal/print_view.

Harrison, Phil. Uranium Radiation Victims Committee. n.d. *Nuclearfreefuture.com*. http://www.nuclear-free-future.com/en/award-presentation/laureates/phil-harrison.

Hendryx, Michael, and Melissa M. Ahern. 2009. "Mortality in Appalachian Coal Mining Regions: The Value of Statistical Life Lost." *Public Health Rep* 124(4): 541–50.

Institute for Southern Studies. 2009. "TRI On-site Surface Impoundment Releases for 100 Top-Polluting US Electric Utility Facilities, 2006." January. https://www.facingsouth.org/2009/01/coals-ticking-timebomb-could-disaster-strike-a-coal-ash-dump-near-you.html.

Jim, Louise. "2009 Uranium Legacy Remembrance and Action Day." Via the EXTRACT archive by Lea Rekow. http://extract.learekow.com/content/uranium action.php.

Klauk, Erin. n.d. "Human Health Impacts on the Navajo Nation from Uranium Mining." DLESE Community Services Project: Integrating Research in Education. Date unknown. http://serc.carleton.edu/research_education/nativelands/navajo/humanhealth.html.

Macmillan, Leslie. 2013. "Peabody mine expansion coincides with Navajo and Hopi artifacts battle." *High Country News*, 20 December.

Minno Bloom, Liza. 2014. "Black Mesa communities continue stand against mine expansion." *Waging non-violence*, 15 December.http://wagingnonviolence.org/feature/black-mesa-communities-continue-stand-mine-expansion.

National Institute for Occupational Safety and Health. 2011. "Coal Mine Dust Exposures and Associated Health Outcomes: A Review of Information Published since 1995." Current Intelligence Bulletin 64, *DHHS (NIOSH)*. Publication no. 2011–172, April. Washington, DC.

Navajo Abandoned Mine Lands Reclamation Program. n.d. http://www.aml.navajo-nsn.gov.

Navajo Nation Division of Economic Development, n.d. http://www.navajobusiness.com/fastFacts/demographics.htm. http://www.navajobusiness.com/fastFacts/Overview.htm.

Nichols, Jeremy. n.d. "Powering Past Coal at the Four Corners Power Plant." WildEarth Guardians. http://pdf.wildearthguardians.org/site/DocServer/Four_Corners_Fact_Sheet.pdf.

New Energy Economy. n.d. http://newenergyeconomy.org/native-power.

Nikolewski, Rob. 2012. "EIB repeals greenhouse gas rule in New Mexico." Blog: *New Mexico Watch Dog*, 16 March. https://www.watchdog.org/newmexico/eib-repeals-greenhouse-gas-rule-in-new-mexico/article_ec67034a-2891-57f8-9a9f-65513e53fbcd.html.

Nuclear Information and Resource Service. n.d. http://nirs.org/home.htm.

Palmer, Margaret. 2010. "Mountaintop Mining Consequences." *Science* 327 (5962): 148–49, 8 January.

Povinelli, Elizabeth A. 2016. *Geontologies: A Requiem to Late Liberalism.* Durham: Duke University Press.

Russia Today News. 2014. "Polluter pays: Record $5bn cleanup settlement for Kerr-McGee legacy." 4 April. http://rt.com/usa/kerr-mcgee-cleanup-settlement-285/.

Sands, Will. 2005. "Area power plant cleans up its act: Polluter to implement $200 million in clean-air fixes." *Durango Telegraph*, 17 March. http://www.durangotelegraph.com/05-03-17/cover_story.htm.

Shuey, Chris. 2007. "Church Rock Uranium Monitoring Project 2003–2007 Report Summary." Southwest Research and Information Center. http://www .sric.org/uranium/docs/CRUMPReportSummary.pdf.

Shuster, Erik. 2007. "Tracking New Coal-Fired Power Plants." National Energy Technology Laboratory, Office of Systems Analyses and Planning, Department of Energy, Pittsburgh, 10 October. http://www.netl.doe.gov/coal/refshelf/ncp .pdf.

Singer, James. 2007. Navajo Nation Government Reform Project, Draft report by the Diné Policy Institute for The Office of the Speaker of the Navajo Nation. 29–31. http://www.dinecollege.edu/institutes/DPI/Docs/GovernmentReform Draft.pdf.

Spangler, Jerry, and Donna Kemp Spangler. 2001. "Uranium mining left a legacy of death." *Deseret News*, 13 February. http://www.deseretnews.com/ article/250010691/Uranium-mining-left-a-legacy-of-death.html?pg=all.

Umberger, Allyson. 2013. "California Public Utilities Commission Applies Utility's Acid Rain Program Credit Sale Proceeds to Renewable Energy Project On Native American Lands." *Center on Urban Environmental Law*, 19 February. http://ggucuel.org/california-public-utilities-commission-applies-utility %E2%80%99s-acid-rain-program-credit-sale-proceeds-to-renewable-energy -projects-on-native-american-lands.

Union of Concerned Scientists. n.d. "Environmental impacts of coal power: Air pollution." http://ucsusa.org/clean_energy/coalvswind/c02c.html.

Uranium Miners Millers and Transporters Benefits Center. n.d. http://www.uranium workers.info.

US Department of Energy. 2014. "Abandoned Uranium Mines Cost and Feasibility Topic Report Draft Final Report." February.

US Department of the Interior Bureau of Reclamation. n.d. http://www.usbr.gov/ WaterSMART/.

US Environmental Protection Agency. n.d. Acid Rain Program. http://www.epa .gov/airmarkt/trading/factsheet.html.

———. Uranium Mills Tailings Information. http://www.epa.gov/radiation/docs/ radwaste/402-k-94-001-umt.html#source.

Waugh, W., Jody, D.E. Miller, S.A. Morris, L.R. Sheader, E.P. Glenn, D. Moore, K.C. Carroll, L. Benally, M. Roanhorse, and R.P. Bush. 2010. "Natural and Enhanced Attenuation of Soil and Groundwater at the Monument Valley, Arizona, DOE Legacy Waste Site—10281." WM2010 Conference, Phoenix, AZ. 7–10 March. http://energy.gov/sites/prod/files/Waugh%20et%20al.%20 WM2010.pdf.

Wise Uranium Project. n.d. "Compensation of Navajo Uranium Miners." http:// www.wise-uranium.org/ureca.html.

Yurth, Cindy. 2011. "Mining depleted N-Aquifer more than predicted." *Navajo Times*, 28 July. http://www.navajotimes.com/news/2011/0711/072811water —.php#.UtAILPbXX-k.

Agency and Time on Active Grounds

A memoir of Bruno Latour
and Gaïa Global Circus

Robert Boschman

Fig. 1. Looking for the K-T: Bruno Latour, Geological Time, and the Geosocial.

In the Fall of 2016 the French philosopher and anthropologist Bruno Latour visited Mount Royal University in Calgary, Alberta, to give a keynote address at Under Western Skies 2016, an interdisciplinary conference on the environment. The biennial gathering was begun at a time when Calgary had come into global view as a point of corporate headquartering for the Athabasca Oil Sands, and the conference has involved a complex confluence of agencies with speakers coming from environmental humanities backgrounds but also from the sciences and business, on a campus sponsored in part by energy interests. Latour was intrigued and brought with him from Sciences Po, Paris, a troupe of actors called Gaïa Global Circus, under the direction of Frédérique Äit-Touati and Chloé Latour, to perform an eponymous drama about Global Climate Change and its increasingly urgent political and ecological realities.[1]

A tragicomedy written by Pierre Daubigny (2018) and performed a final time at Under Western Skies 2016 after a three-year tour, Gaïa Global Circus is a drama rich in allusions that extend across millennia to Aeschylus's *Oresteia*.[2] As the co-founder of an environmental company of actors and writers, Bruno Latour was already heavily invested in the core message of the play: on a global scale, time is running out on changes first wrought by humans a long time ago.

At the core of *Oresteia* is the quest for and development of community justice on the summit of the Parthenon rather than in its subterranean depths where the Furies demand vengeance, and out of which families and communities are caught in endless cyclical destruction. Indeed, it is the Furies, with their horrifying masques, who in *Oresteia* preside over what to the Greeks of Aeschylus's time was an already ancient form of retributive justice. As Agamemnon returns to Mycenae

Fig. 2. Actor Claire Astruc, Gaïa Global Circus.

after the Trojan War with the captive seer Cassandra, she despairingly foretells their demise in a never-to-be-resolved cyclone of archaic politics.

In both plays, the hazards involved in offending and then making reparations to Gaia or Artemis (Nature) are repeatedly demonstrated, although Gaïa Global Circus contains a single and telling difference that heightens the current crisis. In both plays, the broad human community is given voice—fragile, anxious, ignorant—from the Chorus of *Oresteia* to the series of actions in public spaces that Gaïa Global Circus evokes. In both plays, moreover, geopolitical formations are prominently featured, from the Parthenon in *Oresteia* to a United Nations Climate Change Conference somewhere in Europe. Both acutely recognize how human life takes place in what Latour in his keynote address called *phusis,* the biospheric shell of earth, air, fire, and water in which all life exists. The human collective, Latour argues, lives *in* the earth, not *on* it. Also, both dramas explore anarchy and violence, restorative justice, and human relations in the context of weather, climate, and the anger of Artemis, the ancient Greeks' goddess of nature. Finally, both plays portray flawed male protagonists. As a violent, head-strong warrior-king, Agamemnon in his core dramatic act—sacrificing his adult daughter in order to alter the weather in *his* favour—is inherently anthropogenic, a precursor to the blind, violent will-to-power of the capitalist Ted portrayed in Gaïa Global Circus. Agamemnon is Ted and Ted is Agamemnon. While more than 2,000 years separate these plays, the storm that allows Agamemnon to bring his massive Greek fleet to Troy is arguably the same storm with which Gaïa Global Circus begins.

Yet here is the latter play's new and significant addition to the intertext: when the massive helium-infused canopy of Gaïa Global Circus sails out over the audience, it extends the stage at the same time as it portrays both audience and stage as actors—and as actants—thus representing Western human history as agentive and complicit in its own destruction. As the audience witnesses the climate canopy swing overhead, its members understand their own agency in the roiling and turbulent climate events occurring now. At the same time, as Äit-Touati and Latour make clear in their preface to Pierre Daubigny's 2018 publication of Gaia Global Circus, the canopy itself also acts: "During the work of rehearsal and creation, we very concretely experienced the power to act of this immense puppet that we had constructed but whose presence and movements occupied an essential place, that of an actant, and even, literally, of an overwhelming fifth actor" (2018, 100). The play makes clear that climate change is not some detached, transcendent event "out there" but emerges instead from human actions within the earth system itself. Even to say that "we are a part of this" does not fully capture the point that agency per se is deeply, relationally, materially, and temporally part of biospheric phenomena.

⁀

After the conference, an unexpected opening appeared. Latour did not want a cab to the airport for his immediate return flight. Instead, he asked what we could do the next day. So I laid out options along traditional lines for most visitors to Calgary: Banff National Park, an alpine drive or hike, something sublime, a westward day trip.

"No," said Bruno, slightly smiling. "Let's do something else."

It turns out he was interested in *down*, not *up*—in the downward depths of soil, earth, rock, and geological strata inhabited (in Aeschylus) by the ancient

196

Fig. 3. Alberta Badlands, aerial view.

Furies and forming the bases of the *phusis* that Latour had just spoken of in his conference address. I wondered if we could head west to see the Burgess Shales, a site that interested Latour but requires site exploration permits we did not have time to obtain. It occurred to me that eastward from Calgary along an isolated, winding grid road (#848) a traveller will discover a rupture in the prairie, a sudden opening that extends downward multi-directionally and from the air appears like a fractal design along the Red Deer River (see Fig. 3). This extraordinary sight captures the sheer materiality of human and non-human agencies. Descend from the lip of that opening on #848, where wind turbines compete with oil derricks in a complex landscape that, in my experience, has at times invoked the sublime, and you are eventually below the K-T Boundary layer, which signifies the abrupt end of the Cretaceous Period and of dinosaur life 65 million years ago. Here are hoodoos, multi-coloured layers of the earth representing geological time, and the fossils of large animals long extinct. Here too is the Royal Tyrrell Museum of Paleontology. I suggested this as an eastward daytrip, and Bruno, from whom exuberance radiates without interruption, said yes.

We were joined by Bruno's colleague, Olivier Vallet, a member of director Frédérique Äit-Touati's GCC production group, and started driving straight east in my car along the Trans-Canada Highway through Strathmore. In staggered fashion, we made our way north and east to the #848 grid road and began to descend gently to the edge of the Badlands. Where the road departs from the surveyor's rule laid down by the Dominion Survey more than a century ago, we could see our road winding below us until it seemed to disappear into the geological depths of deep time, until I realized that it was actually doing so and that what I was seeing was not a metaphorical movement of matter in time.

I stopped the car and we all got out to look. Bruno and Olivier stood in the middle of the road and gazed east (see Fig. 4). A nearby oil derrick prompted a brief discussion of the sublime, in its conventional sense defined as "an encounter with an object or phenomenon of such overwhelming power, grandeur, and immensity that it is almost beyond comprehension" (Kover 2014, 125). Could this scene, with its extensive, humbling view, but one that also included energy extraction, be considered sublime? For Latour, the oil derrick obviated such an experience and, if anything, now produced guilt and regret. Here, in the middle of the Saskatchewan River Basin, where coal formations had been exploited for a century, before giving way to the search for oil and gas, human agency had, he argued, made the sublime an artifact of history. As he states for the record in *reset MODERNITY!'s* "Procedure 3: Sharing Responsibility: Farewell to the Sublime,"

Fig. 4. Bruno Latour and Olivier Vallet look east.

"[Y]ou realize, at least if you consider the earth, that you, you the human agent, have become so omnipotent that you have been able to inflict definitive changes on its system" (2016a, 169):

> To feel the sublime you needed to remain "distant" from what remained a spectacle; infinitely "inferior" in physical forces to what you were witnessing; infinitely "superior" in moral grandeur. Only then could you test the incommensurability between these two forms of infinity. Bad luck: there is no place where you can hide yourselves; you are now fully "commensurable" with the physical forces that you have unleashed; as to moral superiority, you have lost it too! Infinities today seem to be in short supply! You are now entering an "era of limits." The question is no longer to take delight in the contradiction of infinite matters and infinite soul, but to find a way, at last, to "draw limits," this time voluntarily, because the world is no longer a spectacle to be enjoyed from a secure place. Such is the reset. (170)

Indeed, in addition to the oil derrick, my car and the road we traversed, with its ditches and barbed-wire fencing, also had to be taken into account. And along the bottomlands highway to Drumheller and the Tyrrell Museum, there was infrastructure installed to create a built environment for viewing the expanse of geological time: parking lots, stairs, safety rails, refuse bins (see Fig. 1). These constitute what Latour has called "the social" and has spent decades intricately describing, portraying, and theorizing in its complex relationship to and intersections with "nature."

When we stopped by the side of Highway 570 to see the hoodoos, we found the Social waiting for us in the small plaque explaining this scene of geological

Fig 5. The Social: A plaque stands just below the K-T boundary.

199

forces, just below the exposed K-T Boundary (see Fig. 5). We could look up and pinpoint the iridium-rich layer marking the abrupt end of dinosaur life 65 million years ago. The K-T is, where we stood, only inches high but conspicuously black with what geologists call "shocked minerals." In comparison, the Anthropocene layer—what humans will leave for a distant future—indicated by the residue of our modified environs, will also be both measurable and visible. That is the point of the term's inclusion in current discourse, although perhaps this Anthropocene layer will be not so much "shocked minerals" as "stocked plastics," a refined petroleum compote, slightly irradiated. Writing for *reset MODERNITY!*, Dipesh Chakrabarty calls the Anthropocene "a thought experiment among geologists [that is] based on stratigraphical evidence" (2016, 191): "There is a certain chutzpah and perversity to the concept, no doubt. For geological periods are usually named long after they are gone. Here scientists are trying to convince themselves and other scientists that stratigraphic evidence already exists for us to be able to imagine the geological history of this period from the point of view of both geologists of the present and those who may come—in human terms at least—in the very, very distant and probably posthuman future" (191).

Our next stop was the Royal Tyrrell Museum of Paleontology, just outside Drumheller, an early-twentieth-century coal-mining hub that has found new life in tourism as visitors from around the world are drawn to its rich fossil discoveries, its multi-storey T. Rex, and of course its renowned museum. I was curious to watch Bruno expertly inspect the museum. Just inside the entrance is a very large globe that, as you spin it, turns through geological time. A titan struggle would presently ensue here, as a little girl appeared who insisted on spinning the world counter to Bruno Latour. Their respective pairs of hands brought the globe to a

Fig. 6. Spinning the Devonian, with Bruno Latour and Olivier Vallet.

dramatic stop while diplomacy and her parent resolved the issue. Soon Olivier and I watched as Bruno returned to his examination of the Devonian Period (Fig. 6). I knew from our discussions in the car that he deeply admired James Lovelock, who had posited the Gaia Hypothesis in the 1970s. In Fact, Bruno had just published an article and book that explored and defended Lovelock's hypothesis. An American scholar named, oddly enough, Tyrrell had provoked Latour's article, "Why Gaia Is Not a God of Totality," which sifted and clarified this concept of Gaia. Under the heading "The Same Prefix: Two Opposite Reactions," Latour relates a playful provocation on his meeting geoscientists, asking them why not *Gaia*. After all, "geo- and Gaia share exactly the same etymology: both come from the same entity Gè, actually a chthonic divinity much older than Olympian gods and goddesses" (2016b, 1).

> [T]he somewhat wild proliferation of the prefix "Gaia" exactly parallels the transformation of how the distant presence of the earth has been formatted in public discourse: what, as far as we remember, had constituted a solid but distant and faithful background for various *geosciences*, and for staging the usual drama of *geopolitics*, has now become, no matter which political persuasion you come from, an *actor*, at least an *agent*, let's say an *agency* whose irruption or intrusion upon the foreground modifies what it is for the human actors to present themselves on the stage.
>
> Whereas you could consider "geo" from the outside standpoint of a disinterested observer, with 'Gaia' you are inside it while hearing the loud crashing of outside/inside boundaries. To be a disinterested outside observer becomes slightly more difficult. We are all embarked in the same boat—but of course it's not a boat! (2)

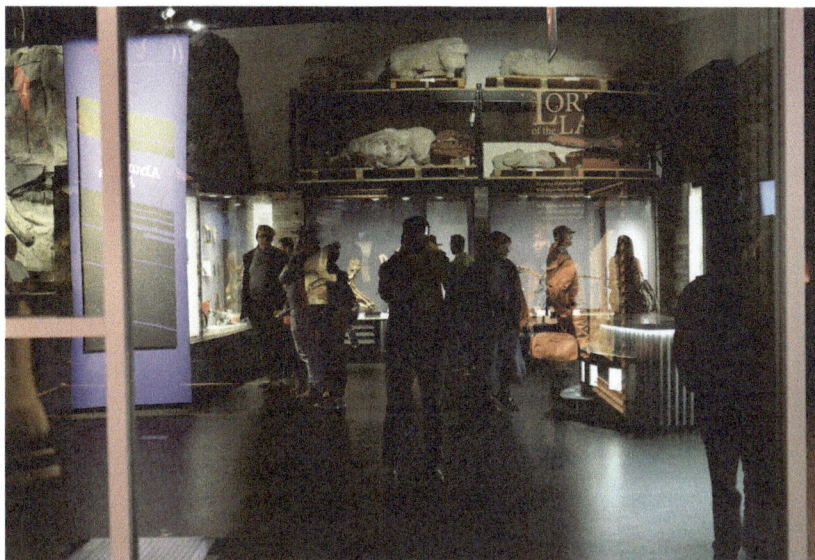

Fig. 7. The Collection: Royal Tyrrell Museum of Paleontology, Drumheller.

Fig. 8. Bruno Latour and the Composition of Facts.

I have visited the Tyrrell Museum many times. As a photographer, I find it a fascinating venue for studying answers to the fatigued and increasingly problematic dichotomy of subjects and objects—of "disinterested outside observers" and "the loud crashing of inside/outside boundaries." This is especially the case, I've noticed, as one approaches the Preparation Lab, where visitors can observe a fossil preparation laboratory in action. My photo instincts were heightened as Latour approached this part of the museum (Fig. 7), and I opened and closed the shutter just as he and Olivier, along with a number of other people, moved toward and then away from the lab. I did this swiftly just as we passed a mirror, in order also to include myself. The resulting image merges subjects and objects in a museum devoted to the spatio-temporal. It suggests the Latourian emphasis on *phusis* and the enmeshment within it of materialities, mentalities, agencies, and temporalities, both on site and, by implication, throughout the biosphere. The image expresses the *at-onceness* of existence that the introduction to this volume suggests is necessarily characteristic of Environmental Humanities research. As a photographer and academic in the environmental humanities, I find that such images themselves carry a form of agency, as kinds of assemblages (as Trono also suggests in his chapter in this volume), with the message of this specific image evoking what Latour means when he uses the term *collective* and speaks of *composing a common world* (2005).

When he finally noticed the lab itself, Latour pulled out his iPad to take a photo. I quickly pressed my shutter button again (Fig. 8). In my own recent visits to the museum with my children, I'd contemplated the Preparation Lab repeatedly as a zone where the nature–culture complex (previously imagined as bipartite) is properly understood as ontologically uniform. As a Latour reader, I had wondered to myself what the founder of science studies would do when encountering this working lab among the museum's exhibits. As it turned out, he also chose to create an image, but I do not know why. Perhaps he had in back of mind work by his compatriot Gilles Deleuze, who writes in *Cinema 2: The Time-Image* about the connections between concepts, their planes of immanence, and the forms of time that fuel conceptual creativity.

Visiting the Alberta Badlands is, I like to think, a kind of time travel. Entering and leaving, I experience the stratigraphic realities of the earth system—Gaia—as a temporal and material complex that is fluid. Ever in motion, it is in its history of movements here in this place both emphatically revealed and intimately accessible. Otherwise to experience so closely the revelation of temporal geological processes, you'd need to live through an earthquake, as the young Charles Darwin did in Chile in 1835: "A bad earthquake at once destroys our oldest associations: the earth, the very emblem of solidity, has moved beneath our feet like a thin crust over a fluid;—one second of time has created in the mind a strange idea of insecurity, which hours of reflection would not have produced" (323).

Driving out of the valley and west toward Calgary on our return, Olivier, Bruno, and I could have stopped at any number of points, gotten out, and crossed the roadside ditch to place a finger on the "shocked minerals" of the K-T Boundary. Instead we drove on till we saw Horseshoe Canyon off to the right and parked in its muddy lot to view this badlands tributary, this last opening in the earth for the traveller heading west.

Fig. 9. Horseshoe Canyon, Alberta.

All around Horseshoe Canyon, farmland extends as far as the eye can see and farmyards dot its edges. In summer, tourists can buy a ride in an orange helicopter and wander in circles over the canyon (see Fig. 3), shaped indeed like the piece of iron that immigrant European farriers attached with nails to their horses' hooves during the colonization of the Americas. Even if we'd had the time to walk down into this canyon, though, a steady rain that day made it difficult to venture farther than the horseshoe's edge (Fig. 9). Our footwear became instantly caked and we spent ten minutes scraping our soles before returning to the car.

Both the Chilean Earthquake of 1835 and, much further back in time, the formation of the K-T Boundary took place very quickly: the former caused by the sudden shifting of tectonic plates, the latter quite likely by an incoming asteroid of sizable proportions. Darwin ([1860]2008) imagined how a large earthquake "in the dead of night" would impact England—it would become "at once bankrupt" and "[i]n every large town famine would go forth, pestilence and death following in its train" (326). The sciences that inform us now about earthquakes and asteroids, which strike quickly, are also warning repeatedly of another event—global climate change—not as an instantaneous shock in time but rather moving slowly in the current of the years, centuries even, largely unnoticed at first but now gathering momentum enough for concerned citizens to witness for themselves, apart from the sciences and their incoming facts. The sciences repeatedly inform us that the cause of these changes, and they are as drastic globally as any earthquake can be locally, is a human one. We have altered the planetary exchange of gases and changed the balance enough to warm Earth, with consequences that are actual and measurable. If global climate change were to occur in an instant, like an

Fig. 10. Gaïa Global Circus: post-finale packing.

earthquake, it wouldn't just be England—or Alberta or Paris—but all the globe's hemispheres that would become "at once bankrupt" (326).

The forces of negotiated resolution and agency are central to Aeschylus's *Oresteia,* the trilogy of ancient plays that begins with *Agamemnon.* The drama's overarching narrative demonstrates how, for the ancient Greek peoples, retributive justice could not be replaced with communal jurisprudence until the chthonic Furies were satisfied—not by ongoing vendetta but instead, in a pivot, by human resolve and jurisprudence. Only the latter may justly foreclose on acts of violence and their consequences. While the force of human agency at a pivotal moment in history constitutes the beating heart of *Oresteia,* it is no less central to the drama of Gaïa Global Circus. When at the close of the play the actor Claire Astruc lies on the stage and pulls the atmosphere down to meet with and enclose her form while she looks directly at the audience, the audience is meant to understand its own material, collaborative role as actor as well (Fig. 2). Represented by a massive tarp floated by heavy-duty black-and-white balloons infused with weather-grade helium, the "atmosphere" created by director Frédérique Äit-Touati blends not only with the human troupe but also with the ground: *phusis.* Äit-Touati's invention was used for the final time on 29 September 2016, at the Bella Conservatory Theatre at Mount Royal University in Calgary, Alberta. Gaïa Global Circus's three-year tour would end here. In the cleanup and packing after the performance, the actors became the crew. They worked expertly and quickly, having done such work many times before. This time, however, they gave all the balloons away.

Notes

All photos in this chapter courtesy of Robert Boschman. The author wishes to thank Frédérique Äit-Touati for her advice and suggestions.

1 *Gaia Global Circus,* a play by Pierre Daubigny based on Bruno Latour's project and idea, was directed by Frédérique Äit-Touati and Chloé Latour, with Claire Astruc, Luigi Cerri, Jade Collinet, Matthieu Protin, and Mathias Marty, with mechanical and optical effects by Olivier Vallet, lights by Benoît Aubry, sound by Christophe Hauser, and costumes by Elsa Blin.

2 Pierre Daubigny's *Gaia Global Circus* was published in Yale's *Theater* by Duke University Press, 1 May 2018.

References

Aït-Touati, Frédérique, and Bruno Latour. 2018. "Performing Gaia." *Theater* 48(2): 95–101. doi: https://doi.org/10.1215/01610775-4163038. Accessed 27 November 2018.

Chakrabarty, Dipesh. 2016. "The Human Significance of the Anthropocene." In *reset MODERNITY!,* edited by Bruno Latour, with Christophe Leclercq, 189–99. Cambridge, MA: MIT Press.

Darwin, Charles. [1860]2008. *The Voyage of the Beagle: A Naturalist's Voyage Round the World.* Project Gutenberg.

Daubigny, Pierre, and Bruno Latour. 2018. "Gaia Global Circus." *Theater* 48(2): 103–23. doi: https://doi.org/10.1215/01610775-4163026. Accessed 25 November 2018.

Kover, T.R. 2014. "Are the Oil Sands Sublime?: Edward Burtynsky and the Vicissitudes of the Sublime." In *Found in Alberta: Environmental Themes for the Anthropocene*, edited by Robert Boschman and Mario Trono. Waterloo: Wilfrid Laurier University Press.

Latour, Bruno. 2005. *Politics of Nature: How to Bring the Sciences into Democracy.* Cambridge, MA: Harvard University Press.

———. 2016a. "Procedure 3: Sharing Responsibility: Farewell to the Sublime." In *reset MODERNITY!*, edited by Bruno Latour, with Christophe Leclercq, 167–71. Cambridge, MA: MIT Press.

———. 2016b. "Why Gaia Is Not a God of Totality." *Theory, Culture, and Society*, Special Issue: *Geosocial Formations and the Anthropocene*. 1–21. PDF.

III.
ANIMAL AGENTS AND
HUMAN / NON-HUMAN
INTERACTIONS

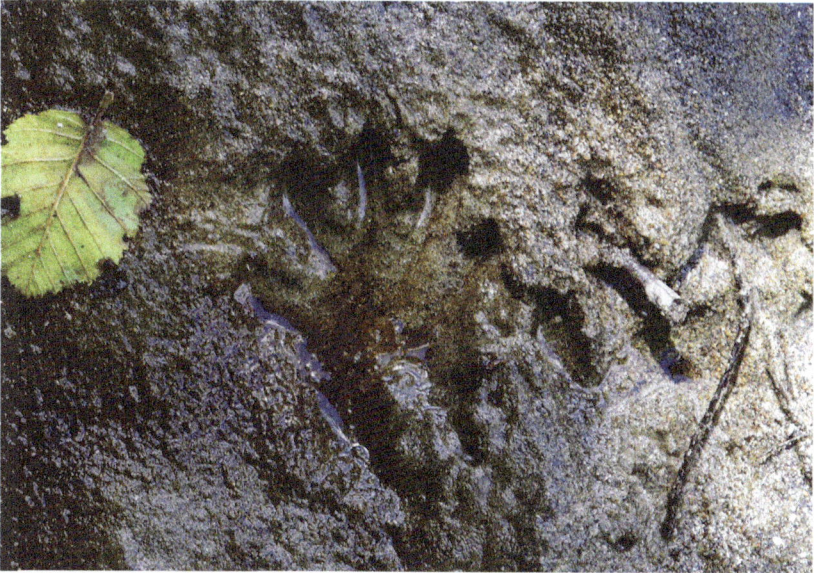

Raccoon print. (Photo courtesy Robert Boschman)

CHAPTER 7

The Gaze of Predators and the Redefinition of the Human

Karla Armbruster

> What would we find if we were once again to step into a space
> where we would willingly become edible, a space that might
> even be called inhumane? What would become of our own
> sense of ourselves as human and humane beings?
>
> —James Hatley, "The Uncanny Goodness of Being Edible to
> Bears"

Though we humans living in wealthier, more industrialized nations certainly suffer from our share of fears and anxieties, one thing we rarely worry about is being watched, stalked, or killed by an animal of another species. When placed in the context of our prehistory and evolution, this complacence is quite remarkable. As wildlife biologist Joel Berger explains in *The Better to Eat You With: Fear in the Animal World*, "*Homo sapiens* did not evolve as armed pursuers of big game, despite our prowess as hunters for the last forty of fifty thousand years or more. To the contrary; our behavior has been molded by half a billion years as prey" (2008, 268). Since that time, though, our sense of agency in relation to our environment has changed dramatically. As most human cultures have worked to gain control of the natural world around us, our instinctive fear of predators has been relegated to the depths of the subconscious, and, as Sy Montgomery puts it in *Spell of the Tiger,* we have "felled the forests and eradicated the predators so that we could pretend we are not made of meat" (1995, 73). Any animal who rips through this illusion by attacking a human for food is almost always eliminated, having committed the ultimate crime against our self-proclaimed position atop the web of predator–prey interactions. We have so convinced ourselves of the rightness of this position that I often hear students and others state as a given that humans are at the top of the food chain. You can see why they think this, given that Western culture's sense of human agency has metastasized into the prevailing assumption that humans are separate from and superior to nature, including the wildlife we lump into this abstract category.

However, as contemporary environmental thought attests, the ecological (and, increasingly, social) consequences of this assumption have been devastating. Consequently, the challenge of how to shift to a more ecological view that understands humanity as inseparable from the rest of nature and extends agency to animals and other natural entities has been one of the preoccupations of contemporary ecocriticism. In *Bodily Natures: Science, Environment, and the Material Self*, Stacy Alaimo promotes such an understanding with her concept of transcorporeality, which emphasizes the ways "the human is always intermeshed with the more-than-human world" (2010, 2). Rather than viewing "nature" as a mere backdrop to human activity, she insists that it "is always as close as one's own skin—perhaps even closer" (2010, 2). Her analysis highlights the way material substances, including biological agents of disease, toxic chemicals, and food, move across and through bodies, with her goal being to "catalyze the recognition that the environment, which is too often imagined as inert, empty space or as a resource for human use, is, in fact, a world of fleshy beings with their own needs, claims, and actions" (2010, 2).

As Alaimo's argument suggests, the risks to human life and health that most concern environmental thinkers today exist at the other end of the spectrum from the possibility of being eaten by another animal. Both Alaimo and Ursula Heise, whose *Sense of Place and Sense of Planet: The Environmental Imagination of the Global* has also had a major impact on ecocriticism, build on Ulrich Beck's concept of risk society, which is characterized by hazards such as radioactivity, toxic chemicals, and global climate change—distinctly modern hazards that are the effects of the modernizing process itself (rather than the type of risk posed by predators, whose elimination is often considered the first step towards modernization). These hazards are also, in some cases, truly planetary in scope (Heise 2008, 146), travelling across bodies and landscapes in ways that are often imperceptible to the ordinary person without the assistance of experts and complex technology (Alaimo 2010, 72) rather than taking the form of another living being that one can see, hear, feel, and possibly avoid.

Given the reality that twenty-first-century humans are far more likely to encounter these invisible, widespread risks rather than the very tangible, local threat of predation, it seems significant that the fear of being eaten by other animals still lingers, surfacing in our nightmares, fairy tales, and horror stories. The fact that in many societies we have gone to such lengths to remove this risk of being eaten, yet quickly jump to worrying about it when it becomes even a remote possibility—a phenomenon that becomes clear after just a few conversations with anyone venturing into grizzly or mountain lion country for the first time—suggests that the possibility of becoming prey retains a powerful hold on the human psyche. In her discussion of transcorporeality, Alaimo suggests that an understanding of "the [human] body as changing and changeable" may help us move past the damaging dualistic construction of nature and culture—one an active agent, the other acted upon—that is at the root of so many environmental problems and arrive instead at a vision of humans in dynamic material interchange with the rest of the world (2010, 5). Viewed a certain way, the notion that we can be eaten, digested, and incorporated into the bodies of other animals can certainly bring home the changeability of the human body. Accordingly, this chapter explores the possibility that the intensity of our feelings about becoming prey can propel us beyond horror and the punitive reaction against potential predators that it so often evokes to a point where we can instead use the idea to help us recalibrate our understanding of the human role in the biosphere. In "The Uncanny Goodness of Being Edible to Bears," philosopher James Hatley asks, "What would ensue if we were to let down our defensive posture in regard to the natural predator—if only for a moment—and desist from our ongoing domestication of predatory space, allowing the animal who eats a chance to address us more fully?" (Hatley 2004, 15). Responding to this question and to what Alaimo calls "the disturbing, unpredictable agency of corporeality itself" (2010, 138), my argument suggests that we might be able to develop an understanding of the human body as temporary, unstable, vulnerable, and belonging to the rest of the world as much as to the individual who claims it.

I. HUMANITY AS CONFIRMED BY OTHER ANIMALS

For most of us, an essential aspect of human identity is the privilege of being off-limits as food for other living things. As noted, one reason we can take this privilege for granted is the eradication of predators, but that eradication has been part of a much more all-encompassing marginalization of non-human animals by most human societies, especially in the urbanizing global North. Before asking what difference it might make to question this privilege, it's worth exploring this relentless pushing down and away of other species (and the possibility of their agency) and related assumptions about what it means to be human. John Berger's influential 1980 essay "Why Look at Animals?," which eloquently describes and mourns the ways in which humans have become ever more alienated from other animals, provides a complex perspective on these issues, evoking the definition of humanity as separate from animality that has arguably fuelled that process of alienation, but also suggesting an alternative way in which animals might help us think differently about what it means to be human.

Berger argues that non-human animals once existed in a powerful, immediate way for humans, standing with us "at the centre of [our] world" (1980, 1) and interceding between us and our origin in nature. However, he claims that this relationship was eroded during the nineteenth and twentieth centuries, when "every tradition which had previously mediated between man and nature was broken" (1980, 1), and that our quest for ever-extending knowledge and power over other animals has only fed the gap between us and them: "The more we know, the further away they are" (1980, 14). The result, he pronounces, is that animals have been rendered "absolutely marginal" (1980, 22). You can quibble with what Jonathan Burt has called Berger's "romanticisation of the non-capitalist relations of [humans and other animals in] the past" (2005, 204), but it's harder to deny Berger's claim that by the end of the twentieth century, other animals had all but disappeared from our cultural lives, persisting (apart from pets) only on the margins as images and in spectacles such as zoos. For Berger, the act of gazing at these marginalized animals—such as the disappointing experiences many of us have in zoos—demonstrates that all potential for any authentic intimacy between humans and other animals has been lost; as Burt suggests, the implied answer to the question that Berger poses in the title of his essay is, more or less, "don't bother." Today, Berger explains, "Animals are always the observed. The fact that they can observe us has lost all significance" (1980, 14).

As this quotation implies, for Berger, the exchange of looks between humans and animals was at the heart of the significance they once held for us—and the power (now lost) of that interaction came largely from their difference from us. Using present tense in a way that enshrines this exchange as mythical and ideal, Berger explains that the human and the animal scrutinize each other across similar (though not identical) "narrow abyss[es] of non-comprehension," and at least in the case of the human, across "ignorance and fear" (3). The abyss between humans can be bridged by language, in his view, but the animal's "lack of common language, its silence, guarantees its distance, its distinctness, its exclusion, from and of man" (4). In his characterization of this unbridgeable abyss, Berger seems to accept the dualistic Western tradition that has defined the human against the animal, elevating the human based on a range of supposedly distinctive traits including the possession of a soul, rationality, the capacity for language, the ability to use tools, or self-consciousness—in other words, the very tradition responsible for the marginalization of animals that he laments. However, Berger ultimately avoids this particular set of assumptions, carefully framing the idealized exchange of looks in a way that unsettles them instead.

First, Berger finds irreplaceable value in the difference of other animals rather than using it to diminish their importance, and he tempers his insistence on that difference with an emphasis on similarity as well. He explains that the look of the animal feels familiar to the human because of the similarity in their two experiences of gazing across an abyss of noncomprehension: "when [the human] is *being seen* by the animal, he is being seen as his surroundings are seen by him" (1980, 3). This familiarity, combined with the incomprehensible difference of the other animal, means that the animal's life "can be seen to run parallel" to the human's, a position that invests the animal with mystery ("the animal ... can still surprise the man") and power (1980, 3). And it is the loss of that mystery and power in our lives that Berger mourns, suggesting that it once made us feel less alone and more at home on the planet: "With their parallel lives, animals offer man a companionship which is different from any offered by human exchange. Different because it is a companionship offered to the loneliness of man as a species" (1980, 4).

This concept of animals' parallel lives once offering companionship to our loneliness as a species is another way Berger approaches, yet deviates from, traditional Western assumptions about the human–animal relation. Although he claims that "[m]an becomes aware of himself returning the [animal's] look" (1980, 3), opening up the possibility that the companionship of other animals traditionally helped us understand human identity, he does not suggest that humans defined themselves *against* other animals in any static or discrete way. Plainly stating that "[n]o animal confirms man, either positively or negatively" (1980, 3), he leaves us with the question of what it means to become aware of oneself without experiencing a confirmation of one's identity. As Michael Emerson suggests in his close reading of Berger's essay, Berger sees the animal's gaze not as providing a coherent human identity but as putting "the human subject into question … creat[ing] a self-reflective puzzle or aporia in which the human being reflects 'My identity seems assured and guaranteed in the look of other human beings, but the animal's look comes across as a question never to be completely answered'" (2015).

So Berger's elegy for human intimacy and meaningful relations with other animals powerfully critiques the cultural developments that he blames for this loss and evokes the related assumptions about what it means to be human; furthermore, it suggests that animals themselves hold—or held—the potential to help us find a less damaging way to think about being human, a way that doesn't so much "confirm" a definition of the human as unsettle that concept and the binary of human versus animal that fuels it. For Berger, tragically, that potential has vanished: "The fact that [animals] can observe us has lost all significance" (1980, 14). However, Berger does not address the gaze of the predator. If the look of any animal still possesses the power simply to *mean something* to a human being, to assert agency, it must be the look of an animal that wants to feed on that person. While we are much less likely than ever before to encounter that look due to the developments Berger discusses, even the idea of another animal stalking, killing, and eating a human may be able to remind us that we are not the only ones who look, who see, who eat, and this chapter will argue that the experience of thus being seen as our surroundings are seen by us, as Berger puts it, can still unsettle and transform our sense of what it means to be human.

II. TIMOTHY TREADWELL AND RESPONSES TO HIS "BECOMING GRIZZLY"

While the visceral horror that most people feel at the idea of becoming prey may hold the potential to shake us up and help us re-envision our human role in the world, the horror itself is an obstacle to that shift in perception. Untangling the various forms this horror takes and the ideological assumptions thus reflected is the first step in understanding how to move beyond it. The case of Timothy Treadwell is instructive, at least with regard to attitudes in the United States and likely most of the global North. Treadwell lived with brown bears along the south-west coast of Alaska, primarily in Katmai National Park and Preserve, for thirteen summers, from 1991 to 2003. Brown bears are the same species, *Ursus arctos*, as the bears we call "grizzly," but they live on the coast, as opposed to the inland grizzly, and they tend to be less aggressive (Jans 2005, 20). Treadwell considered himself the bears' protector and advocate and eventually took more than one hundred hours of film footage of them (and of himself interacting with and talking about them). Eventually, he co-founded a not-for-profit called Grizzly People, wrote the 1997 book *Among Grizzlies: Living with Wild Bears in Alaska*, spoke to numerous groups of schoolchildren about bears, and attracted attention from national media. Then, in 2003, he and his girlfriend Amie Huguenard stayed in the park into the beginning of October, later than usual, and the pilot who came to pick them up found their remains, along with evidence that they had been killed and partly eaten by a bear or bears. Park rangers shot the large, older male bear identified as responsible for the killings (remains were found in the bear after he died) as well as a younger bear that seemed aggressive to them. Perhaps most chilling, Treadwell and Hugenard had audiorecorded their own deaths by somehow turning on a video camera (without removing the lens cap) as the fatal attack began.

Once the news got out, controversy ensued. As Nick Jans writes in *The Grizzly Maze: Timothy Treadwell's Fatal Obsession with Alaskan Bears*, Treadwell became a "lightning rod crackling with emotion," a catalyst for people to express strongly held feelings about the proper human relationship to bears and to the natural world (153). As Julie Kalil Schutten points out in "Chewing on the Grizzly Man: Getting to the Meat of the Matter," a common reaction was—in the words of Sam Egli, a helicopter pilot who helped recover Treadwell and Huguenard's remain—that Treadwell "got what he deserved" (2008, 194). According to Jans, an entire "postmortem lynch mob of actual and self-styled outdoorsmen" quickly emerged to criticize and even mock Treadwell (one column in *Field and Stream* was subtitled "Brown bears were his life; he was their lunch") (149). (Interestingly, Huguenard is rarely blamed, when mentioned at all; instead she is often portrayed as a victim of Treadwell's poor judgment, a tendency that certainly invites an analysis of gender stereotypes!)

And what did Treadwell do to merit this harsh criticism, which effectively casts him out of the category of the inedible human? Jans explains that many Alaskans believe that they are closer to nature and understand it better than people from the Lower 48, like Treadwell, and that they interpreted his actions as a form of trespass on their turf and disrespect to them and the bears. Certainly, Treadwell's attitude towards the bears contrasted with that of most Alaskans. He did not hesitate to declare that he loved them and considered himself their devoted defender. As bear expert and advocate Charlie Russell writes, "the one thing that upsets people the most about Timothy, to the point of loathing him, was that he talked to bears in a kind way. In Alaska this type of behavior is [considered] unforgivably stupid. If Timothy had spent those thirteen years killing bears and guiding others to do the same, eventually being killed by one, he would have been remembered in Alaska with great admiration" (2006).

The reaction that Treadwell deserved his fate extends beyond Alaskans or outdoorsy types, though. When I show *Grizzly Man*, Werner Herzog's film about Treadwell, to students, some of them dismiss his experiences with the bears by saying he did not know what he was doing (perhaps influenced by Herzog's portrayal of Treadwell's troubled history with substance abuse, Southern California surfer image, and garrulous, sometimes childlike persona). Herzog himself takes an extremely didactic approach to the story, arguing that Treadwell "crossed an invisible borderline" between humans and nature and paid the inevitable price (*Grizzly Man* 2005). In fact, Herzog eloquently distills and packages the essence of most of the charges against Treadwell: that he profoundly misunderstood the proper relationship between nature and culture, thinking he could befriend the bears when in reality they were unfeeling killing machines lacking all agency, bound to eat him in the end. It is true that Treadwell was not a trained wildlife biologist. He may have gone too far in his sometimes anthropomorphic and even infantilizing discourse about the bears (for example, he gave them names that seemed more appropriate for stuffed animals, like Mr. Chocolate, Rowdy, and Cupcake, and once described one on film as "a big bear" in the same tone one might use to tell a child he's a "big boy"). It is also true that his justification for his presence among the bears (that they needed him to protect them from poachers) is quite a flimsy one: as Herzog points out, the area where Treadwell lived with bears was part of a national park, where killing bears was prohibited, and Jans indicates that in all the research he conducted for his book on Treadwell, he found little to no evidence that poachers posed any credible threat to the brown bears of Katmai (33–39).

However, dismissing Treadwell as an idiot who deserved to be killed ignores the evidence that he actually knew something about bears. Could he really have coexisted with them for thirteen summers without a realistic, thorough knowledge of bear behaviour and, for that matter, of the natural world? According to Herzog, Treadwell's foundational error was to romanticize nature, naively ignoring its dangerous, ugly, and inharmonious aspects. Herzog is so invested in his critique of Treadwell that he ignores footage included in his own film that contradicts it by showing Treadwell explicitly acknowledging the dark side of life in the wild, as when he describes the loss of a fox cub to wolves as "a sad turn, but ... a real turn" (*Grizzly Man* 2005). When Treadwell's critics ignore his accomplishments and knowledge in their quest to condemn him in this way, they often seem propelled by outrage—specifically, outrage that Treadwell and the bear who killed him violated an imaginary hierarchy in which humans are at the top, removed from the food webs of nature—and by a corresponding sense that justice was served when both were killed. As Schutten argues, the level of hostility and contempt directed towards Treadwell represents more than a disagreement about how to understand the natural world; they are no less than a defence of culture's ideologically privileged position over nature. Treadwell's death made it all too clear that human bodies are vulnerable, that humans can be "'pieces of meat' and, as such, objects rather than subjects" (Schutten 2008, 195), immersed in nature rather than safe on the other side of a nature/culture dualism. For many people, this knowledge is so traumatic that it must be vehemently denied and the ideological boundaries protecting human culture and identity repaired by any means necessary. It's worth noting that the assumption that human bodies must never be food for other animals goes so deep that even Treadwell's harshest critics did not question the decision to kill the bears and recover his and Huguenard's remains.

For Herzog, the gaze of the grizzly conveys nothing but a reflection of the same philosophy that he hammers home throughout the film: in his voice-over narration, he declares that "in all the faces of all the bears that Treadwell ever filmed, I discover no kinship, no understanding, no mercy. I see only the overwhelming indifference of nature" (*Grizzly Man*). The fact that Herzog is reacting to filmed faces rather than describing grizzlies he has seen in person seems unintentionally fitting: rather than encountering their bodies in the flesh, giving the bears the opportunity to see him, return his gaze, and perhaps assert their agency and make him aware of the unique bear perspectives lurking behind their faces, he views them exclusively as objects of a one-way gaze that they don't ever experience—and his certainty that there is nothing worth knowing behind their faces prevents him from even suspecting that there might be a difference between these two modes of looking. There is no possibility of meaningful interaction between humans and other animals, in this view, and being human means accepting that and turning away from the animal gaze to the business of culture.

Interestingly, some of Treadwell's friends and supporters, though devastated, took a profoundly different view of his fate, focusing on the fact that he understood and accepted the risks of living with the bears. He regularly referred to the fact that the bears could kill him if they wished (Herzog's film even starts with footage of him making this point), and at times he stated it would be his preferred way to die. Doug Peacock, author of *Grizzly Years: In Search of the American Wilderness*, recalls Treadwell telling "people he would be honored to 'end up in bear scat'" (2004, 20). While we can't discount the degree to which Treadwell made these statements for dramatic effect, since he was a quintessential performer, it's hard to believe he didn't also know that they were true. As some of his friends have said, perhaps it was in fact the most fitting way for him to die, given his oft-stated desire to become one of the bears. Finding the idea of being eaten by another animal fitting, and accepting (or even embracing) the concept of literally becoming part of another creature's body through the processes of digestion and metabolism, is certainly an unusual perspective, at least in Westernized cultures, and it's easy to dismiss it as part of Treadwell's identity as an outsider marching to the beat of his own drum. However, as I will explore in the following section, it is an idea with profound implications and rewards.

III. EMBRACING OUR EDIBILITY

Ecofeminist philosopher Val Plumwood tells a powerful story of being attacked by a crocodile while canoeing on the East Alligator River in Kakadu National Park in Australia's Northern Territory. In her essay titled "Being Prey," she explains that at a certain point in her journey, she realized how vulnerable she was as "a solitary specimen of a major prey species of the saltwater crocodile" (2000, 2). Shortly after that, a crocodile attacked her canoe, attempting to knock her out. As she steered under a tree and prepared to jump into it, the crocodile rushed alongside her canoe, and they exchanged looks: "its beautiful, flecked golden eyes looked straight into mine" (2000, 2). As she waved her arms in an attempt to drive it away, "the golden eyes glinted with interest" (2000, 2). After being pulled underwater in a terrifying "death roll" three times in a row, she managed to clamber up the bank of the river and escape. Though she was severely injured, she was able to get to the edge of the swamp, where she was rescued by a search party organized by a park ranger who noticed she had not returned from her journey.

In her various accounts of this experience, she makes an eloquent case for responding to the gaze of a predator like the crocodile with an ecological humility that understands humans' place in the food web. As she made her way to the edge of the swamp after she escaped, still unsure she was going to survive, she was "apologizing to the angry crocodile, repenting to this place for [her] intrusion" (2000, 4). Once rescued, she acted as an advocate for the crocodiles, preventing her rescuers from going back into the swamp and killing a crocodile—any crocodile!—in revenge for the attack. She emphasizes that she was the intruder in the crocodile's territory and that the crocodile's attack was ecologically appropriate. Her experience of being prey—of realizing on a visceral level that the crocodile saw her as meat—led her to an acceptance of ecological identity that counters the dominant Western tradition of denial "that we humans are also animals positioned in the food chain" (2000, 7). As she writes, "Crocodiles and other creatures that can take a human life also present a test of our acceptance of our ecological identity. When they're allowed to live freely, these creatures indicate our preparedness to coexist with the otherness of the earth, and to recognize ourselves in mutual, ecological terms, as part of the food chain, eaten as well as eater" (2000, 7).

As an ecofeminist philosopher, Plumwood was no doubt primed to experience a predator attack differently than the average person (think of Treadwell's critics or the people who wanted to kill a random crocodile for revenge after her attack). But even so, coming to terms with the idea of being prey was a struggle, and her account suggests the kinds of obstacles anyone might face in moving from the understandable impulse to value one's own survival over everything else to experiencing the predator's gaze as a reminder of our ecological role as food for other beings to acknowledging the agency of those beings and reconciling ourselves to the limits of our own. She explains that the attack profoundly shook her sense of subjectivity, which she characterizes as a view of the world "'from the inside,' structured to sustain the concept of a continuing, narrative self" (2000, 3). She explains that in order to maintain action and purpose—to go about the daily business of living, in other words—one needs to "remake" the world in this way, "investing it with meaning, reconceiving it as sane, survivable, amenable to hope and resolution" (2000, 3). However, her traumatic experience with the crocodile jolted her outside of the boundaries of the ego so that she could—temporarily—see the world "'from the outside,' as a world no longer [her] own, an unrecognizable bleak landscape composed of raw necessity, indifferent to [her] life or death" (2000, 3).

As Plumwood shows, the instinctive reaction to this view of the world from the outside—to the notion of being food for another—is shock and denial; she recalls thinking during her attack, "This can't be happening to me. I'm a human being. I am more than just food!" (2000, 7). But her experience also demonstrates how the shock of being placed outside one's accustomed sense of (more than edible) self in this way can offer an opportunity to reconceptualize one's identity as a human. In his essay "The Uncanny Goodness of Being Edible to Bears," philosopher James Hatley explores how the realization that another creature sees one's flesh as food can help one understand that "my flesh is not only my own but also a mode of becoming" that other creature (a bear, in his hypothetical example) (2004, 21). In other words, when we are reminded that we can be eaten, it can orient us to the temporary and mutable nature of our bodily forms. As Hatley puts it, "my very body is revealed as the capacity to be the body of a bear, as well as that of a human" (2004, 21).

But an embrace of one's edibility can go beyond an acceptance of the ecological/ontological human role as food for other creatures to a deep sense of fulfillment in the possibility of being reintegrated into the world, as illustrated in Chase Twichell's poem "The Smell of Snow." The poem's title refers to the smell of a fisher, called a fishercat by the "local people" (1995, 61), encountered by the speaker in the forest near her home. For her, this smell marks the fisher as representative of the more-than-human world: "When I was a child I thought / that was the smell of God because / it obscured what was human" (1995, 75–77). Though fishers don't kill and eat humans, the speaker feels the risk of becoming prey when she encounters this fierce predator, thinking "*is this my death?*" (1995, 58). Rather than reacting with fear or outrage, she indulges in an imaginative vision of the ways that becoming prey could provide an entryway into the larger processes of life, yearning

> to be the single
>
> creature of his desire,
>
> the one he would tear open,
> drag off in pieces to devour,
>
> and thus disappear
> in violence into the world of his flesh,
>
> go where his flesh goes,
> even into the coyotes' hunger

when they finally pull him down,
into their scat

with its clots of hair and berry seed,

living on a while longer
in blood, piss, fur, musk,

before my bleached dust is abandoned

to the roots and leaves, and I become
the words the wind says
to the birch tatters, the song

The hawk's shadow sings to the ground ... (1995, 98–115)

By longing for the hunger of a predator and envisioning her integration into the food chain after being consumed, the speaker conveys
the beauty and rightness of participating in the processes of eating and
being eaten, celebrating the endless possibilities of transformation it
offers. And the rightness of this imagined fate evokes more than just
an ecological vision, subtly suggesting what Hatley calls a "plethoric"
sense of the way the body comes into being out of a larger "inter-
twining of flesh with flesh" (1995, 22) in the world. In this view, the
realization that we can be eaten not only reminds us of the other bodies
and physical forms that we can become, but also makes us aware of the
ways that our bodies are *already* interwoven with every other animate
existence that eats; all the other others are implicit in the bodies in
which we come into being. As Twichell's speaker encounters the fisher,
she tells us that "I felt my body long toward his" (1995, 88), almost
as though she recognizes the plethoric sense in which her body and
fisher's arise from the same intertwined "flesh of the world" and thus
have an age-old relationship that she recognizes on a physical level.
Her later reference to the "world" of the fisher's flesh also suggests the
multitude of other beings, both past and present, with whose flesh her
body is already entangled.

For this speaker, who feels the pain of alienation from nature that
pervades so much of Twichell's poetry, the otherness in the fisher's
"small black / God-eyes ... empty of / any language [she] could extract"
(1995, 82–85) aligns it with the divine. The notion that she will "go

where his flesh goes" carries a powerful sense of companionship after the loneliness and emptiness of living cut off from nature, suggesting yet another reason to embrace one's edibility: the knowledge that we can be food for other creatures provides a tremendous reassurance that we are not alone, that we belong here in this world. Even the division between nature and culture represented by the assumption that language is a uniquely human characteristic, which she originally accepts, is eroded away in her vision of becoming the "words" of the wind or the song of the "hawk's shadow." If we accept Gary Snyder's definition of the sacred as that which takes us out of ourselves and out into the universe (1990, 101), this poem suggests that understanding our bodies as food for the world can bring about a great sense of spiritual fulfillment.

But how can we reconcile such an understanding with the more everyday notion that our bodies belong to us? For humans whose bodies are attacked, imprisoned, or otherwise claimed or violated by other humans, the idea of relinquishing one's hard-won claims to bodily autonomy and agency may be especially unappealing or difficult. In *Eye of the Crocodile*, a manuscript unfinished at the time of her death and published posthumously in 2012, Val Plumwood contrasts the "individual justice world" of modernist liberal individualism in which we own our lives and bodies with the "food chain world" she viewed "through the eye of the crocodile," "a Heraclitean universe where everything flows—where we live the other's death, die the other's life" (2012, 35). She explains that this second world seemed to her at first terribly unjust, but that eventually she was able to see it differently, as "a world of radical and startling equality—it is not unfair, it treats all the same way" (2012, 36). In fact, she explains, it is pervaded with a certain kind of generosity: "one in which our bodies flow with the food chain. They do not belong to us; rather they belong to all. A different kind of justice rules the food chain, one of sharing what has been provided by energy and matter and passing it on" (2012, 35).

For Plumwood, this perspective offers a new way to know oneself as human, a way that is both "astounding and utterly dismaying" (2012, 36): as a being that exists for others as well as oneself. The challenge is how to reconcile this view with the perspective of the other world we inhabit—the world in which our individual survival matters above all else. Plumwood argues that despite the radical

incommensurability of these two worlds, we exist in both simultaneously and need to find a way to live fully and consciously in both
rather than consigning some beings (typically humans) to the "individual justice world" and others (typically everything else) to the "food
chain world." The key, she concludes, lies in the realm of narrative,
asking "What sort of story would you have to tell to bring the two
worlds together?" (2012, 42). As I have suggested, narratives like
Plumwood's own story and Twichell's poem suggest ways of doing so,
of experiencing the gaze of the predator as a confirmation that being
human means sharing mortality, mutability, and an essential molecular
interrelatedness with other beings. The consolation for relinquishing
our normally viselike grip on our individuality and autonomy is the
knowledge that not only are we all in this together, but we also *are*
all this, together.

IV. LIVING WITH PREDATORS

While writers and thinkers like Plumwood, Hatley, and Twichell have
embraced the notion that humans are caught up in a web of ongoing change and exchange of matter with other living things and the
rest of our environment, understanding our current bodily forms as
temporary and unstable and not at all a teleological endpoint for the
material we're made of, they are surely exceptions within the Westernized cultures they inhabit. While their stories and visions may have an
impact on others, it's also worth exploring whether an entire culture
can live within the gaze of potential human predators in a way that
produces a more ecological and dynamic sense of human identity. One
place where humans still face the reality of becoming prey on a daily
basis is the Sundarbans archipelago of mangrove islands that spans the
southern edge of Bangladesh and the Indian state of Bengal, home to
infamous "man-eating" tigers. Remarkably, the Sundarbans islanders
have largely come to terms with this reality without demonizing or
trying to eradicate their predators, to a great degree accepting the
tigers as agential partners in an ecological community. Although this
relationship is threatened by Western-style conservation that sometimes prioritizes the conservation of the tigers over the lives of local
residents, it still provides valuable insights into how a unique physical
environment and culture allow (and perhaps even require) residents

to integrate the desire for individual self-preservation with a holistic, other-oriented acceptance of participating in a system of shared matter and energy.

Humans in the Sundarbans lead a precarious existence, often barely subsisting by fishing and collecting "forest products" such as firewood and honey. Mobility is limited by dense mangrove forests and powerful tides, with most travel conducted by boat. The area is subject to frequent cyclones, and residents are at risk of attacks by sharks, crocodiles, and, perhaps most terrifying, tigers. Constituting roughly 3 percent of the world's population of this endangered carnivore, the Sundarbans tigers are unique in more than one way: they have adapted to their watery environment, making fish a large part of their diet, and they are also far more likely to attack and eat humans than tigers in other parts of Asia. It is very difficult to determine exactly how many people Sundarbans tigers have killed in recent years, though, in part because attacks often occur when residents illegally enter protected tiger conservation areas in search of forest products and so go unreported. For the West Bengal side of the Sundarbans, estimates range from ten people a year killed by tigers (Alexander 2008) to 150 people lost to crocodiles and tigers annually, with a similar number presumed for the Bangladesh side as well (Jalais 2010, 44). The threat of tiger predation is so woven into the fabric of life in the Sundarbans that wives dress as widows whenever their husbands go into the forest (Ghosh 2005, 67–68; Alexander 2008).

Though my focus is on the ways the islanders have adapted to tiger predation, it's important to note the very real devastation it causes. Many of the families who lose someone to the tigers are already desperately poor, and the person who is lost is often a man, who dies in the process of supporting his family by fishing, or more likely by collecting honey or gathering wood, since these activities necessarily bring humans into the tigers' territory. Families who lose a male "breadwinner" are especially vulnerable to economic devastation, since women marry quite young and usually have no skills outside the home (Hagler-Geard 2012). One might think people would lash out against such an ever-present threat and seek vengeance and safety as they do in one memorable scene in Amitav Ghosh's novel *The Hungry Tide* (set in the Indian portion of the Sundarbans): a tiger that has killed two people and numerous livestock becomes trapped in a livestock pen, where the villagers surround him, jab and blind him with pointed sticks, and then burn him alive. And killings like this occur in real life, especially when tigers enter villages. But they are surprisingly rare, and Sundarbans residents generally accept the tiger's presence despite the risks it entails. In her study *Spell of the Tiger: The Man-Eaters of Sundarbans*, Sy Montgomery reports: "Some wish there were fewer tigers here; some—particularly tiger widows, who have never worked in the forest, wish that the Forest Department would shoot known man-eaters rather than attempt to capture and relocate them. But no one who has grown up around the mangroves and worked in the forest calls for the eradication of the tiger. No matter how many men are killed, no matter how deeply the man-eaters are feared, the tiger is not hated" (1995, 195).

How is it that the islanders are able to maintain such an extraordinary relationship with a dangerous predator, valuing the tiger's presence while acknowledging and managing the risks of living on the fringes of its territory? To some degree, this achievement reflects the islanders' understanding of the role the tiger plays in their local ecology and political economy. In "Tigerland," Caroline Alexander explains that many residents feel "[t]here would be no Sundarbans if there were no tiger," believing that the tigers' presence protects the forest from over-logging (2008). Sy Montgomery quotes a tiger reserve official who characterizes the tiger as a kind of god, "'silently doing the work of ecodiscipline … looking after the forest, and the forest is looking after it'" (1995, 222). The residents seem to understand that they and the tigers are all part of a community and need each other even if they also pose risks to each other. But the residents' relationship with tigers is far more complex than this, growing out of long-standing cultural traditions and a cosmology with deep roots in the very particular physical environment of the Sundarbans.

Certainly, Sundarbans culture is not exempt from a sense of dualism between culture and nature, which surfaces in the different sets of rules for behaviour that apply in the villages and in "the forest." However, the residents of this region can't completely exclude themselves from what they see as nature, since they directly depend on it for survival, and they demonstrate a remarkably high tolerance for ambiguity, change, and porous boundaries, no doubt reinforced by their physical environment. Situated where the fresh waters of the Ganges,

Meghna, and Brahmaputra Rivers flow into the saltwater tides of the Indian Ocean, the Sundarbans is an ever-changing hybrid land/water-scape, described in *The Hungry Tide* as possessing "no borders ... to divide fresh water from salt, river from sea. The tides reach as far as two hundred miles inland and every day thousands of acres of forest disappear under water, only to reemerge hours later. The currents are so powerful as to reshape the islands almost daily" (Ghosh 2005, 6–7). In fact, the people who live there call it not the Sundarbans—which means "forest of beautiful trees"—but *bhatir desh*, country of the ebb tide, since many islands appear only when the tide is out (2005, 7). Ghosh's novel also conveys the powerful effect of the mangroves, which quickly take over when a new island is formed to create a dense forest "utterly unlike other woodlands or jungles," with such poor visibility that no one could have any doubt of "the terrain's hostility to their presence, of its cunning and resourcefulness, of its determination to destroy or expel them" (2005, 7).

Despite this sense of the natural environment's hostile agency, the islanders believe that humans have the right to interact with the forest itself, the realm of "nature," as long as they behave properly. In *Forest of Tigers: People, Politics and Environment in the Sundarbans*, anthropologist Annu Jalais explains that they see the forest as a realm of equality and sharing, in contrast to the dry land, which carries connotations of hierarchy and violence because landownership carries status and makes people greedy (2010, 51). Given Plumwood's emphasis on the role of narratives in integrating individual and holistic perspectives, it's significant that this ethic of sharing, which extends to tigers and the entire non-human realm, is intimately connected to the story and rituals surrounding the forest deity Bonbibi. Bonbibi is a paragon of boundary-straddling and hybridity, originally abandoned by her mother in the forest and raised by a deer and ultimately mediating "between Allah and humans, between village and forest, and between the world of humans and that of tigers" (Jalais 2010, 69). Even her story, *Bonbibi Johuranamah* (The Glory of Bonbibi), is a syncretic mix of Islam and Hindu.

The story begins with Dokkhin Rai, a Brahmin sage whose greed prompted him to claim the entire mangrove forest for himself, a claim he enforced by transforming himself into a tiger in order to kill and eat humans who were "stealing" the products of the forest. This led the other tigers to follow suit and begin preying on humans as well. However, Allah took pity on the humans and called on Bonbibi to help them. The conflict ended in a truce and friendship after Bonbibi gained the upper hand over Dokkhin Rai. But the story does not end there: Dukhe, a young boy living with his impoverished, widowed mother, joined his uncle on an expedition collecting honey in the forest. Dokkhin Rai promised the uncle great wealth in exchange for Dukhe, and the greedy uncle left Dukhe on the shore. As Dokkhin Rai was about to devour him, Dukhe called out to Bonbibi, who rescued him and sent Dukhe home with enough treasure that he never needed to work in the forest again.

This complex story establishes and reinforces important principles of the Sundarbans island culture. As presented in *The Hungry Tide*, Bonbibi's actions show the world "the law of the forest, which is that the rich and greedy would be punished while the poor and righteous were rewarded" (Ghosh 2005, 88), sending a message that forest products belong to all and should be shared. However, as Jalais notes, the story includes non-humans in the community with rights to the forest. After Dukhe was rescued, Dokkhin Rai took the advice of the Gazi (an Islamic holy man) and pleaded for Bonbibi's forgiveness. Accepting Dokkhin Rai as her "son," she listened to his fears that humans would destroy the forest if given free rein and brokered an agreement among Dukhe, Rai, and the Gazi that humans and non-humans would treat one another as brothers in the future. This sense of kinship tempers the hostility that might otherwise result from competing for the same forest resources, instead establishing an expectation that humans and non-humans can "depend on the forest and yet respect the others' needs" (Jalais 2010, 72).

As they walk a narrow line between the drive for survival and the realization that humans are necessarily part of the material flux of the natural world, the islanders can turn to rituals and rules that build on this story in order to cope with conflict between these two modes of perception. Villagers believe they must enter the forest with pure hearts and minds, which means taking no more of the products of the forest than they need for survival, curbing greedy wishes and deceitful motives, and avoiding arguments or any sort of violence. The forest is viewed as a place where divisions based on caste and religion must be stringently disregarded (2010, 68). Thus, "entering the forest ... can be seen as a sort of 'liminal phenomenon' that brings people who are in its realm to partake of a sense of community feeling ... with each other" (2010, 86)—and even with a fear-inspiring predator like the tiger. Simultaneously, they can retain some sense of agency and control over their fates in the face of what would otherwise seem like unbearable risk, believing that following prescribed codes of behaviour will protect them and that they can call out to Bonbibi for help in moments of danger.

But there may be one more ingredient in the recipe for this relationship: the tiger's relative invisibility. As the narrator of *The Hungry Tide* remarks, "[t]he great cats of the tide country were like ghosts, never revealing their presence except through marks, sounds and smells. They were so rarely seen that to behold one, it was said, was to be as good as dead" (Ghosh 2005, 91). This lack of visibility stems both from the tigers' physical nature and from the dense foliage of the forests in which they live. As Laura A. White points out in an analysis of *The Hungry Tide*, more than many other environments, the Sundarbans encourages and even requires non-visual ways of knowing. Needless to say, a reliance on the visual is a markedly human tendency, one that has grown out of our biology but that is also freighted with a powerful cultural history of looking as an act of domination, control, and distancing. In the Sundarbans, roles are reversed, with the tigers looking and the humans aware that they are much more likely to be seen than to see. Awareness of being the potential object of a predator's gaze—an awareness that many in the Sundarbans experience regularly—must be profoundly humbling, reminding humans of the limits of our knowledge and power and encouraging a sense of our vulnerability and mutability as embodied beings—qualities we share with all other living beings. Animal studies theorist Cary Wolfe suggests that when this shared vulnerability and mutability become part of our core definition of what it means to be human, we may finally be able to move beyond the damaging dualisms of the Western traditions of individualistic liberal humanism.

V. CONCLUSION

Even with the tremendous potential for reconceptualizing what it means to be human offered by the gaze of the predator, it seems naive to imagine that most people will seek out that gaze, even imaginatively or vicariously through the stories of others. And even if they do willingly consume stories such as Plumwood's, Twichell's, or those of the people of the Sundarbans, the effect may be limited: Plumwood warns that stories that allow us to simply "visit" the "food chain world" will not be enough to integrate an understanding of the world into the "individual justice world" of our lived experience. She suggests that we bring the two worlds "into harmony at the point where they touch," which is also "the point where what we eat morally touches our lives" (2012, 39). In striving to embrace the world's claim on us in a spirit of generosity, she argues, we can begin by responding to the generosity of the beings who become our food by "be[ing] generous in our turn and relinquish[ing] the desire for a mean-spirited self-maximisation that takes as much from each life we use as we possibly can" (2012, 39).

In addition to practising more conscious, generous eating, I wonder if the path to developing culturally influential new stories that dethrone the human from an imagined point outside and above the pulsating webs of energy and matter that make up the universe could begin with reinterpreting stories that already possess ideological heft. What stories do we tell in Westernized societies that celebrate a human dying and relinquishing his or her body to others to eat? I can't help but think of my Catholic education, attending mass with my grade school class and my family and being instructed that Christ had died for our sins. When we received communion, we were told we were receiving and eating his blood and flesh—literally, according to the doctrine of transubstantiation. This was presented as a great gift to us ordinary humans. When Christians model their lives on Christ's, could it include the idea of following in his footsteps and giving themselves to the rest of creation? We are told that Christ took human form in order to suffer, to experience the vulnerability and weakness of being fully human. Taking this interpretation of what it means to be human seriously could put quite a chink in the wall we tend to erect between ourselves and the natural world. Like any cosmology, Christianity is what you make of it. But the foundations for a humbler, more vulnerable and interconnected view of ourselves as human arguably already exist within one of the most powerful stories in the Western world.

In the end, though, what most of us need is practice. In the Sundarbans, they don't have the luxury of ignoring what Hatley calls "the world's many hungry mouths" (2004, 23). Those of us in the global North generally do, and so it's unlikely we will dwell on the unlikely horror of being hunted and eaten by a large carnivore. What if instead we shifted our focus and made a point of paying attention to all the much smaller, less charismatic creatures taking a proprietary interest in our flesh—creatures most or all of us encounter every day, such as the mosquitoes and ticks thirsting for our blood, the dust mites congregating to feast on our dead skin cells, the beneficial gut bacteria that feed off the sustenance we intend for our "selves" alone, and even the worms and other decomposers waiting to usher us into our next material incarnations (if they get the chance, since our conventional burial practices are designed to thwart this process)? If we also move away from our anthropocentric privileging of the visual and ask how those creatures experience us, we might learn that in addition to (or instead of) being seen, we are heard, we are smelled, we are felt—and yes, we are tasted. And not just by mammal mouths similar to ours, but (probably more often) by orifices very different than ours: the needle-like "mouthparts" of the mosquito, the suckers of the tapeworm, the almost invisible but muscular slit-like mouth of the earthworm. The next step in developing a more integrated, generous sense of our relationship to the animate, hungry world is to ask yourself, on every occasion possible, how much you need to preserve yourself and how much you can give. Accept that as you go through life, you will get a bit "nibbled," as Annie Dillard puts it in *Pilgrim at Tinker Creek,* and that the world will take you apart in the end. There are limits to human agency, so why not embrace them?

Please don't misunderstand: I am not suggesting we should let everything that wants a bite of our bodies have its way; striving to survive and to avoid injury and illness is the right of every living thing. But we can check the outrage when other creatures look to us for a meal and stifle the desire to eradicate every threat to the (ultimately inevitable) breaking apart of our bodies. It's popular to say everything is connected, and the idea that we are all part of the "circle of life" can be empty and unthreatening enough to serve as a theme in a Disney movie. But when we peer into the sometimes shocking truth of *how* we are connected with everything else, being human does not mean perching atop the rest of creation, riding it like surfers on a wave, propelled forward and upward, exhilarated, but ultimately left intact, only temporarily dampened by a momentary, voluntary connection with the otherness of nature. Instead, we are always swimming, surrounded and literally made of the same stuff as the rest of creation. We have never been "intact." We can't have it both ways—wanting that warm and fuzzy connection with the world evoked by the circle of life but treasuring our human uniqueness and superiority as a rationale for vilifying and exterminating any other being that threatens not just our survival but even our sense of bodily integrity. That fight for integrity was lost before it started. We are already made of everything else, symbiotic with bacteria, constantly losing and gaining different molecules from our environment. And we don't have to wait for the gaze of a huge, terrifying predator to remind us that we are not only eaters but also food, not only beings who are agents but also raw material for other beings: those beings are already around and within us, confirming our humanity through our mutual dependence, if only we can learn to perceive it.

References

Alaimo, Stacy. 2010. *Bodily Natures: Science, Environment, and the Material Self.* Bloomington: Indiana University Press.

Alexander, Caroline. 2008. "Tigerland: A Journey through the Mangrove Forests of Bengal." *The New Yorker*, 21 April, 66–77.

Berger, Joel. *The Better to Eat You With: Fear in the Animal World.* Chicago: University of Chicago Press, 2008.

———. 1980. "Why Look at Animals?" In *About Looking*, 1–26. New York: Pantheon Books.

Burt, Jonathan. 2005. "John Berger's 'Why Look at Animals?': A Close Reading." *Worldviews* 9(2): 203–18.

Emerson, Michael. 2015. "Looking Again at Looking at Animals: Berger's 'Why Look at Animals?' and the Abyssal Limit of the Gaze." Paper presented at the biennial meeting of the Association for the Study of Literature and Environment, Moscow, Idaho, 23–27 June.

Ghosh, Amitav. 2005. *The Hungry Tide.* Boston: Houghton Mifflin.

Grizzly Man. 2005. Directed by Werner Herzog. Santa Monica, Lionsgate Home Entertainment.

Hagler-Geard, Tiffany. 2012. "Tiger Widows: Life in the Sundarbans." *Picture This: ABC News*, 29 February. http://abcnews.go.com/blogs/headlines/2012/02/tiger-widows-life-in-the-sundarbans.

Hatley, James. 2004. "The Uncanny Goodness of Being Edible to Bears." In *Rethinking Nature: Essays in Environmental Philosophy*, edited by Bruce V. Foltz and Robert Frodeman, 13–31. Bloomington: Indiana University Press.

Heise, Ursula. 2008. *Sense of Place and Sense of Planet: The Environmental Imagination of the Global*. New York: Oxford University Press.

Jalais, Annu. 2010. *Forest of Tigers: People, Politics and Environment in the Sundarbans*. New Delhi: Routledge.

Jans, Nick. 2005. *The Grizzly Maze: Timothy Treadwell's Fatal Obsession with Alaskan Bears*. New York: Dutton.

Montgomery, Sy. 1995. *Spell of the Tiger: The Man-Eaters of Sundarbans*. Boston: Houghton Mifflin.

Peacock, Doug. 2004. "Blood Brothers." *Outside*, January, 18–22.

Plumwood, Val. 2000. "Being Prey." *Utne Reader*, July–August. Accessed 5 May 2009. http://www.utne.com/2000-07-01/being-prey.aspx.

———. 2012. *Eye of the Crocodile*, edited by Lorraine Shannon. Canberra: Australian National University Press.

Russell, Charlie. 2006. "Timothy Treadwell—*Grizzly Man*." 21 February 2006. http://charlierussellbears.com/treadwell.html.

Schutten, Julie Kalil. 2008. "Chewing on the Grizzly Man: Getting to the Meat of the Matter." *Environmental Communication* 2(2): 193–211.

Snyder, Gary. 1990. *The Practice of the Wild*. San Francisco: North Point Press.

Twichell, Chase. 1995. "The Smell of Snow." In *The Ghost of Eden: Poems*, 61–66. Princeton: Ontario Review Press.

White, Laura A. 2013. "Novel Vision: Seeing the Sunderbans through Amitav Ghosh's *The Hungry Tide*." *ISLE: Interdisciplinary Studies in Literature and Environment* 20(3): 512–30.

CHAPTER 8

Anim-oils: Wild Animals in Petro-Cultural Landscapes

Pamela Banting

> "Here's the truth," Margaret says, now emotional. "Where are the animals? There's no too-da-loos, the little one-armed fiddler crabs. Ya' don't hear birds. From Amelia to Alabama, Kevin never saw a fish jump, never heard a bird sing. This is their nestin' season. Those babies, they're not goin' nowhere. We had a very small pod of sperm whales in the Gulf, nobody's seen 'em. Guys on the water say they died in the spill and their bodies were hacked up and taken away. BP and our government don't want nobody to see the bodies of dead sea mammals. Dolphins are choking on the surface. Fish are swimming in circles, gasping. It's ugly, I'm tellin' you. And nobody's talkin' about it. You're not hearing nothin' about it. As far as the media is reportin', everythin's being cleaned up and it's not a problem.
>
> —*Margaret Curole, chef and Cajun shrimper, qtd. in Williams (2010, 40)*

> ... political economists have long recognized that capitalism is predicated upon our metabolism of nature ...
>
> —*Bakker 2010, 726*

A political cartoon in the 12 January 2012 issue of the *Times Colonist* (Victoria) depicts several animals gathered along British Columbia's ocean shoreline—an eagle, an otter, some clams and mussels on land, and a seal, three waterfowl, and two fish in the water. The caption reads, "Major stakeholders on the B.C. Coast who will not be invited to the Enbridge Pipeline hearings." The cartoon encapsulates the fact that if environmental groups and the individuals who support them are silenced, or if the hearings are disrupted, discredited, terminated, or if the public input of the hearings were simply ignored, there would be virtually no organized voices left to speak on behalf of animals.

Animals—both human and other-than-human—are profoundly affected by oil, bitumen, and gas in countless ways—through spills, roadification, loss of habitat, air pollution, oil and chemical contaminants in soil and water, seismic testing, dramatically lower estuarine flows, unmanaged human garbage at work camps, and other problems associated with energy mining, transport, processing, and use. As Andrew Nikiforuk notes in his seminal, award-winning book *Tar Sands: Dirty Oil and the Future of a Continent*, the Alberta tar sands tailings ponds alone literally consume hundreds of other-than-human animals, particularly birds:

> Every year the ponds quietly swallow thousands of ducks, geese, and shorebirds as well as moose, deer, and beaver ... To a bird's eye, the toxic ponds look like a nice bit of ice-free real estate in the spring and fall. The ponds also lie under a major migratory flyway for birds travelling to the Peace–Athabasca Delta, which Environment Canada calls "one of the most important waterfowl nesting and staging areas in North America." A trumpeter swan or snow goose coated in bitumen often dies of hypothermia. Dene and Métis hunters have found slimed, near-dead ducks 135 miles north of the ponds in the bush. (Nikiforuk 2008, 81)

Images of ducks covered in sludge from the ponds have circulated around the world. Eventually a court case was won that required the Syncrude corporation to pay a fine of $3 million for the deaths of 1,600 ducks at the ponds in 2008.[1] Downstream from the tar sands in Lake Athabasca, Indigenous fishers have caught some fish bearing massive tumours and others that exhibit genetic mutations involving, for example, the appearance of two mouths.[2] Caribou numbers are rapidly declining, due almost certainly to loss of habitat in the region of the tar sands. Instead of dealing with lost habitat by ceding some territory to the animals, the then Conservative Alberta Government had been addressing the crash in the caribou population by shooting, sterilizing, and poisoning literally hundreds of wolves (Linnitt 2012, n.p.) (along with other animals that also consume the baits or that consume animals that have eaten them). Poor and illegal garbage management practices at the tar sands have also been linked to provincial government fish and wildlife officers shooting 145 black bears in the tar sands region.[3]

Ironically, oil *is* animals. The remains of ancient plants[4] and animals (zooplankton) are the very constituents of fossil fuels.[5] Summarizing the geological history of Alberta and the processes that form coal and oil, writer Sid Marty observes: "Without the animate, there would be much less inanimate mass" (Marty 1995, 16). Moreover, animals have also been used as sources of heat, light, energy, and human dietary fat for aeons. Even a partial list of animal oils would include bear grease, duck fat, beaver oil, buffalo and beef tallow, sheep's lanolin, whale oil, turtle oil, shark oil, the oil of oolichan or eulachon (also known as candlefish), and many more.[6] Pemmican—or grease as it was called—which propelled the men who propelled the fur trade canoes during the fur trade era in Canada, is largely fat. As Patricia Yaeger writes in "Literature in the Ages of Wood, Tallow, Coal, Whale Oil, Gasoline, Atomic Power, and Other Energy Sources," "energy sources have varied wildly over time and space and include almost anything that burns: palm oil, cow dung, random animal carcasses mounted on sticks" (Yaeger 2011, 307).

Many articles and books in the arts and humanities have demonstrated that anthropocentrism is responsible for the marginalization, denigration, abuse, and killing of other animals. Yet little if any scholarly work in these disciplines has been done on the effects of fossil fuel production and use on contemporary wild animals. While the exploitation and destruction of ecosystems by the fossil fuel industry is, of course, an instance of anthropocentric activity, I propose to consider the effects upon wild animals of industrial resource extraction and the North American petro-culture that enables it. In doing so, I will draw upon three works of literary nonfiction—Eden Robinson's *The Sasquatch at Home: Traditional Protocols and Modern Storytelling*, Farley Mowat's *A Whale for the Killing*, and Sid Marty's *The Black Grizzly of Whiskey Creek*—and the two emergent fields of animal studies and petro-cultural studies to examine how the lives of certain animals—oolichan,[7] whales, and bears respectively—have been affected by oil cultures and how, by contrast, adherence to traditional accords between humans and other animals (such as are recorded in traditional stories and, in many cases, supported by scientific studies) could help us resist the worst effects of these catastrophic industrial processes on animals.

Given Matthew T. Huber's argument that it is necessary to understand oil not only as a thing-in-itself but also as something produced through a set of historico-temporal social relations (Huber 2011, 44), I examine oil in terms of social relations that cross the species divide. As Huber writes, oil

> is increasingly framed not as the [societal] lifeblood, but as an *addiction*—an unnatural fluid uncontrollably entrenched within the American bloodstream. It seems no matter how much oil is linked directly to war, climate change, or catastrophic oil spills ... its embeddedness within everyday social reproduction persists. Thus, whether framed as the lifeblood or addiction, oil's singular power over us is naturalized. (Huber 2011, 33)

Like Huber, Marion Grau figures the North American relationship with oil in terms of addiction: "Substances we depend upon can quickly devolve into objects of dependency—the result is substance abuse. This takes different forms in different contexts. In urban contexts, oil-dependency contributes to apathy and inaction, creating large numbers of moral bystanders, silent beneficiaries of the addiction, if not outright dealers and pimps" (Grau 2007, 449). It is this naturalization of oil's power over us and other animals, its figurative and literal incorporation or corporealization—as blood flow, drug, and, as I shall argue, fat—and its tendency to rob both humans and more-than-human animals of our respective agencies that I shall address by examining oil's role in encounters between humans and other species.

If oil is both blood and drug, then what sorts of tests might we administer in order to come up with a way of sensing the coursing and pulsing of this "lifeblood" within us and our imaginaries? How can we diagnose its presence, effects, and affects within us? Can the addict heal oneself? Given the contemporary neoliberal extension of capitalism into zones and life worlds previously uncommodified, can we now begin to detect the symptoms of our oil addiction via narratives that are in part about other-than-human animals and oil? Might narratives of encounter between us and other animals help illuminate our own corporeality? As the cartoon I alluded to above suggests, other animals are always already stakeholders in oil and gas and their movement to market. However, it is not just a question of who can speak for these stakeholders and endow them by proxy with a form of agency. Instead we should ask, who can hear them? How can we begin to hear, see, and understand what we already know but what lies repressed in our collective petro-unconscious? What might narratives of encounter and relationship with wild animals across petro-landscapes tell us about our need and greed for oil? If oil is blood, bodily fluid, and drug, a drug that metabolizes both in our individual bodies and in the body politic, then what do we make of narratives of encounter with wild animals and specifically of scenes in which there may be—and sometimes is—blood, both ours and theirs?

As we typically deploy the term, "wild" connotes that which is outside of or extraneous to human social, cultural, and political domains. Naturalist and bear expert Charlie Russell succinctly encapsulates the culturally predominant notion of the wild animal: "By definition, a wild animal is one that is fearful of humans" (Russell and Enns 2002, 12), a notion his life's work radically problematizes. While there has been broad acceptance and adoption of the notion of the social construction of nature, when we attempt to think about or with wild animals we tend nevertheless to default to older pre-constructivist notions of them as largely outside our domain.[8] However, in the same way as the wild is not born free of our cultural constructions, the wild, wilderness, and wild creatures are not free of our obsession with maintaining and enforcing our regimes of energy and consumption. The image of the pitiful, oil-enshrouded animal—duck, pelican, turtle, otter, beaver, dolphin—is not only "us" in some senses[9] but also a representation of our "metabolism of nature" (Bakker 2010, 726), our petro-addiction[10] and the socio-political economies that underwrite and police it.

ꙅ

Dad's house was broken into the year before and the big screen TV and stereo systems and DVD collection were untouched. The only thing taken was a gallon jar of grease.
—Robinson (2011, 22)

In her Henry Kreisel lectures titled *The Sasquatch at Home: Traditional Protocols and Modern Storytelling*, Haisla/Heiltsuk writer Eden

Robinson illuminates how closely the diet, stories, and culture of her people are interwoven. One of the fish that figures prominently in her work is the oil-rich oolichan, which "arrived at the end of winter when most stored food supplies were depleted and, in harsh winters, people were facing starvation" (Robinson 2011, 19). In their article "Sharing Resources on the North Pacific Coast of North America: The Case of the Eulachon Fishery," Donald Mitchell and Leland Donald quote a 1929 source that "for the Tsimshian, 'It was the first fish caught and was very often styled the starvation fish, as it came just when the people were on the verge of starvation before the appearance of salmon and always in great quantities and always easy to get.' The Nisga'a still refer to eulachon as /ha'liimootkw/ 'means of saving, salvation' or 'saviour'" (Mitchell and Donald 2001, 22). Mitchell and Donald provide a snapshot of the dietary role and health benefits of the oolichan for the coastal peoples:

> Early historic claims for exceptional health-giving and curative properties have only in part been confirmed by more recent nutritional analyses. Both fish and processed grease are rich in vitamins A and E and the fish, fresh, smoked, or dried, are a source of modest amounts of calcium, phosphorus, iron, and zinc. Perhaps most important, unsaturated fat content, at 65 percent, is double that of saturated (32 percent). Further, the grease contains significant levels of a beneficial fatty acid which acts to reduce blood cholesterol and triglycerides. Oils, like that derived from eulachon, have additional or other desirable properties. They are in demand as a condiment, important to make more palatable a diet that for so much of the year consisted of dried foods and no doubt aiding in passage of such fare through the alimentary canal. Another use was as a preserving medium for fruit. Drucker notes that from the Nuxalk (Bella Coola) and Heiltsuk north, crabapples and berries were stored in a frothy eulachon oil emulsion, in which medium "they could be kept for a long time." Included in a long list of plant foods and materials exchanged by inhabitants of Northwestern North America are Pacific crabapples, bog cranberries, all of which, fresh or preserved in water or eulachon oil, were "widely traded" along the coast "and probably among interior peoples." (Mitchell and Donald 2001, 21)

Clearly though, as the authors go on to elaborate, oolichans provide more than nutrition and a means of food preservation. In her article on the fish, anthropologist Gloria Cranmer Webster points out that its importance within Kwakwaka'wakw culture is reflected in the fact that even some place, tribe, and personal names relate to oolichan and the grease made from them. She writes:

> Kliniklini is an Anglicized form of *tli'na*, the word in our language for the rich oil, commonly called "grease," which is processed from eulachons. The name *Dzawadi* means "place of *dzaxwan*," or eulachon. The name of the tribe occupying Kingcome Inlet is *Dzawada'enuxw*, meaning "having eulachons" … *Maṇtla* describes a state of satiation, in particular that caused by consuming the oil with other foods, such as dried or smoked salmon. So, the name, *Maṇl'ida'as* refers to the place where you become satiated, that is, the house of a generous host, who feeds his guests well at feasts. *Maḷidi* is the name of someone who makes you satiated. These are names that are still passed on today. (Cranmer Webster 2001, 37–38)

These oily fish structured in part, lubricated, and maintained the social order, functioned as indicators of prestige, and provided cultural linkages between coastal and interior peoples through trade along well-travelled "grease trails" (Cranmer Webster 2001, 20).[11]

Today, however, the oolichan runs are already severely "compromised by effluent from the town of Kitimat, Eurocan Pulp & Paper and Alcan Aluminum smelters" (Robinson 2011, 21-22).[12] Moreover, only "a couple of families in the Village still know how to render the oolichan grease" (22). Given the time and commitment required, no grease has been made recently, and its price has skyrocketed to $300 per gallon. During a fishing trip with her father, Robinson cannot help but notice the ironies involved in fishing for one's own food today: "[Y]ou have to [be] fairly well-off to eat traditional Haisla cuisine. Sure, the fish and game are free, but after factoring in fuel, time, equipment, and maintenance of various vehicles, it's cheaper to buy frozen fish from the grocery store than it is to physically go out and get it" (22). On the other side of the ledger, however, are tradition, protocol, family, community, memory, and story—things that elude quantification. As Robinson remarks, "I hate to think of thousands of years of tradition dying with my generation. If the oolichans don't return to our rivers, we lose more than a species. We lose a connection with our history, a thread of tradition that ties us to this particular piece of the Earth, that ties our ancestors to our children" (23).[13]

A colonialist, instrumentalist, or capitalist mentality might relegate such statements as Robinson's about the connection between oolichans and tradition, nature and culture, to the zone of sentiment or nostalgia for a bygone world. However, when read against the background of the recreational killing we witness in Farley Mowat's *A Whale for the Killing*, or when considered against the backdrop of petro-cultural thinking and the neoliberal commodification of everything, one begins to see that, to the contrary, it is only through the sensation of the exuberant vitality of life forms and of linkages between ecosystems and culture that we can hope to survive as a viable society. Cranmer Webster concludes her article by drawing attention to the importance of the fish not only to humans but to other animals as well: "Not only are eulachon important to human beings, but they are also a vital link in the food chain including sea mammals, eagles, seagulls, dogfish, salmon, cod, halibut, and bears. If the eulachon were to disappear completely, the people of the Northwest Coast would lose a precious part of our heritage and other forms of life would share our loss" (Cranmer Webster 2001, 41). In other words, culture is not just a top-down invention or collective creation by human beings alone, nor is it just the story of humans and our tools and resources.[14] It supersedes the "patterns, traits, and products considered as the expression of a particular period, class, community, or population," as it is defined in *The American Heritage Dictionary of the English Language*. Culture is, essentially, an ecological relationship. Culture is the result of trans-species flows. What we call culture begins with metabolism: food, eating, energy.

～

Remote from universal nature, and living by complicated artifice, man in civilization surveys the creature through the glass of his knowledge and sees thereby a feather magnified and the whole image in distortion. We patronize them for their incompleteness, for their tragic fate in having taken form so far below ourselves. And therein we err, and greatly err. For the animal shall not be measured by man. In a world older and more complete than ours they move finished and complete, gifted with extensions of the senses we have lost or never attained, living by voices we shall never hear. They are not brethren; they are not underlings; they are other nations, caught with ourselves in the net of life and time, fellow prisoners of the splendour and travail of the earth.

—from *The Outermost House*, by Henry Beston; epigraph to Farley Mowat's *A Whale for the Killing*

In one of his many literary nonfiction books, *A Whale for the Killing*, Farley Mowat depicts the series of events he witnessed in early 1967 when an endangered fin whale was trapped by receding tides and the shallow mouth of a cove on the southwest coast of the province of Newfoundland. In Chapters 4 and 5 he provides an overview of the history of whaling and whaling techniques and draws attention to the historical moment when the practice of hunting whales for food metamorphosed into hunting them for their blubber and baleen. Mowat writes:

> Only the oil and baleen were wanted now; the oil to fuel the lamps of an increasingly urbanized European society, and the baleen for the manufacture of "horn" windows and utensils. Thus the whale had been transformed from edible game into an article of commerce. When that happened man ceased to be a pin-prick irritant to the whale nation and became a deadly enemy. From this time forward whales were slaughtered without quarter and with every weapon and by every method the planet's most accomplished killers could devise. (Mowat 1972, 49–50)

Mowat's concise history of whaling and the transition from consumption to consumables is supported by an article by Eric J. Ziegelmayer titled "Whales for Margarine: Commodification and Neoliberal Nature in the Antarctic" in which he examines the history of whaling and the processes by which whales came to be commodified. One crucial development in this history of the transformation of whales from meat to lamp fuel to tubs of margarine[15] was the creation of "Blue Whale Units" (BWU), a system of measurement that served both to quantify the catch and to despecify one species of whale from another. As Zieglemayer writes:

> The mechanism was a quota system expressed in Blue Whale Units (BWU). Since the Blue Whale was the largest and most profitable whale to kill, it suffered accordingly, and the whalers [whaling cartel], with scientific assistance, devised a system to regulate the death of whales in light of this fact. Under the system, one blue whale would be the equivalent of two fin whales, two-and-a-half humpbacks, or six sei whales. This device had nothing to do with the whales; it simply expressed their utility in oil production.
>
> Each company was allotted a fixed quota of BWU to limit production, and efficiency was encouraged by the stipulation that each BWU produced 110 barrels of oil. (2008, 73–74)

This system, devised by the cartel itself for the season of 1930–31 ostensibly for purposes of conservation for the sake of continued exploitation, is not unlike contemporary practices of quantifying and trading carbon-offset credits or "ecosystem services."[16]

Mowat recounts the series of incidents in which several local men shot the whale repeatedly with .22-calibre rifles and even ran over it with outboard motor boats. The focus of Mowat's text is the killing of the trapped fin whale neither for food nor for illuminant, nor for any other consumable product. That is, the fin whale is killed not *for* fuel but rather *by* a surplus of boat fuel and ammunition (gunpowder being a particularly dramatic form of fuel) available for recreational purposes. Mowat, who with his wife Claire owned a summer home near the cove where the whale was destroyed over the course of many days, notes that most of the men shooting, harassing, and driving over the whale were ones who had left their villages and small-scale traditional economies for salaried, labouring jobs in other places. Here is his characterization of the group of them:

> Although these men had all been born on the Sou'west Coast, they had all spent some years away, either in Canada or the United States. Returning home for one reason or another, they had rejected the vocations of their fishermen forefathers and had instead sought wage employment at the plant as mechanics, tradesmen and supervisors. They were representative of the new Newfoundlanders envisaged by Premier Smallwood—progressive, modern men who were only too anxious to deny their

outport heritage in favour of adopting the manners and mores of
20th-century industrial society. (Mowat 1972, 105–6)

Mowat refrains from identifying the individuals in question and analyzing their motives, preferring instead to place their behaviour, as
Ralph Lutts writes, "in the larger context of Canadian government policies that had reshaped and destroyed the region's economy and way
of life" (Lutts 1998, 14). It is worth noting that their brutal actions
in tormenting the whale were a strange mixture of gestures otherwise
associated with whaling, hunting on land, and military exercises in
Newfoundland during the Second World War during which whales
were metaphorically construed as enemy submarines and used for target practice.[17] Mowat, an aficionado of outport life and an admirer of
the folk who live there, implies that it was, in part, the repudiation of
their forefathers' hard and dangerous work and their own exposure to
wage employment of a technical-managerial kind that brutalized the
men who in turn brutalized the whale. Having earned decent wages
in the outside world but perhaps missing the elements of home and
community (unable to establish them elsewhere, suspended between
the two spheres of home and away), they acted out a hideous and
prolonged skit bordering on a parody of their forefathers' labour as
fishers and whalers.

꒜

In order to examine narratives of animals and oil as if the individual
lives of individual animals matter, let me turn now to automobility,
one of the most common ways in which our lives and those of animals
are imbricated with oil and gas. Just as North American human lives
are thoroughly imbued with oil and its manifold by-products, so too
are those of other animals. In her article on "Intimate Bureaucracies:
Roadkill, Policy, and Fieldwork on the Shoulder," Alexandra Koelle
lists some of the ways in which one aspect of automobility—roads and
highways—complicates the nature/culture binary, with severe consequences for wild animals:

> Snakes bask on paved roads. Some get run over, and their bod-
> ies attract scavengers, who find a steady food source along the
> asphalt. Eagles gorge themselves on carrion, becoming so heavy

that they need a running start before they can fly. Cars often hit them before they are airborne. Bighorn sheep glean salt from de-icing agents stuck in the cracks of the pavement. Bats roost under highway bridges, and the mowing that keeps grasses short along medians can increase forage for small mammals … Busy roads create a barrier effect and isolate populations, noise from cars impacts bird communication, and particulates from exhaust harm the respiration of animals and the plants on which they depend for food and shelter. Roadkill—direct mortality—is only the most apparent effect to us: the olfactory onslaught of a skunk's final, ineffective defense; the tissue-thin butterfly wings you find stuck in the grille when stopped to get gas. Yet roads' impact on megafauna alone is staggering: collisions are the number one source of mortality for moose on the Kenai Peninsula; of barn owls in the U.K., and of the endangered Florida panther. In the United States, roughly a million deer-vehicle collisions occur annually, resulting in 200 human fatalities, 29,000 human injuries, $1 billion in property damage, and, not least, the deaths of a million deer. (Koelle 2012, 652–53)

During the summer of 1980—the summer during which the events of Sid Marty's literary nonfiction book *The Black Grizzly of Whiskey Creek* take place—when garbage management practices in Banff National Park[18] were still very lax and underenforced, an anomalously black grizzly and an uncharacteristically large, brown black bear were trolling the town's restaurant refuse and gorging on what Marty refers to (as if from the bears' perspective) as "cowfish," that is, a heady blend of scraps of steak and lobster meals and used cooking oil.[19] The bears' need for nourishment, for quick recovery of their body weight after hibernation, and for fat storage for next winter's hibernation and reproduction[20] draws them into confrontations with humans at a site just below the Trans-Canada Highway. Threats to them include collision with oil- and gas-powered motor vehicles; human intrusion into their traditional territories for recreational activities such as hiking, fishing, and the proliferation of restaurant waste, including used cooking oils.

In other words, interspecies encounters in the park are conditioned, in large part, by energy and by a variety of different kinds of oil. In the following passage from *The Black Grizzly of Whiskey*

Creek, Marty describes the grizzly and the various urgencies and life forces that compel him:

> Every day he is growing fatter and more powerful and every day he feels the need to be bigger still ... His weight and his strength are his glory; they secure his mating rights and the passing on of genetic material ... But a thick layer of fat under his hide is what promises his survival in the winter den; that was the fuel his body burned all winter while he slept. But now that fat is diminished and so finding food is his first concern. (63)

While we store our various fuels, albeit temporarily, in our cupboards and pantries, gas stations, refineries' holding facilities, and on the stock market, animals store theirs under their hide or fur. The bear's body is fuelled in both present and future tenses by a thick layer of insulating and nourishing fat. While obviously body fat is in no way the same substance as crude oil or bitumen, as forms of energy and fuel for propulsion through space and time their functions and histories overlap.[21]

As Marty's text demonstrates, the bear's need for fatty fuel places him on a potentially lethal and literal collision course with automotive culture. The narrative introduces the grizzly at the point when he emerges from hibernation in the spring of 1980, comes down a mountain in the Canadian Rockies an hour west of the city of Calgary, just outside the town of Banff in Banff National Park, and crosses what the bear (in Marty's formulation) thinks of as "the Meatmaker," the Trans-Canada Highway. The bear recalls that that highway had in a previous year yielded a road-killed bighorn sheep, but its asphalt stink and unnatural heat on his footpads offends his sense of propriety. Like many other-than-human animals, this bear has a conflicted history with highways and automobiles:

> He crosses the ditch with a crunching of gravel underfoot and stops at the asphalt margin of the highway. He remembers this Meatmaker; that is, his stomach reminds him that it had fed him once before, long ago. He bends his great head down to examine it with his nose; he smells the foreign odour of gasoline, of spilled oil, the sweet poison of antifreeze, and he sniffs loudly, for he detects the tentative nose-candy richness of pooled cruor. The odour tells him that a male bighorn sheep was killed here a

few days earlier. He can almost taste the musty tang of its scent glands that lingers in the roadside grasses and on the pavement. It rouses a picture in his head of a meat-beast with massive horns. He prowls back and forth in the ditch for a few moments, looking for the carcass, but there is nothing there but scent. (63)

The Meatmaker had given him road-killed bighorn sheep in the past, but this time its odoriferous provender is missing, having likely been picked up by parks or public works employees so as to discourage other animals from lingering near the kill-site to feed on it and thereby causing another collision and mortality. All that lingers is the scent of the sheep, the signifiers and not the signified. In addition to these insults to his olfactory sense and his grizzly semiotics, this contradiction of the evidence of his senses and rupture of the link between scent and meat, the unnaturalness of the pavement also offends him:

> The smell of blood makes him salivate, and he tests the asphalt with his pink tongue. He is not prepared for the heat it still contains from the previous day of sun, and he recoils from the warmth. He rears up on his hind legs and stares, front feet in the air, feeling the heat waves radiating still in the darkness. He inhales sharply, as a breath of cool air from the river floats to his nostrils. On all fours now, he steps onto the warm pavement, then breaks into a rough gallop for the other side. But the tacky warmth of the road on his padded feet puzzles him, and so he hates it: the earth should be cool and firm at night and give some relief from summer's heat. Reaching the graveled verge he rounds on it, slams a forepaw down on the pavement with a rattle of great yellow claws, lifts and strikes again, then pauses, paw raised as if in admonition, and emits a low, querulous moan. (The thin grey scars he leaves in the pavement might have been understood by a good tracker, the kind that has been dead a hundred years now, but only from sun-up to noon, after which the heat of full day anneals the scar to black again.) (64)

The grizzly burns his tongue on the surface of the road. Bears use their tongues for gathering information, as well as for foraging, prey detection, guiding foods such as berries into their mouths, to lick and heal wounds, to bond with their young, and to show affection; so, as anyone who has ever scorched their bare feet on pavement or caught their out-stretched tongue on a metal object on a cold winter day can imagine, this sudden heat on his tongue might feel to the bear like yet another assault or insult.

The pavement's unnatural heat and its foreign odours of tar, gas, oil, and antifreeze offend the bear's senses and sensibilities; more than that, the traffic on the Trans-Canada Highway is a genuine threat to his life. In the next moment, just as he slowly turns away from the road, "a shaft of brilliant light makes the rock cut above the highway jump out of the darkness, full of brooding stone faces. The grizzly gallops in among the trees before the oncoming headlights illuminate his retreating form" (64). The headlights turn night instantly and unnaturally into day and create near-hallucinatory visual effects that shock the bear, troubling his sense of physics as rock cuts metamorphose into "brooding stone faces" and unsettling his sense of the nature of things. He flees the scene before he too, like the rock face, is turned into a mere apparition or image—before, if I may say so, both his epistemology and his ontology can be further shaken. The grizzly's two swipes at the pavement protest its unnatural heat and tackiness and the way it both gives and takes away its beneficence in the form of roadkill. The highway's offensive tarry malleability under the hot sun adds a further, deferred insult insofar as the marks of his protest against it are soon annealed to black. Following his nose and his memory to the nearby stream, the bear's thought-balloon, as it were, is, in Marty's words, "The old trail has been abraded by machines and lost" (64). The grizzly's annoyance or ire in this scene is aroused by a combination of factors: the loss of that to which he, as the top predator of the area, feels entitled (the road-killed mountain sheep); the way in which the removal of the bighorn confounds the evidence of his senses and possibly even makes him feel foolish; the sudden illumination of the rock face and the close call with an oncoming vehicle, which together reinforce his sense of epistemological confusion and ontological dismay; the interruption and erasure of bears' own traditional trails by roads and the traffic on them; and the erasure of his marks of agitation, which, had they been inscribed in the bush, would have remained legible by many creatures for days and weeks.[22]

As we can see, animal agency is compromised or eroded by more aspects of oil than oil spills alone. Nicole Shukin's analysis of the logic of automobility is useful here. She writes:

While roadkill is perhaps most emblematic of the violence at material intersections of animal and automobile, car culture materially displaces animals in far more systematic ways as well, through the infrastructure of roads and highways that transect animal habitat and through the incalculable costs of fossil fuel extraction. Moreover, if automobiles emerge, in part, out of a desire to replace the animal traction of the horse, across the twentieth and early twenty-first centuries they have also worked to outmimic animal life and to symbolically occupy the place of animal life. (Shukin 2009, 126)

It is not impossible that, taken together, these factors add up to a level of frustration on the bear's part that could provisionally be ascribed to having been "out-mimicked" by us and our machinic inventions and prostheses.[23] I am not suggesting that bears only like things to be "natural"—after all, they do not object to feasting on our garbage—or that they "rage against the machine" as if they were members of the sixties counterculture. My point is that through our fossil-fuelled and other technologies, we have—if we think as if from the point of view of bears—not only out-horsed the horse in terms of physical speed and traction but also become un-"bear"-able, so to speak. Bite me,[24] as the popular saying goes: I am entitled. Indeed we seem to be saying exactly that, at the level of the culture as a whole. We have upset traditional codes of mutual respect and long-standing ecological relationships between other animals and ourselves and in so doing have robbed both parties of agency.

By stark contrast, the Stoney people of the area consider the bear to be the earth's ear (Marty 2008, 1). From his conversations with Stoney elders about their beliefs, Marty learns that

when we upset the community of bears, the effect is to unbalance, in a sense, the people who are spiritually connected to them. There appears to be what we might call an "ecology of spirituality," a spiritual interconnectedness between animals and between people and animals that mirrors the physical ecology in which all life forms are connected. When we mess with the bear [e.g., by tranquilizing or trapping them], we mess with the bear's medicine. The bear cannot help us anymore, and it might elect to hurt us instead. (1)

For their part, bears are much stronger and faster than we are, and they cannot understand why we thwart them or why we so jealously guard our "meat-hoards"—Marty's term for the restaurant discards that to the bears are delectable, especially during a hard summer when the berry crops have largely failed. They cannot understand why we have turned our backs on ancient systems of reciprocity.[25] As Marty writes, bears cannot understand why we fail to understand when occasionally one of them warns or bluff-charges us or "decides to eat one of us either literally, which is rare, or figuratively in the form of a bear mauling" (19). Instead we, in effect, charge *them* (something that in a bear's *umwelt* is simply not done!) in our gas-powered cars and trucks, motorcycles, trains, and helicopters as we barrel along, almost totally unconscious of the links between eating food and burning fuel.

When we think about motor vehicles and highways, automotive culture and oil, while simultaneously taking into account the perspective of another species such as grizzly bears, surprising insights come to light: bears are good to think with. Marty states that although bears are deservedly famous for their gargantuan appetites, their tendency to eat "an Everest of grub" (18) pales when compared to our own level of consumption:

> Filling the stomach is the least of humanity's appetites. Not only do we consume water, flesh, grain, and fruit in great quantity [just as bears do], we reach into the earth to gnaw out valuable ores and minerals, to feed an insatiable appetite for iron ore and coal, or to pump out fossil fuels to feed other machines that slave away to sate our greed for endless variety in plastic gimcracks, and to feed our need for bigger houses, faster cars, and wider-screen TVs. What is a car, or an ATV, or a motor home, if not a device for eating up space and time by displacing the natural world from out of our path as we speed from A to B? We are not savouring our time on this earth, we are cramming it down our craws at a ravenous pace. Our days vanish in the wake of our furious assault on our own bodies and spirits. (19)

Because we read an expression like "hungry as a bear" as mere metaphor, even a dead metaphor, and "consumption" as merely another metaphor based on eating, the above passage rewards serious reflection. Such metaphors are not mere figures of speech. Driving is akin to "eating up" space and time (and often not in the mode of savouring or feasting on either, but rather of abolishing them, disposing of them quickly with little or no visible trace). Especially if we keep in mind the commonality between oil and fat, driving not only compares with but constitutes a kind of eating. If all we have, fundamentally, are space and time, then, as Marty suggests, in attempting to vanquish both, we are in effect eating our lives alive.

We have forgotten that bears are the top predators. Compounding this amnesia, as Marty says, given that a bear is so much stronger than any of us, the bear does not realize that "humans, not bears, are the gods of eating." "Our flesh," he writes, "is not for the eating" (19). We drive; therefore we are human; therefore we are not for the eating. When we drive, we deny or disavow our mammalian nature, the speed and energy limitations of our legs and feet, and our place in historical, socio-political, and ecological systems. In driving we lay claim to space and time and to the world. When we drive, our feet, not unlike those of the old gods of Western civilization, do not touch the ground.[26]

Even human food "drives," in effect: our food is even more well travelled than we are. The steak (the cow in the bear's "cowfish") in the Banff restaurants may or may not be locally raised, but the lobster (fish) has certainly come from away. Bears and other wild animals, by contrast, eat locally. As Marty reminds us, early humans may well have learned what to eat from bears:

> Scholars suggest that ancient men probably learned the uses of some plant foods by following the tracks of the great bear and observing his digs, his foraging habits, and his feces. No doubt they learned from observing other omnivores and herbivores as well. But here was a beast unlike the others, whose footprint was eerily similar to our own, a beast that suddenly stood up on its hind legs in the middle of a berry patch, mimicking our human posture, and examined us as if expecting a password; one who, stripped of its hide by a flint blade or a Buck knife, looks weirdly man-shaped, as if a man had taken refuge in the skin of a beast. Little wonder that neolithic man attributed supernatural wisdom and occult powers to the bear. (20)

Marty's reference to the dietary and physical similarities of bears and humans invites us to return to the question of mimicry raised by the quotation from Shukin above. However, here we see that although in automobility we may have out-mimicked the horse,[27] when the bear rises up she or he reminds us forcefully of our own mammalian existence and of extant questions of our own edibility; for us, a bear is a unique kind of Meatmaker. Though we may have outpaced and replaced the horse and forgotten our ancient debt to bears for having shown us what to eat and where to find it, when we encounter a bear on foot we are returned immediately and impressively to the fact that we have not out-mimicked the bear. The bear is our double, our Other, in uncanny ways: we eat many of the same foods, their footprints are like ours, they can stand on two legs as we do, their skinned carcasses resemble our naked bodies, and, most importantly in such encounters, they do not have to yield to us in the berry patch or on the trail when we meet them on foot. Moreover, although bears only very rarely eat human flesh, there is always the possibility such that, if eaten, we would not only mimic the bear or they us: we would *become* bear, bear flesh, bear life. Just as we have, as Shukin shows, rendered other animals' bodies into commodities such as factory belts, film stock, and advertisements, bears have the ability to render our flesh theirs, our cars and our gasoline notwithstanding. A bear encounter is the point where a mutual potential for mimicry folds over into the possibility of our becoming other-than-human, becoming bear, becoming, even, fat, bear grease. Bite me.

It is not only or perhaps not at all that, as popular belief and some parks management practices would have it, once bears get a "taste" of what we call human food they are "ruined" as wild creatures and have to be deported to the back country or shot. It is rather that our bulimic culture in some ways requires the generation and bestrewal of garbage and waste. As Julie Guthman and Melanie DuPuis argue in their article "Embodying Neoliberalism: Economy, Culture, and the Politics of Fat," about the problematics of fat and obesity under the crisis of accumulation associated with neoliberalism,

> neoliberalism's commodification of everything ensures that getting rid of food—whether in bodies [in the form of over-eating, which leads to obesity], municipal dumps, or food aid, for that

matter, which has been shown to open up new markets—is as central to capitalist accumulation as is producing and eating it. Notwithstanding the laments of those who problematize the costs of obesity, the dieting, health-care costs, and waste management that accompany US food surpluses are internal to the logic of neoliberal capital and are not externalities. In other words, bulimia is not simply a way to read bodies; it is a way to read the neoliberal economy itself. (Guthman and DuPuis 2006, 442)

Guthman and DuPuis quote K. LeBosco's statement that "'bulimics, inasmuch as they satisfy the insatiable needs of the capitalist machine and at the same time please the thin-obsessed society, are the perfect citizens'" (444) and suggest that "[b]ulimia is a metaphor not only for consumption but also for production" (445). Our individual and collective reluctance to reduce or take responsibility for our discards—as Marty illustrates in his account of the ongoing poor to non-existent management of garbage in Banff National Park in the years leading up to the events of 1980—is in effect the perfect illustration of a bulimic culture at large exemplified by a local situation rooted in cheap gas, good highways, and tourist infrastructure.

Within capitalist and especially neoliberal capitalist economies, in the senses sketched here, oil (crude or bitumen) and fat are structurally the same "substance" or commodity. The expenditure of fuel via wanton driving and transportation practices and the manufacture of endless plastic doodads on the one hand and the excess consumption of food and generation of food waste on the other are structurally similar gestures. Guthman and DuPuis themselves draw a similar comparison to mine when they write that admonitions to the public to eat less "may well be construed as a threat to capitalist growth much like 'drive less'—a concept that many consider to be laughable these days" (445). It is not only that we are domesticated fat cats, living off the proceeds and ill-gotten gains of fossil fuels obtained at the cost of the lives, health, and habitats of other animals: it is that fossil fuel and fat serve the same societal function. Whether we strew our food garbage around in the parks or at the labour camps serving the tar sands (thereby attracting bears), we make a nuisance of ourselves, a behaviour we then project onto bears: a bear that comes back repeatedly to one of our refuse sites—sometimes referred to as "nuisance grounds"—acquires the label "nuisance bear." When bears come to be figured as commodities—as trophy objects to be hunted or symbols of wilderness through which to attract tourists to a park—or nuisances, or both, they have acquired a completely different status than that accorded to them by the Stoneys, for whom the bear is a subject rather than an object.

I hope this analysis of aspects of three literary nonfiction texts has shown that we can think of bitumen not only as low-grade, toxic, high-carbon, dangerous fossil fuel but also as fat in many different senses of the term. By taking into account other-than-human animals, *The Sasquatch at Home, A Whale for the Killing,* and *The Black Grizzly of Whiskey Creek* illuminate the fact that fossil fuels too are fat.[28] While they are not the same physical or chemical substance, bitumen and fat are substitutable, and not only in a metaphorical sense. Oil is, as Matthew Huber points out, an "incredible thing" (32) that borders on possessing shape-shifting powers: it can take on the qualities of blood, drug, sacrament,[29] and food on the table. And when you live downwind or downstream—for example, of the Alberta tar sands enterprises—this substitutability of bitumen and animal fat becomes entirely non-metaphorical.[30] Eating the meat from moose or deer that have been drinking the contaminated water or consuming fish who swim in it is tantamount to eating oil and the chemicals used in processing it. For the First Nations people who live along the Athabasca River downstream of the tar sands, to say that oil is "lifeblood"—of the economy, society, or culture—is more than metaphorical. They are literally drinking, cooking with, and bathing in a mixture of water, oil, and the various chemicals used in its extraction and processing, and ingesting it via the wild animals that are part of their diet.[31]

By learning to think with animals and by speaking up for them in debates about energy, we can simultaneously speak for the future of healthier habitats and "humanimal" cultures and communities. Texts like those by Robinson, Mowat, and Marty illustrate how adherence to indigenous and other long-standing accords between humans and other animals such as are recorded in First Nations' stories, and in some aspects of traditional outport and village life, and as are sometimes reflected in scientific studies, has the potential to help us resist catastrophic industrial processes. Instead of viewing damage to other-than-human animals as the unfortunate but inevitable collateral consequence of oil and gas projects—fish with two mouths, or birds in news footage struggling in vain to fly with oil-coated wings—we can begin to recognize not only that other animals are stakeholders in discussions about energy but also that we too are "anim-oils," and that if we speak for them, we speak in the same breath for our own habitats. They are the same habitats, the same places. The eagles, otters, clams, mussels, whales, seals, ducks, and fish—those "[m]ajor stakeholders on the B.C. Coast who will not be invited to the Enbridge Pipeline hearings"—live in the same places and rely on the same ecosystems as many people do.

We will also begin to see the nature that capital cannot see and to understand the connections between the eating and the spiritual economies. In the context of debates about energy, the dialectics between corporations and incorporation understood as ingestion, eating, and consumption is a crucial nexus for thinking with animals and restoring to them—in our own minds—the agency that has been theirs all along if only we would see and acknowledge it. After all, what is agency other than a powerful sensation of the vitality and exuberance of life, of the mesh of ecosystems and ways of being, and of one's immersion and active participation in all of it?

Notes

1   A portion of the funds ($900,000) from the payout has purportedly gone to the restoration of fish and waterfowl habitat close to Edmonton at Golden Ranches Conservation Area.

2   Statistically aberrant numbers of rare cancers have been found among the human residents of the Indigenous communities downstream of the tar sands on the Athabasca River.

3   According to Darcy Henton (2012), sixty-eight bears were shot near the work camps, and fifty-one in residential areas.

4   In his book on energies derived from the sun, Alfred W. Crosby observes that "fossil fuels are the tiny residue of immense quantities of plant matter. An American gallon of gasoline corresponds to about 90 tons of plant matter, the equivalent of 40 acres of wheat—seeds, roots, stalks, and all. Coal, oil, and natural gas are the end products of an immensity of exploitation of sunshine via photosynthesis over periods of time measured by the same calendars used for the tectonic shuffling of continental plates. We are living off a bequest of fossil fuel from epochs before there were humans and even before there were dinosaurs" (2006, 62).

5   In keeping with their notion that the planet is only a couple of thousand years old, however, some right-wing creationists claim that oil does not derive from fossil creatures. As James Howard Kunstler writes: "Some cornucopians believe that oil is not fossilized, liquefied organic matter but rather a naturally occurring

mineral substance that exists in endless abundance at the earth's deep interior like the creamy nougat center of a bonbon" (2005, 4).

6 For a lengthy but still non-comprehensive list, see the entry "animal oil" in the Free Dictionary at http://www.thefreedictionary.com/animal+oil (accessed 1 May 2012). An interesting point about the entry on that page is that, at least on the date I accessed the site, on the right-hand side of the page were several links to jobs in the petroleum industry.

7 There are a number of variant spellings of "oolichan," including "ooligan," "hooligan," "eulachon," "Ullachan," and "ulichan," a word derived from the coastal trading language of Chinook (see, for instance, Donald Mitchell and Leland Donald [2001] and Robinson [2011, 44]). I shall follow Eden Robinson's lead and use "oolichan," which she says is the variant most commonly used among the Haisla, and also because I like it best visually. Within quoted material, however, I shall use whichever variant the author uses.

8 This may not be an entirely bad thing: it may be strategically useful to keep traditional ideas of 'wildness' in play alongside those associated with theories of the social construction of nature.

9 In her article "There Will Be Birds: Images of Oil Disasters in the Nineteenth and Twentieth Centuries," Kathryn Morse collects journalistic examples in which an equation is made between the fate of the imperilled birds and our own human situation.

10 Alfred W. Crosby states that "[f]ossil fuels in some ways remind me of amphetamines ("speed" in the vernacular). Amphetamine pills, like the fossil fuels, are enormously stimulating. The taker doesn't tire, seems to think and in general to function better and faster, and is happier. But people who take the pills can become dependent on them, which is unfortunate because their supply cannot be guaranteed. Furthermore, people who take them may suffer distorted comprehensions of reality. Lastly, amphetamines, taken in large doses for too long, are poisonous" (2006, 164). Elsewhere in the same book he states that "[m]odern civilization is the product of an energy binge" and that humankind has an "unappeasable appetite for energy" (xiv).

11 See "Dzawadi" by Gloria Cranmer Webster for more on the physical, medicinal, economic, and cultural sustenance of the oolichan to the Kwakwaka'wakw people of the Pacific Northwest.

12 Marine life near Robinson's coastal community of Kitimat, BC, would have been under even greater threat from tanker spills had Enbridge's Gateway pipeline been built with Kitimat as its terminus and port.

13 For an analysis of the role of the oolichan in Eden Robinson's novel *Monkey Beach*, see the articles by Ella Soper-Jones (2009) and Megan Gannett (2012).

14 Gary Paul Nabhan refers to Western society's extraordinary distance from the natural world as a state of "ecological illiteracy" (Nabhan 1997, 164). Jim Cheney goes so far as to configure our isolation from our environment as a descent into "species autism" (Cheney 2005, 111). Val Plumwood writes that "an unfortunate and unnecessary side-effect of the long overdue recognition of the creativity of indigenous humans has been a denial of creativity to nonhuman species and ecosystems—*nature scepticism*" (Plumwood 2006, 120, emphasis added).

15 Eric J. Ziegelmayer remarks that "[f]ortuitously, the process of hydrogenation arrived just as demand for these commodities [edible oils, lubricants, and soap] accelerated. In 1907 chemist Wilhelm Norman hydrogenated whale oil, converting it into an edible solid fat for margarine" (Ziegelmayer 2008, 70).

16 See Morgan M. Robertson's 2006 article "The Nature That Capital Can See: Science, State, and Market in the Commodification of Ecosystem Services" for elaboration on the costing of ecosystem services.

17 Mowat also describes how during the 1940s, when US naval aircraft began flying out of Argentia, Newfoundland, whales became "a useful addition to the Navy's anti-submarine training": "Aircraft crews engaged in practice patrol work had been instructed to pretend that any whales they spotted were Russian submarines. The whales became targets for cannon fire, rockets, bombs and depth charges!" (Mowat 1972, 59). As he points out, what is a submarine but "a man-made imitation of a whale, in form" (Mowat 1972, 53).

18 Or, as Marty refers to it in *Leaning on the Wind*, Banff National Car Park. He writes that while he was a park warden a bit of his youthful idealism with regard to the park's mandate—that nature be left unimpaired for future generations—"died week by week with every game animal crushed under the wheels of progress in Banff National Car Park" (Marty 1995, 3).

19 *The Black Grizzly of Whiskey Creek* contains several extended passages in which Marty writes as if from the perspective of a bear. He fully admits in the author's note that prefaces the book that "[n]o one can say with any certainty what goes on in the mind of a bear" (Marty 2008, n.p.); even so, as a former park warden, he is better qualified than most to make his best guess as to how they think. He writes: "Wildlife management, including the live trapping, tranquilizing and relocation of black and grizzly bears, was one of several functions I performed first as a seasonal and later as a full-time national park warden, from 1966 to 1978" (Marty 2008, 2–3).

20 Whereas a female bear may have mated, her fetus(es) will implant in her womb only if she is sufficiently well nourished.

21 As Laurie Shannon writes, in Shakespeare's time for instance, fat in the forms of tallow and grease was far less invisible to the perception of ordinary citizens. For Hamlet, fat is fuel, and fat is us (Shannon 2011, 311).

22 Near the end of the text, Marty writes that the bear scratched his back on a rubbing tree and then rose on his hind legs "to his full extension and majesty to bite his mark into the bark higher than any bear has bitten there before" (Marty 2008, 252–53).

23 We not only inhale carbon monoxide and other air pollutants associated with oil and gas production and consumption from the air. We also identify with our cars and trucks to such an extent that it could almost be said that we too come to embody or represent fuel: our vehicles are our avatars. Ivan Illich writes that "[t]he typical American male devotes more than 1,600 hours a year to his car ... The model American puts in 1,600 hours to get 7,500 miles: less than five miles per hour" (Illich 2012, 53).

24 The online *Urban Dictionary* defines the expression "bite me" as "a taunting phrase, essentially meaning 'I don't care,' used to defend oneself's [*sic*] actions, characteristics or values following an accusation," http://www.urbandictionary.com/define.php?term=bite%20me (accessed 27 May 2012).

25 In his essay "Putting Out the Cows," Marty details how Charlie Russell attempts to enter back into such systems of reciprocity with bears by towing the carcasses of winter-killed cattle up towards the mountains of Waterton Lakes National Park so that when the bears emerge from hibernation and come down slope looking for winter-killed bison they have something to eat (Marty 1999, 276–89).

26 The figure of the driver is perhaps the perfect emblem of the neoliberal subject: often solo, isolated, detached from the natural world, and functioning within the illusion of control and self-control.

27 It goes without saying that cars are the new animal kingdom: Pony, Taurus (the bull), Tiburon (shark), Bronco, Cobra, Jaguar, Mustang, Pinto, and so on. When we drive, I would suggest, we individually out-mimic the horse or feel we have at some subliminal level. That is part of the "pleasure" of driving. In her article "Automotive Emotions: Feeling the Car," Mimi Sheller writes: "We not only feel the car, but we feel through the car and with the car" (Sheller 2004, 228).

28 In her article "Blubber Capitalism," Laura Saunders notes that "[t]he [US whaling] industry's real death-knell was the discovery of petroleum in Titusville, Pa. in 1859. This was an epochal shift. A three-year whaling voyage produced at most 4,000 barrels of oil, whereas one well could pump 3,000 barrels a day" (Saunders 2004, n.p.).

29 As Marion Grau observes in her article "Caribou and Carbon Colonialism," on the caribou of the Arctic National Wildlife Refuge and the Gwich'in people's relationship with them, "[i]ronically, for contemporary consumer societies oil has a sacramental quality that competes with the sacredness of other life-grounding substances. In this industrial society, oil fuels and greases the wheels of the system" (Grau 2007, 449).

30 The biofuel industry, whereby agricultural land is used to grow crops to be converted to fuel, adds another equation between oil and food.

31 Statistically abnormal levels of a rare form of bile duct cancer have appeared among the people of Fort Chipewyan, a community downstream of the tar sands.

References

Bakker, Karen. 2010. "The Limits of 'Neoliberal Natures': Debating Green Neoliberalism." *Progress in Human Geography* 34(6): 715–35.

Cheney, Jim. 2005. "Truth, Knowledge, and the Wild World." *Ethics and the Environment* 10(2): 101–35.

Cranmer Webster, Gloria. 2001. "Dzawadi." *Anthropologica* 43(1): 37–41.

Crosby, Alfred W. 2006. *Children of the Sun: A History of Humanity's Unappeasable Appetite for Energy.* New York and London: W.W. Norton.

Gannett, Megan. 2012. "Dreams in Double Exposure: *Monkey Beach* in Dialogue With Enbridge." *The Goose* 11: 30–42. http://www.alecc.ca/uploads/goose/The_Goose_Issue_11_Summer_2012.pdf. Accessed 19 August 2015.

Grau, Marion. 2007. "Caribou and Carbon Colonialism: Toward a Theology of Arctic Place." In *Ecospirit: Religions and Philosophies for the Earth*, edited by Laurel Kearns and Catherine Keller, 433–53. New York: Fordham University Press.

Guthman, Julie, and Melanie DuPuis. 2006. "Embodying Neoliberalism: Economy, Culture, and the Politics of Fat." *Environment and Planning D: Society and Space* 24: 427–48.

Henton, Darcy. 2012. "Wildlife Officers Shoot 145 Black Bears in Oilsands Region." *Calgary Herald*, 22 February, 1–2. http://www2.canada.com/calgary herald/news/story.html?id=323ebac3-1629-4cbc-8c32-18ef6cd8486f&p=1. Accessed 18 June 2013.

Huber, Matthew T. 2011. "Oil, Life, and the Fetishism of Geopolitics." *Capitalism Nature Socialism* 22(3): 32–48.

Illich, Ivan. 2012. "Less Energy, More Equity, More Time." In *The Localization Reader: Adapting to the Coming Downshift*, edited by Raymond De Young and Thomas Princen, 47–53. Cambridge, MA: MIT Press.

Koelle, Alexandra. 2012. "Intimate Bureaucracies: Roadkill, Policy, and Fieldwork on the Shoulder." *Hypatia* 27(3): 651–69.

Kunstler, James Howard. 2005. *The Long Emergency: Surviving the End of Oil, Climate Change, and Other Converging Catastrophes of the Twenty-First Century*. New York: Grove Press.

Linnitt, Carol. 2012. Blog: "Unethical Oil: Why Is Canada Killing Wolves and Muzzling Scientists to Protect Tar Sands Interests?" *Desmog*, 14 February. http://www.desmogblog.com/crywolf. Accessed 11 May 2012.

Lutts, Ralph H., ed. 1998. "The Wild Animal Story: Animals and Ideas." In *The Wild Animal Story*, 1–24. Philadelphia: Temple University Press. Ser. Animals, Culture, and Society, general eds. Clinton R. Sanders and Arnold Arluke.

Marty, Sid. 1995. *Leaning on the Wind: Under the Spell of the Great Chinook*. Toronto: HarperCollins.

———. 1999. *Switchbacks: True Stories from the Canadian Rockies*. Toronto: McClelland and Stewart.

———. 2008. *The Black Grizzly of Whiskey Creek*. Toronto: McClelland and Stewart.

Mitchell, Donald, and Leland Donald. 2001. "Sharing Resources on the North Pacific Coast of North America: The Case of the Eulachon Fishery." *Anthropologica* 43(1): 19–35.

Morse, Kathryn. 2012. "There Will Be Birds: Images of Oil Disasters in the Nineteenth and Twentieth Centuries." *Journal of American History* 99(1): 124–34.

Mowat, Farley. 1972. *A Whale for the Killing*. New York: Penguin.

Nabhan, Gary Paul. 1997. *Cultures of Habitat: On Nature, Culture, and Story*. Washington, DC: Counterpoint Press.

Nikiforuk, Andrew. 2008. *Tar Sands: Dirty Oil and the Future of a Continent*. Vancouver: Greystone Books—David Suzuki Foundation.

Plumwood, Val. 2006. "The Concept of a Cultural Landscape: Nature, Culture, and Agency in the Land." *Ethics and the Environment* 11(2): 115–50.

Robertson, Morgan M. 2006. "The Nature That Capital Can See: Science, State, and Market in the Commodification of Ecosystem Services." *Environment and Planning D: Society and Space* 24: 367–87.

Robinson, Eden. 2011. *The Sasquatch at Home: Traditional Protocols and Modern Storytelling*. Henry Kreisel Lecture Series. Edmonton: University of Alberta Press and Canadian Literature Centre.

Russell, Charlie, and Maureen Enns, with Fred Stenson. 2002. *Grizzly Heart: Living without Fear among the Brown Bears of Kamchatka*. Toronto: Random House Canada.

Saunders, Laura. 2004. "Blubber Capitalism." *Forbes Magazine*, 11 October 2004. http://www.forbes.com/forbes/2004/1011/096.html Accessed 19 August 2015.

Shannon, Laurie, Vin Nardizzi, Ken Hiltner, Saree Makdisi, Michael Ziser, and Imre Szeman. 2011. "Editor's Column: Literature in the Ages of Wood, Tallow, Coal, Whale Oil, Gasoline, Atomic Power, and Other Energy Sources." *PMLA* 126(2): 305–26.

Sheller, Mimi. 2004. "Automotive Emotions: Feeling the Car." *Theory, Culture, and Society* 21(4–5): 221–42.

Shukin, Nicole. 2009. *Animal Capital: Rendering Life in Biopolitical Times*. Minneapolis: University of Minnesota Press.

Soper-Jones, Ella. 2009. "The Fate of the Oolichan: Prospects of Eco-Cultural Restoration in Eden Robinson's *Monkey Beach*." *Journal of Commonwealth Literature* 44(2): 15–33.

Williams, Terry Tempest. 2010. "The Gulf between Us: Stories of Terror and Beauty from the World's Largest Accidental Offshore Oil Disaster." *Orion* 29(6): 34–53.

Yaeger, Patricia. 2011. "Editor's Column: Literature in the Ages of Wood, Tallow, Coal, Whale Oil, Gasoline, Atomic Power, and Other Energy Sources." *PMLA* 126(2): 305–10.

Ziegelmayer, Eric J. 2008. "Whales for Margarine: Commodification and Neoliberal Nature in the Antarctic." *Capitalism Nature Socialism* 19(3): 65–93.

CHAPTER 9

Reacting to Wolves: The Historical Construction of Identity and Value

Morgan Zedalis and Sean Gould

INTRODUCTION

The success of environmental policy plans depends on more than ecological considerations and, as such, policy plans must always navigate the social, human component of conservation. Because the implementation and planning processes are both situated among various networks of people who often see themselves as already active agents in a particular place at a particular time, it is no surprise that contextualization is critical for understanding the outcomes of environmental policy. How various stakeholders receive a policy is an essential factor regarding the policy's implementation. In this chapter we analyze stakeholder reception of a particular environmental action, the reintroduction of wolves into Idaho, and attempt to methodically identify the operative reasons that support the need for historic contextualization.

Very often, efforts to integrate human dimensions of conservation include economic strategies—for example, cost/benefit analyses and monetary compensation—to offset the negative impacts that conservation measures have on stakeholders. These methods are prevalent in contemporary conservation policies and were used during Idaho's gray wolf reintroduction to mitigate impacts on the livestock industry.

Efforts to appease stakeholders with compensation policies risk oversimplifying the context in which environmental policy is carried out. Implementing environmental policy often involves acknowledging the presence of particular social and political groups that have declared themselves proxy agents for their locale. The historical context in which these stakeholders are embedded complicates the human dimension of environmental policy. Conservation policy must work from, with, around, or through the self-ascribed identities of communities and their value systems.

In this chapter we examine how compensatory strategies have been used during Idaho's wolf reintroduction and argue that such approaches can be only part of a sustainable solution when it comes to mitigating stakeholder concerns about conservation efforts. To address these issues, we draw from environmental anthropology, environmental philosophy, and moral psychology. We analyze how the ranching community's identity and values affected its response to the wolf reintroduction. The research behind this paper (described in more detail below) included a series of one- to four-hour semi-structured interviews that Morgan Zedalis conducted with ten Long Valley livestock producers in rural central Idaho.

Morgan actively listened to ranchers' views on the introduction, often meeting them in their homes. However, while we have taken steps to ensure an accurate representation of their views in this chapter (see the methodology section below), by the very act of *listening* we were inserting ourselves as agents into the overall development of their views. To recall a point made by Sabine Feist (see Chapter 3 of this volume), an inherent sadness is evoked when voices of the earth go unheard. We would add that listening to others discuss their relationship to the earth is equally an environmental act. To listen is, to some degree, to empower, and this empowerment draws out some of the disenfranchisement some people feel when other groups act as environmental agents. One of the participants pointed out that the indifference of "outsiders" and urban academics to their situation was a thorn in the foot of the ranching community. Morgan and I grew up in Valley County, where the interviews took place, but during the time of the research we were living in England, which meant that Morgan was entering the participants' homes, in part, as a listening representative of a distant academic institution.

Interviews were recorded, transcribed, and reread through a process of textual narrative analyses against a backdrop of political philosophy and research in moral psychology. Utilizing a philosophical context for interpreting participant interviews makes it possible to fold those participants' concerns into preliminary suggestions for mitigating some of the negatively perceived effects of wolf reintroduction.

In this chapter we show that livestock producers' views of wolf reintroduction are bounded by their projects, with all the historical and political dimensions and interactions such projects encapsulate. Thus, conflicts relating to environmental value and group identity must be sufficiently resolved before a monetary compensation model can successfully address political tensions over environmental policy.

In investigating the relationship between community values and environmental policy, we reviewed interviews with cattle and sheep ranchers conducted in 2009 with regard to gray wolf reintroduction in Idaho. Our analysis is based on those interviews. That fieldwork vindicates our approach by illustrating that instead of expressing a direct evaluation of wolves per se, the reactions to wolves can be understood as a communicative act that serves an instrumental purpose of establishing each individual's tie to the broader ranching community and to the identity to which they aspire. To a large extent, this communicative act creates a contrastive identity vis-à-vis those *other* human agents whom ranchers view as responsible for the reintroduction. Those views then influence how stakeholder groups receive outsiders' offers of monetary compensation.

Addressing values and identity in a conservation initiative often involves identifying stakeholder groups and framing a policy that is compatible with their social values. When this approach is taken, values play a foundational role regarding how policy decisions are received. One possible way to address the issue of values is to apply an acceptable cost/benefit tradeoff (O'Neill, Holland, and Light 2008; Pearce, Markandya, and Barlier 1989; Hicks 1981). A good example of this method is the compensation program organized by the Defenders of Wildlife that offered monetary compensation for documented wolf attacks. This program ran from 1987 to 2010 and delivered over $433,000 in the state of Idaho alone (Defenders of Wildlife n.d., "Wolf Compensation Trust"). This sort of economic approach assumes that values exist as a fixed, antecedent variable tied to various demographics that can be accommodated by trading one value for others, or by using money as an evaluative standard.

Measuring values in economic terms—which is what monetary compensation does—has achieved mixed results when it comes to how predator reintroduction programs are received (Naughton-Treves, Grossberg, and Treves 2003, 1500–1511). Compensation efforts, like wildlife management policies more generally, are carried out against a backdrop of political and social issues that influence their viability (Montag, Patterson, and Sutton 2003). We acknowledge, as does Geneviève Susemihl (see Chapter 5 in this volume), that heritage itself is a project necessarily carried out from an emic perspective. Heritage is maintained and projected from within the community it represents. So when a stakeholder group values its heritage and sees it as threatened, that threat must be addressed from a position internal to the stakeholders' prior position. We contend that viewing the value of livestock in economic terms fails to situate the reception of compensation among important factors such as the self-ascribed identity held by important stakeholders, in this case especially the ranching community.

In this chapter we propose an alternative presupposition regarding the nature of values and identity. We reject the notion that a group of people base their identity and perception of their situation on a set of pre-established values. Instead of presuming that one's values are formative to one's identity and world view, we suggest working with the notion that one's values stand in reciprocal relation to one's practical identity. In other words, a group's projects and practices determine how it adopts certain ideas as core values. Undoubtedly, this working interpretation of values and identity was influenced by continual rereadings of Sartre's *Being and Nothingness* (1956) between sessions of interview transcriptions. If existence precedes essence, and if we continually create ourselves, then this process must be visible in a variety of contexts, including stakeholder reactions to environmental policy.

In the case of environmental policy, our working, existentialist view suggests that a stakeholder's response is a product of his or her aims, goals, and efforts towards adopting and sustaining various social roles. We argue that goals and projects cognitively structure how people see and evaluate the world. Through engaging in certain practices and maintaining ties with specific social identities, we come to evaluate the world in certain ways.

How policies are evaluated reflects the political and historical contexts of the policy's inception to a much greater degree than is assumed by the economic postulation of antecedent values. Rancher and shepherd identity is shaped by historical perceptions of the land

and wildlife. As anthropologist H. Kawamura explains, "[h]umans attach meanings to the environment, which shape the ways in which they relate themselves to the environment" (Kawamura 2004, 157). As such, cultural identity is dependent on the relationship between people and their places (Thornton 2008). Individuals' historical perceptions of their environment influence not only their contemporary relationship with that environment but also their relations with others (Sciama 2006). Therefore, tensions over conservation and environmental management approaches relate strongly to the identity of those who take stock in the condition of the surrounding environment (Kosek 2006). This chapter addresses influential socio-political interactions in analyzing how participants' views regarding wolf reintroduction in Idaho have been shaped by their relationship with the federal government and outside NGOs.

BACKGROUND

The reintroduction of wolves presupposes that particular individuals and groups exercise varying degrees of agency in proxy for natural processes. So if we are to listen fully to the participants and understand where they are coming from, it is critical that we understand the historical context of that reintroduction.

In 1973 the US Congress passed the Endangered Species Act. The following year, 1974, the gray wolf was placed on the endangered species list, marking a policy shift from eradication towards conservation. As part of its effort to "rebound" the gray wolf population in the lower 48, the US Fish and Wildlife Service (USFWS) presented the Idaho legislature with the 1987 Northern Rocky Mountain Wolf Recovery Plan.[1] For all USFWS's efforts, the Idaho legislature rejected the plan; this blocked the Idaho Department of Fish and Game's (F&G) participation in the recovery, even though it had helped design the plan (Idaho Department of Fish and Game 2009). The rejection of the plan was in part due to a new state governor: Cecil Andrus, who had supported the reintroduction, had been replaced by Phil Batt, whose anti-environmentalist views included opposition to wolf reintroduction (Maughan 2005).

So the state legislature nixed the plan; however, the Nez Perce Tribe, a sovereign nation whose tribal government seat is in Lapwai,

Idaho, offered to lead wolf management responsibilities in the state. This enabled the initial reintroduction, in 1995, of four gray wolves at Corn Creek in the Frank Church–River of No Return Wilderness. That release point was the same location where, in 1805, Meriwether Lewis, William Clark, and the Corps of Discovery had been forced to reroute their expedition owing to the rugged nature of the landscape. After the Corn Creek release, eleven more wolves were released near the Middle Fork of the Salmon River. The Nez Perce Tribe managed the growing wolf populations, both those relocated from Canada to Idaho and those established naturally in the state.

Later, in promulgating a new rule under the Endangered Species Act in 2005, the USFWS granted wolf management responsibilities to the State of Idaho and Idaho F&G. Once Idaho F&G took over management, lobbying began for the gray wolf to be struck from the Endangered Species List. Those in favour of the delisting argued that genetic integrity was already sufficiently established among existing wolf packs. Environmental groups, including the Defenders of Wildlife, disagreed and filed a lawsuit over the delisting. While the court case edged forward, an unsteady situation developed: wolves were delisted, then relisted, then delisted again. As of 2009, wolves were delisted in Idaho and there was an annual hunting season for gray wolves.

In 2011, Idaho F&G reported 101 wolf packs in the state, with an estimated total population of 746. The presence of wolves had already been noticed by the livestock industry, though there had been few financial implications. By 2003, wolf kills in the northwestern United States accounted for less than 0.01 percent of the total annual gross income from cattle ranching (Muhly and Musiani 2009). However, these costs were not necessarily spread equally within the industry. A 2011 report documented that 90 cattle and 147 sheep had died due to wolf depredations; Idaho produced 2,200,000 million cattle and 185,000 sheep that same year (Idaho Department of Fish and Game and Nez Perce Tribe 2012; Idaho State Department of Agriculture 2011). Table 9.1 indicates the percentage of wolf depredations among unplanned cattle deaths for 2011 in Idaho.

Wolf kills, according to this table, constituted 30 percent of all predator kills. From this we estimate that wolf predation accounted for 1.5 percent of adult cattle losses and 2.5 percent of calf losses in 2011.

Table 9.1 Cattle Death Loss (USDA 2011)

	Cattle (adults)	Calves
Total deaths	42,000	51,000
Deaths due to predators	1,900	4,200
Percentage of predator deaths by wolves	30%	30%
Deaths by wolves (approx.)	570	1,260

METHODOLOGY

Zedalis conducted her fieldwork in 2009. She took a qualitative ethnographic approach. Her fieldwork consisted of ten informal interviews with livestock producers, mainly in Valley County in central Idaho. The region is rural. Its principal settlement, McCall, had a population of 2,991 in 2010 (City of McCall 2011). The livestock producers interviewed ran sheep and cattle on either private lands or public allotments. Grazing on public lands was mostly in Payette National Forest, whose 2.3 million acres include the Frank Church–River of No Return Wilderness, the largest wilderness in the lower 48 (Payette National Forest n.d.). Payette's landscape consists largely of mountains, dense conifer forests, open meadows, and alpine lakes.

Zedalis interviewed both men and women, who were selected using both purposive and chain referral sampling techniques. Purposive sampling was used to select sheep and cattle ranchers who had experienced wolf depredations on their livestock. Chain referral sampling was also used—that is, the participants recommended other potential interviewees. Purposive and chain referral sampling techniques allow participants to take an active role in selecting new participants. These techniques facilitate participants' ability to construct a group identity. The participants were asked to select individuals with both similar and divergent views. The result was a group of participants who both had and had not experienced depredations (Bailey 2007; Bernard 2011; Creswell 2009).

The participants identified themselves as full-time ranchers. In age they ranged from 30 to 70. Most of them were continuing a family enterprise that was several generations old. Cattle ranching in Idaho began around 1866 (Jordan 1993, 230). One family interviewed had been living on their land for seven generations dating back to 1889 (Willey 2009). Whatever their scale of production (among other variables), many of them belonged to local and national associations. They viewed these affiliations as providing general support and political representation. They mentioned belonging to organizations such as the Idaho Cattle Association, the Idaho Sheep Growers Association, and on a national level the National Livestock Producers Association.

The qualitative interviews were semi-structured and consisted of between 15 and 20 questions. Those questions touched on the nature of their livestock operation, the presence of wolves and their views towards them, and interactions with outside groups like the Defenders of Wildlife and government agencies. Specific questions included these:

- Have the Defenders of Wildlife compensated you for any livestock lost to wolves? Why or Why not? What were the circumstances? Were you satisfied with the outcome?

- How have the authorities reacted toward the wolf activities on your land in general and do you agree with their reactions?

These questions were prepared prior to the interviews. During each interview, these questions were asked through a conversational process of participant-directed inquiry. Interviews lasted from one to four hours and were conducted in the participants' homes. Ten interviews were completed, at which point data saturation occurred—no new information was being gained, and a repetition of themes had emerged. Interviews were documented with notes and were audiorecorded and later transcribed. Secondary sources, such as newspaper articles and government and organization documents, gathered during the time frame of the research, were also used to inform the research.

Analysis and interpretation of interviews was done using textual and narrative analysis, in the course of which themes emerging from the interviews were coded. Here, textual analysis involved documenting "the text, context, and texture" of interviews and narratives. Documenting the text involved recording what was actually said while taking notes about and photo-documenting the context and the environment of the interviews. Audiorecording and reflective notes also enable researchers to keep track of the "texture" of information; here, "texture" relates to *how* things are said. Narrative analysis helps interpret how participants tell stories, construct narratives, and recount events such as, in this study, wolf kills. From all of this, themes and reoccurring structures emerge (Bernard 2011).

Qualitative interviews and narrative and textual analysis allow us to present aspects of this research in the form of block quotes. These provide an unmediated view of the participants' positions regarding wolf reintroduction.

Several steps were taken throughout the process to ensure that the conclusions we drew from the interviews reflected the views of the participants. Every participant was given an opportunity to judge whether the themes and viewpoints taken down by the researcher accurately represented his or her own position. Survey methods have a single direction of information—from the participant to the researcher—whereas qualitative interviews enable the researcher to check her understanding of participant views directly with the participants as those views are being formed. When, as is the case with this research, the main goal is exposition of participants' views regarding an issue, this sort of dialectic process allows for "member checking" of research findings and helps ensure the credibility of the results. Member checking of findings occurs within this research's methodology and findings again through post-research contact with participants. Zedalis presented each participant with a written copy of the research and a request that participants voice any comments, concerns, or criticisms with the methods or results. Several livestock producers replied with comments on the research, but there were no objections to how their views were presented. The member-checking strategies used for this research support the authenticity and trustworthiness of the findings, affirming their credibility and confirmability. The use of semi-structured interviews within a group of participants who played an informal role in both the selection and the initial analysis of the collected research material enables a close representation of the views and issues the participants wished to express regarding the effect of wolves in Valley County in central Idaho.

RESULTS

All of the livestock producers whom we interviewed opposed the wolf reintroduction. Many cited wolf depredation and economic loss; most, though, expressed greater concern about how the wolf introduction amounted to a disruption of their way of life by outsiders. Wolves' impact on capital and land use was presented as relevant primarily insofar as it affected the capacity of ranching families to continue passing on their practices and identity.

For the participants interviewed, the agency of the wolves themselves played a surprisingly marginal role in the reception of their presence. When wolves themselves were discussed, rather than their *presence*, they were objectified more often than they were represented as agents in themselves. Wolves were primarily depicted as pesky and somewhat gruesome carnivores. It was the wolves' newly acquired *presence*, their actual co-occupation of the here and now in central Idaho, that framed the wolves and gave them symbolic stature. Like the fictional vampires discussed by Robert Boschman (see Chapter 4 in this volume), the wolves' intrusion into the ranchers' world produced a confabulation of the uncannily wild predator with the familiar institution of the federal government. It was this latter aspect, the human agency of the reintroduction, that was prefigured in the disclosure of the wolves.

Dissatisfaction with federal interference was often expressed in terms of a lack of faith in the agencies' knowledge, understanding, and experience of the ranching lifestyle. A political division between livestock producers and the agencies responsible for the reintroduction was expressed, as exemplified in the following:

> These are cattlemen; these are people that have lived here all their life ... They know what they are talkin' about. And to have some government official come in that is some college, pardon my, but he doesn't know. He hasn't lived the life. It is like New York and Washington DC is deciding what to do with our land and our stuff out here and it is wrong.

Livestock producers saw themselves as separate from the reintroduction groups. Furthermore, most of the the interviewees did not trust the government's epistemic standing regarding the effects of wolves.

For example, the latest report (Nadeau et al. 2009) stating that there were around 850 wolves was considered inaccurate. Many participants believed the population ranged between 1,500 and 3,000 wolves. The following came from a cowhand, whose distrust wase sparked when a federal employee incorrectly stated, according to him, that it could be hard to prove wolf depredations of cattle on open-range lands:

> If they would let an individual state that actually knows what the situation is, that actually lives here, just handle their own stuff and not only the wolves, but practically everything, the timber and everything else ... A little bit too much intervention of the federal government.

Other cattle ranchers interviewed indicated similar distrust regarding the efficiency and effectiveness of government programs and employees.

Attempts to form a united front through membership in local or regional associations are common among livestock producers. Association membership offers protection, not only from federal agencies but also from interest groups such as property developers and conservationists. Projecting themselves as a unified group heightens this sense of protection. One participant lamented the lack of solidarity between the Cattle Association and hunting groups—solidarity that could perhaps have blocked the initial reintroduction. Wolf advocates, he felt, were simply better organized. Most livestock producers view proponents of wolf reintroduction as naive urban outsiders from places like New York City. One livestock producer suggested that the reintroduction of wolves was a ploy by the Wild Lands Project to drive farmers off property so that wildlife corridors could be established. Given how common this sentiment was, we found a key result from the fieldwork to be that participating livestock producers framed wolf reintroduction in political terms, with different groups leading different ways of life at odds over control of the land.

Antagonism towards outside interference plays a role how compensation measures are perceived. More broadly, compensation has been criticized for failing to cover non-quantifiable environmental goods. There are some things one cannot put a price on and for which no financial tradeoff is possible. One participant talked about how compensation failed to address the cultural value of husbandry. He felt that his opportunity to display this virtue was being sacrificed to the wolves, which were being allowed to jeopardize the lives of livestock:

> Well, I think the people that love wolves ought to compensate the rest of us that don't, that are affecting our livelihoods. That part is alright. My true opinion is that it is not as much the money as it is the husbandry side of it. That is what really bugs me. And the wildlife, I am bugged about the elk herd we don't have anymore. The elk probably cost us more than the wolves to be honest; they eat a lot of grass, mineral and tore down a lot of fence. But that is just part of the deal. I don't have any respect for a wolf. There is nothing majestic about a wolf in my opinion. I thought elk were.

In other words, it isn't possible for compensation to cover environmental goods and their associated virtues. The participants regularly expressed the importance of horses, chivalry, honour, and a love of husbandry as important aspects of their culture and as things they desired to pass on to their children and grandchildren. In financial terms, this participant viewed elk as more harmful than wolves to his bottom line, yet he embraced interactions with elk as a valuable part of the ranching way of life. In his view, the compensation on offer was incomplete, for it failed to address the diminishing elk population, and furthermore, that monetary compensation was incommensurable with the impact that the wolf reintroduction was having on the participants' desired way of life.

In addition to all this, some participants found it difficult to accept compensation as a matter of political principle. Many livestock producers viewed association with environmental agencies as undermining their own cultural identity. For those who see the federal government, NGOs, and environmental groups as outside threats to their way of life, accepting compensation amounts to colluding with the undermining one's own identity. One prominent sheep producer raised both the unquantifiability and the identity-based objections to compensation:

> I think it was a good faith effort on the front end that bringing wolves in would impact people and helping them with those costs. There are a lot of associated costs that are not compensated for, but there is no way to quantify them either. I think it is true that they [cattle and sheep] will not gain [in weight] as well if the

wolves are in and upsetting the bands, but how do you quantify that. There is that attitude out there locally, in where they will not take compensation because they do not want to be associated with NGO/environmental groups. And there is the attitude that, I am not going to take compensation, they cannot compensate for what they have done ... And it is like well you know at the end of the day you're in business because of your bottom line, just ask GM [General Motors, an automobile company that was experiencing financial fallout], I mean you, it is like when people say with the farm bill programs if you don't participate in the program because you don't believe in a government subsidy, but at the end of the day you have to be a competitor. It does come down to making dollars and cents and I have to make enough dollars to compensate my mens' well-being. Because my employees are very important to me. People use all kinds of rationale. Our industry is very diverse.

This participant is describing how a pragmatic economic concern, a desire to at least be able to pay employees, interacts with an unwillingness to associate with interfering outsiders. Ranchers want to be seen as stakeholders, but this is not solely for the sake of economic security—they are acting to represent and protect their identity as ranchers.

To summarize, the primary complaint against the wolf reintroduction was not presented in economic terms; rather, participants objected to it because it represented a harmful intrusion by outsiders upon a valued set of practices and a particular way of interacting with the landscape.

DISCUSSION

Results from the 2009 fieldwork have implications for assessing public policy decisions with regard to environmental issues. Throughout the interviews, participants identified themselves with a way of life and with membership in a community of practice. Furthermore, they demarcated their membership by contrasting it with outside groups. Call this contrastive identity. Such forms of identity construction are well documented and argued for in other literature (Sciama 2006, Kosek 2006). The participant interviews indicated that contentious policy decisions galvanize identity attributions, thereby exacerbating the political complications involved in policy implementation.

The participants' adoption of a historical reflective perspective regarding rancher identity, and their relationships with outside institutions, must not be underemphasized. Many participants saw themselves as carrying on a specific tradition passed down largely through family ties. At one interview, four generations of one family all identified with ranching practices. They perceived federal interference as a personal hindrance to the passing on of their tradition. This view was further influenced by westerners' ideals of self-sufficiency, as well as by isolationist themes within Idaho politics more generally; for example, all but one participant explicitly referenced "state's rights" in condemning federal wolf policy. Given the historical roots of rancher identity and the increased salience of identity, politics further complicates attempts to use compensation as a method for achieving cooperation, consensus, and fairness regarding policy decisions.

Ranchers' self-identification as "ranchers" is bound to participation in a set of ranching practices; in other words, a "rancher" is someone who "ranches." However, what we include in "ranching" extends beyond the purely pragmatic activities of livestock husbandry. Publicly avowing one's identity and positioning oneself within a specific group, while contrasting oneself to outsiders, do just as much to foster self-identification. This research found that ranchers self-identify by contrasting themselves not only with wolves but also with a federal government that has frustrated their practices through the wolf reintroduction.

Interference with what participants *do*, economically and politically, entails interference with what the participants perceive themselves as *being*. Our results indicate that identity plays a more prominent role than economics in influencing reactions to wolves. On a purely monetary level, the cost of wolves fails to account for the level of opposition to their presence. The percentage of cattle deaths by wolves (1.5% for adults, 2.5% for calves) is far below the reported 5 to 15 percent of cattle lost due to larkspur (delphinium) poisoning (USDA 2011). Moreover, many participants acknowledged that elk impose greater economic costs than wolves. As it happened, discussions of the economic impact of wolves were overshadowed by expressions of resentment towards the federal government. The participants expressed the sentiment that economic compensation was beside the point—mainly, they were irritated by the perceived foolishness of out-side groups.

We suggest that reluctance to accept compensation and the juxta-position of in-group and out-group politics expressed in participant interviews is best explained by positing a three-way reciprocal rela-tionship between identity, values, and practices. To a large extent, issues with wolves are not a result of antecedent values that comprise identity; instead, adopting a specific, historically developed attitude towards outside groups is a necessary step towards manifesting rancher identity, and this political attitude is best expressed by adopting a certain position towards wolves. The opposite (rejected) linear view that having specific values precedes one's practices and resultant iden-tity would predict a stronger emphasis on the economic and practical consequences of wolf reintroduction than is present in the findings. This is not to say that economics and the violence of wolf kills did not together influence the participants' views; rather, it is to flag an important, complicating, and pervasive aspect regarding how wolves, their effects, and their champions are perceived.

By accepting compensation, ranchers would be weakening their defining relationship with outsiders. Ranchers can cooperate with the wolf reintroduction only at the cost of their identity. One informant encapsulated this sentiment. Wolf advocates offered to establish a range-rider system whereby they would pay half the wages for a live-stock labourer to remain in the field with the stock, thereby deterring wolf predation. The informant objected to this, claiming that it would be nearly impossible to find a labourer who had the ranching expe-rience to do the job *and* was able to refrain from shooting wolves on sight. For this informant, being a rancher entailed being against wolves.

Assume that environmental policies must be perceived as fair if they are to fully succeed. Other fieldwork has shown that policies perceived as unjust typically result in failure (Kosek 2006; Knight 2000). To define "fairness," we can follow political philosopher John Rawls's highly influential description of "justice as fairness" (Rawls 1971). Rawls provides two criteria for the fairness of any policy deci-sion in which one stakeholder group perceives itself as not benefiting directly. First, the group must retain political access to the procedure and institutions that implement the change in policy. This is "a right to participation" requirement. Second, no change is acceptable if it constitutes a disagreeable net loss, fiscal or otherwise, for the party least well served by the change. In other words, it is "unfair" to harm any group or person for the benefit of others. Applying these criteria to the wolf reintroduction yields the point that any inequality in control over reintroduction and its effects is fair only if every stakeholder has an equal opportunity for involvement in the policy decision, and if those who suffer from that decision are compensated, thereby offsetting any potential harm brought about by the decision.

Something like Rawls's description of "justice as fairness" is what explains the moral demand that one group be reimbursed for the losses it has suffered due to management decisions. Without compensation there is a loss, and the imposition of a loss is "unfair." One common way that policy-makers determine whether compensation can facilitate desirable, and just, outcomes is by applying something like the Kaldor–Hicks compensation test. Under that sort of test, a possible future situation is an acceptable improvement over the current state of affairs if those who benefit from the change benefit enough that they can and do compensate the losers so that the losers end up better off than they would have been otherwise (O'Neill, Holland, and Light 2008; Hicks 1981). Regarding wolf reintroduction, if one followed Kaldor–Hicks–style reasoning, one might conclude that the policy would be socially acceptable if wolf advocates could amply reimburse livestock producers for their losses.

This research indicates that ranchers viewed the wolf reintroduction efforts as unjust, even though they *were* compensated. The participants as a group felt alienated from the reintroduction and deprived of agency. Also, reintroduction was seen as a failure because the compensation was too low and because some ranchers were unwilling to accept it. If ranchers' alienation from decision-making and dissatisfaction with compensation were a result of resentment, then issues regarding the second principle would be moot. According to many participants, the very act of accepting compensation carried its own costs. That being so, the Kaldor–Hicks–style compensation tests had failed to address issues of fairness.

Two considerations advise against dismissing the participants' positions. Those in favour of the reintroduction cannot disregard the position voiced by this research's participants as, for example, ideo-logical stubbornness. First, charity of understanding is necessary for democratic dialogue and honest listening. Simply disregarding the posi-tion of one group as erroneous tends to result in political stalemates. Second, successful implementation of environmental policy requires not only that the policy be fair but also that it be transparently per-ceived as such. Thus, policy must aim at working with and from the positions of *all* stakeholders. We recommend reading ranchers' dis-satisfaction with the wolf reintroduction as a reasonable response to the dilemma the reintroduction poses to the producers. That dilemma arises because for the reintroduction to be just, there must be coopera-tion between groups. In this situation, cooperation involved a sacrifice of participants' identity, which worked against the political fairness of the reintroduction.

There is also the matter of contrastive identity. By accepting com-pensation, the ranchers would inevitably have been "selling out." Whatever efforts wolf advocates made to offer compensation, the ranchers could maintain their sense of identity and group belonging only by expressing dissatisfaction with the reintroduction.

To summarize: Because the participants focused more on issues of identity and non-monetary value, it is inappropriate to interpret the costs and benefits of the wolf reintroduction in economic terms. It is more appropriate to approach the wolf reintroduction in terms of identity politics.

RECOMMENDATIONS

Wolf reintroduction, as it relates to peoples' identity, exposes issues concerning the historical and political aspects of human–environment and social interactions. An analysis of Idaho wolf reintroduction con-tributes to insights that are applicable worldwide regarding people's perceptions of wildlife and how those perceptions affect human–wild-life interactions and conservation efforts. Wolf protection projects, along with many other reintroduction programs, have expanded globally, including in Sweden, France, and Italy (Linquist 2006; Bath 2000; Boitani and Ciucci 1993). Additionally, many countries, such as Scotland and Japan, are considering wolf reintroduction plans to help re-establish and maintain biodiversity (Nilsen et al. 2006; Knight

2006). This chapter argues that if these efforts are to succeed, anteced-ent public attitudes towards conservation must be understood and addressed.

Aspects of our research support present-day efforts to address the historical socio-political aspects of wolf reintroduction. We encourage work that aims at sharing information regarding the ecological role of wolves. We encourage efforts to collaborate with livestock produc-ers to actively address human–wolf interactions—efforts such as the Defenders of Wildlife's "coexist" program, which includes the range-rider option referred to earlier. Moreover, notwithstanding difficulties with compensation, we feel that to the extent that livestock loss is a significant cost to some stakeholders, compensation methods can and do help foster an atmosphere more receptive to the presence of wolves.

Our research also suggests that a practical way for environmental policy architects to reconcile the presence of wolves with the poli-tics of livestock production is to articulate ways in which acceptance of wolves can be viewed as a *positive* aspect of livestock producers' identity. Here, what is required is to identify some potential aspect of that identity that is (a) self-attributed by the group, (b) valued by the group, and (c) amplified through the wolf reintroduction. Environmen-tal policy, we suggest, is easier to implement by increasing the degree to which it capitalizes on historically present aspects of agency that the various stakeholders attribute to themselves. The positive aspect must exist as a self-attribution for its amplification to be consistent with the preservation of the stakeholder's overall identity. Likewise, the stakeholder must view this self-attribute as a positive one; otherwise its promotion will constitute a harm to the group. So we recommend enabling Idaho pastoralists to explore ways in which wolves can help sustain their identity.

Livestock producers in central Idaho today are proud of upholding of the frontier tradition. Nash, in *Wilderness and the American Mind*, offers a thorough historical narrative of how Americans embraced the hardships of the wilderness as they forged their national identity (Nash 2001). Stewardship of the wilderness is an important aspect of the American tradition. Some participants expressed this regarding their felt stewardship for elk populations (Montag, Patterson, and Sutton 2003).

To succeed as an inclusive policy, wolf management should be presented in a way that emphasizes how stewardship for wolves is a way to help support the environment that first inspired a pastoralist identity. Defenders of Wildlife's Wood River Wolf Project provides one possible approach. The project encourages preventative measures against wolf kills—measures that are implemented, in part, by the livestock producers themselves (Defenders of Wildlife n.d., "Living with Widlife"). In providing an opportunity to participate actively in the process of coexistence with wolves, the Wood River Wolf Project resolves the issues of justice and alienation that can hamstring other wildlife policy enactments.

Encouraging existing stewardship practices can amplify that stewardship among ranchers. Many studies in psychology and sociology have shown that attributions of various characteristics often turn into self-fulfilling prophecies.[2] Cooperative stewardship projects allow for just participation in continued management.

Given the political overtones of the wolf reintroduction, through working with outside, "interfering" groups, livestock producers can act as useful liaisons when it comes to state versus federal environmental discourse. The wolf reintroduction requires a political solution at least as much as it requires economic amelioration through compensation. Instead of disparaging compensation efforts, livestock producers, by interacting with federal and NGO agencies, can view themselves as important representatives of local groups and traditions who maintain their contrastive identity precisely through collaborative efforts.[3]

Notes

1 For information on the Northern Rocky Mountain Wolf Recovery Plan, see US Fish and Wildlife Service et al. (1987), "Northern Rocky Mountain Wolf Recovery Plan," http://www.fws.gov/mountain-prairie/species/mammals/wolf/NorthernRockyMountainWolfRecoveryPlan.pdf.
2 For a survey of relevant literature, see Alfano (2012).
3 Gould and Zedalis would like to thank the participants in the 2009 fieldwork, as well as Alan Thomas, John van Houdt, the Tilburg Institute for Logic and Philosophy of Science, the Tilburg Hub for Ethics and Social Philosophy, participants in the Under Western Skies II conference, Suzanne Stone from Defenders of Wildlife, Robert Boschman, Mario Trono, and the anonymous commentators for their input on this chapter.

References

Alfano, M. 2012. *Character as Moral Fiction*. Cambridge: Cambridge University Press.

Bailey, C.A. 2007. *A Guide to Qualitative Field Research*, 2nd ed. Thousand Oaks: Pine Forge Press.

Bath, A. 2000. "Human Dimensions in Wolf Management in Savoie and Des Alpes Maritimes, France: Results targeted toward designing a more effective communication campaign and building better public awareness materials." https://www.researchgate.net/profile/Alistair_Bath/publication/238096165_Human_Dimensions_in_Wolf_Management_in_Savoie_and_Des_Alpes_Maritimes_France/links/549032390cf214269f2660b7/Human-Dimensions-in-Wolf-Management-in-Savoie-and-Des-Alpes-Maritimes-France.pdf. Accessed 15 April 2009.

Bernard, R. 2011. *Research Methods in Anthropology: Qualitative and Quantitative Approaches*, 5th ed. New York: Altamira Press.

Boitani, L., and P. Ciucci. 1993. "Wolves in Italy: Critical Issues for Their Conservation." In *Wolves in Europe: Status and Perspectives*, ed. C. Promberger and W. Schroder. Munich: Munich Wildlife Society.

City of McCall. 2011. "McCall Place, Idaho Population: Census 2010 and 2000 Interactive Map, Demographics, Statics, Quick Facts." http://censusviewer.com/city/ID/McCall. Accessed 30 August 2012.

Creswell, J.W. 2009. *Research Design: Qualitative, Quantitative, and Mixed Methods Approaches*. Thousand Oaks: Sage.

Defenders of Wildlife. n.d. "Frequently Asked Questions: Transitioning Wolf Compensation." http://www.defenders.org/sites/default/files/publications. faq_transitioning_wolf_compensation.pdf. Accessed April 28, 2013.

———. n.d. "Living with Wildlife in the Northern Rockies: Coexisting with Wolves in Idaho's Wood River Valley." http://www.defenders.org/sites/default/files/publications/coexisting-with-wolves-in-idahos-wood-river-valley.pdf. Accessed March 8, 2014.

———. n.d. "Wolf Compensation Trust." https://defenders.org/publications/statistics_on_payments_from_the_defenders_wildlife_foundation_wolf_compensation_trust.pdf. Accessed 28 April 2013.

Hicks, J. 1981. *Wealth and Welfare*. Oxford: Blackwell.

Idaho State Department of Agriculture. 2011. "Idaho Agriculture Facts," http://www.agri.idaho.gov/Categories/Marketing/Documents/English%20Final%02011%20-%20for%20emailing.pdf. Accessed 26 May 2012.

Idaho Department of Fish and Game. 2009. "Wolf Reintroduction and Recovery Timeline."

Idaho Department of Fish and Game and Nez Perce Tribe. 2012. "2011 Idaho Wolf Monitoring Progress Report." Boise: Idaho Department of Fish and Game; Lapwai: Nez Perce Tribe Wolf Recovery Project, https://idfg.idaho.gov/old-web/docs/wolves/reportAnnual11.pdf. Accessed 26 May 2012.

Jordan, T. 1993. *North American Cattle-Ranching Frontiers: Origins, Diffusion, and Differentiation*. Albuquerque: University of New Mexico Press.

Kawamura, H. 2004. "Symbolic and Political Ecology among Contemporary Nez Perce Indians in Idaho, USA: Functions and Meanings of Hunting, Fishing, and Gathering Practices." *Agriculture and Human Values* 21(2–3): 157.

Knight, J. 2006. *Waiting for Wolves in Japan: An Anthropological Study of People–Wildlife Relations*. Honolulu: University of Hawai'i Press.

Knight, J. ed. 2000. *Natural Enemies: People–Wildlife Conflicts in Anthropological Perspective*. London: Routledge.

Kosek, J. 2006. *Understories: The Political Life of Forests in Northern New Mexico*. Durham: Duke University Press.

Lindquist, G. 2000. "The Wolf, the Saami, and the Urban Shaman: Predator Symbolism in Sweden." In *Natural Enemies: People–Wildlife Conflicts in Anthropological Perspectives*, edited by J. Knight. New York and London: Routledge.

Maughan, R. 2005. "Overview and history of the central Idaho wolf reintroductions." http://www.forwolves.org/ralph/wpages/idaho-o.htm. Accessed 4 September 2009.

Montag, J., M.E. Patterson, and B. Sutton. 2003. "Political and Social Viability of Predator Compensation Programs in the West. Final Report." Wildlife Biology Program, School of Forestry, University of Montana, Missoula. http://wildlife.state.co.us/SiteCollectionDocuments/DOW/WildlifeSpecies/SpeciesOf Concern/Wolf/postfinalreport.pdf. Accessed 28 April 2013.

Muhly, T.B., and M. Musiani. 2009. "Livestock Depredation by Wolves and the Ranching Economy in the Northwestern U.S." *Ecological Economics* 68: 2439–50.

Nadeau, M.S., C. Mack, J. Holyan, J. Husseman, M. Lucid, D. Spicer, and B. Thomas. 2009. "Wolf Conservation and Management in Idaho: Progress Report 2008." Idaho Department of Fish and Game and Nez Perce Tribe. https://idfg.idaho.gov/old-web/docs/wolves/reportAnnual08.pdf. Accessed 13 August 2009.

Nash, R.F. 2001. *Wilderness and the American Mind*, 4th ed. New Haven: Yale University Press.

Naughton-Treves, L., Grossberg, R., and A. Treves. 2003. "Paying for Tolerance: The Impact of Depredation and Compensation Payments of Rural Citizens' Attitudes toward Wolves." *Conservation Biology* 17: 1500–1511.

Nilsen, E., E. Milner-Gulland, L. Schofield, A. Mysterud, N. Stenseth, and T. Coulson. 2006. "Wolf Reintroduction to Scotland: Public Attitudes and Consequences for Red Deer Management." *Proceedings from the Royal Society*. http://www.wolvesandhumans.org/pdf-documents/wolf_scotland_reintro duction.pdf. Accessed 31 August 2012.

O'Neill, J., A. Holland, and A. Light. 2008. *Environmental Values*. New York: Routledge.

Payette National Forest. n.d. "Deep Canyons, Deep Wilderness, Deep Snow." US Department of Agriculture, Forest Service. http://www.fs.usda.gov/main/ payette/home. Accessed 31 August 2012.

Pearce, D., A. Markandya, and E. Barlier. 1989. *Blueprint for a Green Economy*. London: Earthscan.

Rawls, J. 1971. *A Theory of Justice*. Cambridge, MA: Harvard University Press.

Sartre, J.P. [1956]1966. *Being and Nothingness*, translated by Hazel E. Barnes. New York: Washington Square Press.

Sciama, L.D. 2006. *A Venetian Island: Environment, History, and Change in Burano*. New York: Berghahn Books.

Thornton, T. 2008. *Being and Place among the Tlingit*. Seattle: University of Washington Press.

US Department of Agriculture. 2001. "Cattle Death Loss." http://usda.mannlib .cornell.edu/usda/current/CattDeath/CattDeath-05-12-2011.pdf. Accessed 28 April 2013.

———. 2011. "Poisoning of Livestock by Various Larkspur Species (Delphinium): 2011 Annual Report."

Willey, J. 2009. "Bray and Mary Willey Family." In *Free Land! Hopes and Hardships of Pioneers of Valley County, Idaho*. McCall: Valley County History Project.

IV.
SYSTEMS
CHANGE
IN TIME

Under a crumbling bridge on the Salmon River, British Columbia, a chain anchors the structure to a tree. (Photo courtesy of Robert Boschman)

CHAPTER 10

Declarations of Interdependence: Unexpected Human–Animal Conflict and Bhutanese Non-Linear Policy

Randy Schroeder and Kent Schroeder

INTERDEPENDENCE THEN: CONFUSION

Just over halfway through Charles Fisher's wildly interdisciplinary *Dismantling Discontent: Buddha's Way Through Darwin's World*, a single sentence erupts: "Life is incredibly complex and awe-inspiring, but it is neither as efficient nor as reasonable as we sometimes imagine it to be" (2007, 274). On one hand, Fisher's insight is so familiar it is almost banal. On the other, it is startling, because it refreshes a perennial suspicion about the efficacies and agencies of evolved human intelligence. How so? In many academic contexts, we take Fisher's insight as obvious, because we imagine we can acknowledge, confront, grasp, or even model the wildness of nature, while somehow leaving it wild. Ironically, we imagine ourselves possessed of reasonable and efficient models to explain the unreasonable and inefficient: theories of complexity, non-linearity, emergence, ecosystem. As Religious Studies scholar David L. McMahan argues, almost every discipline today declares some version of interdependence, the basic admission that everything results from a "complex multiplicity of factors that extend out into an ever-widening causal web" (2008, 149). The decision to celebrate this "interwoven world," in which one might be subsumed into an intimacy with the "living fabric of life" (2008, 151), is not limited to academics—it is widespread among Eurocentric "spiritual" types, who mix any number of religious traditions into New Age cocktails, and also among "many of the most prominent Asian leaders of

contemporary Buddhism" (2008, 152). But is it possible to delude ourselves that we understand interdependence, while failing to acknowledge—*really* acknowledge—the inexorable truth of interdependence?
We think so. A deep recognition of process and relation will demand
concessions that no conventional Western policy-maker will make: the
acceptance of a virtual infinitude of variables in a virtual infinitude of
relations, and, even more, the acceptance of those relations in evolving
patterns, where they reconnect, redistribute, and reconstruct in every
instant. Implacable time and incalculable agency are the iridescence
of interdependence: gazed at briefly, sidelong, through the mist and
variegations, when the light is at the perfect angle and intensity—just
so, then gone. Humility: the iridescence of deep recognition.

One obvious field that declares for interdependence is the study
of natural systems. As McMahan notes, commonsensical critiques of
anthropocentric and mechanical "Western" paradigms—both inside
and outside the University—entwine with the celebrations of more
interconnective world views, such as that of the shaman, the Buddhist, the hunter-gatherer, the ecologist: all those who understand that
"because everything depends on everything else, altering the balance
of the web of life can be—and has been—catastrophic" (2008, 151).
The view is undoubtedly correct in its general contours, at a gross
level of scale: our inherited atomism, reductionism, and "substantialism"[1] are all demonstrably false and noxious. One needs only to
recall images of factory cows, beached whales, shocked monkeys, and
homeless cheetahs. But as McMahan argues throughout his book—
with the obligatory nod to Foucault—the view, as currently inflected,
is also undoubtedly historically determinate: it is a contemporary
hybrid of various indigenous Buddhist concepts, German Romanti-
cism, American Transcendentalism, systems theory, recent ecology, and
pop science. An emblematic example would be the bestselling work
of physicist Fritjof Capra, who notoriously swishes "new sciences"
together with loose interpretations of "Eastern wisdom." In *The Web
of Life: A New Scientific Understanding of Living Systems* (1996), he
makes the now familiar case that recent insights from all the usual
suspects—chaos theory, quantum physics, network theory, systems
thought—have overturned reductionist models of science in favour
of a new world view that sees the world as holistic, integrated, and
"ecological." The paradigm shift, for Capra, is filled with potential

to refresh our understanding of and relations with ecology. The shift would, for example, help us understand how to index animal extinction with poverty and other seemingly remote phenomena.

The Himalayan country of Bhutan—famous now for its Gross National Happiness development model—has installed, as a constitutional directive, an integrative policy approach that acknowledges the dense feedback loops among ecology, economy, culture, social equity, and good governance. But Gross National Happiness also understands that unexpected consequences for any program of action are an integral part of an integrative world view, and that part of the unexpected is how a consequence will be felt and understood from any particular standpoint. It is not clear that Capra's version of interdependence is as balanced, sophisticated, or true as any version that springs from deeply rooted Himalayan Buddhist values and knowledge. Again, McMahan clarifies: the genealogy of "interdependence," as deployed by thinkers like Capra, is hardly that of Bhutan, since the Buddhist tradition traces back to a set of classical formulations whereby the phenomenal world of interdependence is not a "web of wonderment" but a "binding chain" (2008, 153). So we might ask, with McMahan: how did what was once conceptualized as the grim web of life morph into a contemporary version that "celebrates this-worldly life and promotes activist engagement?" (2008, 153).

The issue has been noticed by scholars in many idioms. In the late 1990s, in *Prisoners of Shangri- La: Tibetan Buddhism and the West*, Donald S. Lopez, Jr. demonstrated that the Himalayas have long been subject to Orientalizing distortions of the Western imaginary. In his introductory inventory of Western representations of Tibet, Lopez includes a scene from the 1995 film *Ace Ventura: When Nature Calls*, in which Ace performs penance in a monastery for failing to rescue a raccoon, and another from the 1983 film *Return of the Jedi*, in which the cuddly, bearlike Ewoks supposedly speak in high-speed Tibetan (1998, 1–2). So perhaps we have a long tradition of romanticizing interdependence and its putative implications for ecology. Do we love our animals too much? Feminist epistemologist Lorraine Code, in 2006's *Ecological Thinking: The Politics of Epistemic Location*, concedes early in her argument that "interconnective modes of thought and being are as *likely to activate retrograde romantic fantasies as moral social interventions*" (2006, 6). Philosopher Rosi Braidotti, in *The*

Posthuman, qualifies more forcefully with regard to animal–human interactions. For Braidotti, a basic understanding of interdependence *may well result* in a healthy displacement of *anthropos*, and we are right to hope so (2013, 75). But, as she acknowledges throughout her book, interdependent "posthuman" thought is always a hair shy of recuperation, into innovative yet residual patterns of instrumentalism and exploitation, into the endlessly adaptive and necro-political networks of hypercapitalism itself. For Braidotti, the naturalized celebration of management and its panoply of techno-solutions leads easily to the bloody mistreatment of whatever species or *bios* happens to be currently useful, marketable, and disposable within the interdependent web of commerce. Elsewhere, Braidotti has argued that many existing declarations of interdependence, such as Deep Ecology, quite naturally regress to an "excessive degree of anthropomorphism," almost as if interdependent thought is humanism's back door of choice (2006, 115–18). The fundamental point is that one cannot conveniently snip out all the unwanted causes and undesirable effects in an interdependent world, leaving only the goodies. Nor can one erase or ameliorate the non-linear complexity and unexpected consequences that comprise the signal features of interdependence. One cannot outwit time and evolution. One cannot imagine that the idea of "interdependence" will somehow escape interdependence itself.

Back to Charles Fisher, who offers a highly useful concept to begin sorting out the accidents of interdependent thought and being. Throughout *Dismantling Discontent: Buddha's Way through Darwin's World* (2007), Fisher returns to a slightly romanticized argument that hunter-gatherers are highly "interdependent" in their navigations of the world. As such, they possess attention skills that are healthier than those of your average scholar, and more attuned to a genuine web of life. But Fisher qualifies deeply, using recurrent incantations of an old evolutionary concept called *antagonistic pleiotropy*. What is antagonistic pleiotropy? It is, first, a simple term from studies of senescence that describes how adaptive traits can become maladaptive as organisms age. But in Fisher's hands, antagonistic pleiotropy becomes an active figure for how interdependence is far more interdependent than the most enthusiastic Western cheerleader for interdependence could ever admit. Antagonistic pleiotropy is a deep metaphor that confronts the difficulty in assigning permanent value to any given consequence in any given context. Antagonistic pleiotropy suggests that one may not be able to construct an interdependent theory of unexpected consequences that does not finally outwit itself. It may also suggest, paradoxically, that one has to try anyway, as the current and ongoing Bhutanese experiment with "multi-dimensional" policy tools demonstrates. Time and agency: the iridescence of interdependence.

INTERDEPENDENCE NOW: HAPPINESS

If anyone could sort out the knotty and naughty vexations of inter-dependence, as they manifest themselveson the ground, one would expect it to be Bhutanese policy actors and citizens. The Bhutanese, moreover, might be best equipped to confront the prevalence, inexo-rability, distribution, and velocity of unexpected consequences. As we have argued elsewhere, Bhutan's Gross National Happiness (GNH) model presents as a paradox for the West, in that it aims for nuanced understanding of an interconnected world that must be confronted, simultaneously, with both radical acceptance *and* radical intervention (Schroeder and Schroeder 2014). Put another way, GNH seems to recognize the unexpected while also attempting to manage it.

So, from a policy perspective, what does Bhutan's GNH experience tell us about interdependence? Does GNH navigate a policy path that recognizes, even embraces, emergent and unexpected consequences while still effectively managing policy goals? Or is policy paralysis the inevitable outcome as economy, ecology, and society entangle, inter-connect, and overlap in emergent and unpredictable ways that drive both desirable and undesirable effects?

Gross National Happiness was initiated by the fourth King of Bhutan after he ascended the throne in the early 1970s. It is rooted in the assumption that happiness, not the accumulation of wealth, is the end goal of development. Happiness in this sense is understood not as immediate satisfaction or pleasure but as deep-seated well-being grounded in a balance of material, mental, emotional, and spiritual concerns. Moreover, happiness is not merely a private matter. Its real-ization requires that it be shared. True happiness has a reciprocal link between the individual and society; increased happiness in one increases happiness in the other. According to the Bhutanese government, the happiness that underlies GNH is ultimately about the achievement of the "full and innate potential" of being human (GNH Commission 2009, 17). GNH as a development strategy is therefore holistic and multi-dimensional. It seeks to balance the material and non-material needs of individuals and society in the pursuit of happiness as the ultimate end of development. Such an understanding of happiness diverges from our Western understanding of the term. Indeed, the nature of happiness within Bhutan's GNH is intimately tied to the Mahayana Buddhist tradition (Hewavitharana 2004; Lokamitra 2004; Tashi in McDonald 2010; Tideman 2011). The Buddhist values of the sanctity of life, interdependence among all sentient beings, compassion, respect for nature, social harmony, and compromise are the foundation of GNH. From a policy perspective, this Buddhist-infused development approach takes shape in a framework of four GNH pillars. Those pillars are equitable socio-economic development, environmental sustainability, cultural preservation and promotion, and good governance. All policy-making in Bhutan must be fed through the four pillars and take each into account. Ultimately, promoting the four pillars creates the enabling conditions necessary for happiness in the Bhutanese context.

Perhaps the most critical characteristic of the GNH framework is the interdependent and integrated nature of the four GNH pillars. It is not merely a framework of multiple pillars but a framework where the pillars are understood to be interdependent, engaging, and interacting with one another. Their interdependence recognizes the complexity and interrelationships within and across social, economic, ecological, cultural, and governance systems. Official Bhutanese policy documents make this clear. The four-pillared GNH framework is described using such terms as "synergistic," "harmonious balance," and "interwoven in reality" (GNH Commission 2009, 17; 2011, 8; Planning Commission Secretariat 2000, 6; RGoB 2005, 15; Thinley 2007).

The recognition within GNH of the complex interdependencies across economic, ecological, social, and governance systems further requires a deep and pervasive recognition of the unexpected and emergent outcomes such interacting systems can generate. Accordingly, GNH itself is constructed to be open to evolution and change (Mathou 1999, 618; Planning Commission 1999, 10). Implementing GNH requires an ongoing orchestration of the pillars (Rinzin 2006, 30), particularly as emergent outcomes arise. As Bhutan has gained experience implementing GNH, the policy framework has evolved with a set of GNH policy instruments, recently developed. The instruments shape policy design, implementation, and evaluation in a way that explicitly incorporates interdependence: the GNH Index, created in 2008, expands the four pillars of GNH into nine domains that measure the achievement of happiness.

How, then, does Bhutan's GNH framework—with its holistic recognition of interdependencies and the unexpected outcomes they create—play out in the actual policy process? Does Bhutan's GNH framework provide a practical means to both accept *and* manage the unpredictable interdependence that may drive both desirable policy outcomes and undesirable counter-effects? Analyzing the implementation of GNH policy on the ground provides some clues. In particular, exploring Bhutan's response to the issue of human–wildlife conflict is instructive. Do we love our animals too much? Human–wildlife conflict (HWC) unfolds in agricultural fields across Bhutan as increasing numbers of wild animals destroy crops and livestock. The problem is a significant challenge for a Buddhist-inspired strategy like GNH that values the sanctity of life and harmony among all sentient beings. Interdependencies within the HWC problem also generate conflicting and unexpected outcomes.[2]

The issue of HWC is often traced back to Bhutan's historical record of conservation. The country was almost entirely closed to the outside world until 1960. As it emerged from its global isolation, Bhutan initiated a process of planned development that would enable the country to shape its future with its own values in the context of new global influences. As a result, Bhutan often privileged environmental issues as a means to counterbalance the global dominance of economic growth and consumption as the primary focus of development. This privileging was clearly evident with natural resource management. The Bhutanese government played a minimal role in natural resource management prior to 1960 as traditional and community-based approaches predominated. As Bhutan initiated its process of planned development, however, the government centralized natural resource management and formalized a reliance on scientific management approaches led by the state. This approach continued with the inauguration of Gross National Happiness in the early 1970s. In order to balance the GNH pillars in a manner that protected Bhutan's environment from unrestrained economic growth, tight restrictions were placed around the economic uses of natural resources, accompanied by strong conservation strategies, including Bhutan's commitment to keep 60 percent of the country forested in perpetuity, a commitment now enshrined in the constitution.

The 1995 Forest and Nature Conservation Act of Bhutan (FNCA) replaced previous conservation legislation but continues Bhutan's historical attempt to balance economic growth with sound conservation. The FNCA continues to emphasize the notion of scientific management of natural resources but recognizes the need for greater community participation (RGoB 1995). It also continues to maintain strict prohibitions and restrictions around economic activities such as hunting, fishing, grazing, and forestry. A framework for the conservation of animals designates 23 kinds of wild animals that are "totally protected" and cannot be killed or captured anywhere unless human life is threatened. All remaining animals are "protected" and can be killed or captured only in cases of imminent attack on humans or livestock or to defend against crop damage on private land. The killing of animals beyond 200 metres of private land is prohibited. The FNCA is bolstered by other initiatives such as the Biological Conservation Complex, which designates protected areas and biological corridors. These protected areas and corridors currently make up approximately 51 percent of Bhutan's total area. No other country in the world has a higher proportion of protected land (MoA 2009, 29–30).

Bhutan's policy focus on conservation has had dramatic results. Forest cover, which is targeted at 60 percent of the country, is actually over 70% (WCD 2010, 8). Substantial biodiversity is evident, with more than 200 species of mammals, 770 species of birds, and 7,000 species of vascular plants (8–9). The country is recognized as part of one of the world's biodiversity hot spots and has won UNEP's inaugural "Champions of the Earth" award and the WWF's J. Paul Getty Award for Conservation Leadership.

Bhutan's historical policy emphasis on conservation as a counterbalance to unchecked economic growth has not, as one might expect, come at the expense of such growth. Blessed with abundant water fed by the glaciers of the Himalayas, Bhutan has developed a hydro-electricity industry that has been a key driver of 7.8 percent annual economic growth since the 1990s and 8.7 percent between 2005 and

2010—this, even while heeding the country's strict environmental regulations (GNH Commission and UNDP 2011, 31). Clearly, Bhutan's GNH-based policy approach has, therefore, experienced significant success in balancing conservation and economic growth without allowing the latter to dominate. Yet that approach also appears to have driven a negative and unexpected policy outcome. Hydroelectricity may generate significant economic growth, but it is not a labour-intensive industry. Much of the Bhutanese labour force continues to engage in small-scale agriculture. In the 2000s, the Bhutanese government outlined what it perceived to be the unintended consequence for farmers of its years of strong conservation policy. The centralized and scientific approach to conservation was seen to have undermined rural livelihoods (RGoB 2011, 60; NCD 2008, x). Years of strict protection of wildlife have increased the numbers of animals, and increased forest cover and protected areas have provided them with more space to roam. The result has been growth in the populations of wild pigs, monkeys, deer, and elephants, which are encroaching on agricultural land and destroying crops. Meanwhile leopards, tigers, and bears are killing more and more livestock (NCD 2008).

The implications of this situation for rural livelihoods are stark. Entire crops can be wiped out in a single night, dramatically reducing rural incomes and contributing to food insecurity. Moreover, given the restrictions on killing and trapping animals, farmers are having to work in the fields all day and then guard their crops and livestock at night. This has had implications for family life, mental health, and risk of injury or death from large predators.

As a multi-dimensional and integrated development framework, Bhutan's GNH approach has engaged in a policy pivot in response to the unexpected emergence of HWC. The pivot recognizes HWC as an emergent outcome that needs to be addressed through a policy evolution that maintains the original conservation focus while accounting for its unexpected and negative outcomes. Promoting a rebalancing of the GNH pillars has been a necessary response. The government's resulting human–wildlife conflict strategy more explicitly links conservation and rural livelihoods as two interdependent components. It assumes that effective conservation can improve rural livelihood opportunities while appropriate livelihood opportunities can strengthen conservation. A self-reinforcing and virtuous GNH circle is the intention. The policy takes a twofold approach. First, it strives to improve rural livelihoods by reducing incidence of HWC and promoting alternative livelihood opportunities rooted in conservation. Second, it strives to do so while maintaining current strict conservation efforts (NCD 2008).

The policy pivot is an attempt to rebalance the socio-economic and environmental pillars of GNH in the context of an unexpected policy outcome. It is also explicitly linked to the cultural and good governance pillars of GNH. By promoting alternative livelihoods while continuing strong conservation measures, the HWC strategy intends to prevent a shift of cultural attitudes away from valuing the sanctity of life of all beings, including those that destroy livelihoods. The strategy also promotes a decentralized approach to HWC and governance. The centralized scientific-management approach of past Bhutanese conservation practices thus requires an evolution towards greater local participation, in order to better understand the nature of HWC and distribute the benefits of conservation more equitably. Overall, the policy pivot attempts to account for the interdependence of all four GNH pillars, as it co-evolves with the emergence of human–wildlife conflict.

A key component of the HWC strategy is an integrated conservation and development program (ICDP) intended to mitigate crop and livestock destruction while empowering self-sufficiency among rural communities (NDC 2008). The ICDP strategy initially took place in nine model sites with the intention of scaling them up across the country. Multiple strategies make up the ICDP approach (NCD 2008: 14–21):

1   crop protection through fencing, alarms, and alternative crop cultivation;

2   intensification of existing agricultural land in HWC hotspots through enhanced production, such as the use of improved seeds and increased production of organic products;

3   intensification of livestock production in HWC hotspots through the introduction of improved livestock breeds and their management;

4   exploration of community-driven alternative revenue generating opportunities such as high value crops and backyard farming with pigs and poultry;

5   development of crop and livestock insurance schemes aimed at community self-sufficiency;

6   promotion of community management of non-wood forest products;

7   increased patrolling of HWC hotspots involving local communities; and

8   implementation of socio-economic surveys specific to problem area species.

Implementation of the ICDP strategies demonstrates that Bhutan's GNH approach is able to pivot and address HWC as an emergent and unexpected policy outcome, even while maintaining strong conservation measures. However, given the hypercomplexity of interdependency, it is not surprising that the policy pivot itself has driven further unexpected and potentially unwanted outcomes. On the one hand, the HWC strategy has achieved notable success in decreasing crop destruction and promoting alternative livelihood opportunities without watering down conservation policy. First, there is evidence that the use of fencing and alarms has decreased localized incidences of HWC. Solar and electric fencing provided to farmers through the HWC strategy has kept marauding wildlife away from crops. Some unsurprising challenges remain; for example, the use of fences has shifted the HWC problem elsewhere, and there have been funding and communication problems related to the availability of fences and alarms. Even so, arranging for appropriate fencing has proved to be a viable alternative to weakening conservation policy.

Second, some alternative livelihood activities are achieving significant success. Livestock intensification is perhaps the best example. Through the HWC strategy, under a cost-sharing program, the government is providing farmers with new breeds of cattle that produce significantly more milk. These new breeds are not grazed in the forests (the traditional method of husbandry); instead they are stall-fed to keep them away from predators in the forest. In addition, farmers are establishing collectives to market the milk; this is increasing household revenues, and the profits that go to those collectives provide low-interest loans to group members. The reported results have been significant. It is reported that the revenues generated by increased milk production are increasing considerably among those farmers raising improved breeds. In addition, stall-feeding smaller numbers of more productive cattle instead of grazing large herds in the forest takes less time, so farmers can spend more time with their crops. Also, the quality of forest cover has reportedly improved with the shift away from grazing, although this is disputed by some.

Insurance schemes have had mixed success. A significant amount of money has been paid out to those who have lost livestock, and this has shielded rural households from dramatic losses of income. At the same time, challenges related to sufficient funding and a lack of knowledge among some farmers of the kinds of insurance available have hampered the effectiveness of those schemes. A recently established endowment fund for community-based insurance schemes is trying to address these challenges.

So as a whole, the policy outcomes generated by the HWC strategy are demonstrating notable success, with some ongoing challenges related to insufficient funding and faulty communications. Yet a new and unexpected issue now looms. Bhutan's policy position is that HWC emerged as an unexpected consequence of past conservation policy (NCD 2008; Planning Commission 2002, 117). However, as the HWC strategy has been implemented, often with notable success, the policy actors responsible for its implementation on the ground have developed divergent understandings of the cause of HWC itself. Cause and effect have been blurred because the very nature of the interdependence between livelihoods and conservation is interpreted differently by different implementation actors. As a result, the foundational assumption of the HWC strategy is now subject to significant disagreement. A curious situation has resulted: successful policy outcomes are addressing the problem of HWC even as an unexpected ambiguity grows around the ontological status of the problem itself.

Four competing positions exist among policy actors as the cause of HWC. One position is consistent with the official view: past conservation policy overemphasized the environmental pillar of GNH, and the success of that policy has led to increased wildlife and forest cover, both of which are now encroaching on agricultural land, leading in turn to increased contact (and conflict) between animals and farmers. Other policy actors view the problem in polar opposite terms: human–wildlife conflict is a result not of strong conservation policy but of rollbacks to the environment pillar of GNH. On this view, the fault lies with increased agricultural production and development activities (such as road construction) that are encroaching on forests, not the other way around. In other words, overemphasis on the socio-economic pillar of GNH is the cause of HWC. Still others suggest that both conservation success and expanding agricultural production—two seemingly opposing processes—have generated HWC in different ways in different contexts at precisely the same time. Complicating these three views is a fourth assumption among a small group of policy actors: HWC is hardly a problem at all, and the issue has been blown out of proportion by the Bhutanese media. Thus, as the HWC strategy has been rolled out, the interdependency between environmental, economic, and social systems has blurred understanding of the HWC problem among those actually doing the strategizing. Yet the HWC strategy appears to be successfully confronting the results of the problem, at least for now, from the perspectives of those most committed to the strategy.

Overall, Bhutan's policy experience with HWC demonstrates an apparent cocktail of interdependency that defies any definitive conclusions about cause and effect. Yet Bhutan's GNH approach, which recognizes and accepts interdependencies and the unexpected, demonstrates an ability to evolve while still managing the problem within an overall framework of ambiguity. Definitive understandings of the problem may not exist, but policy paralysis has not been the result. It's almost as if successful management requires a measure of surrender.

INTERDEPENDENCE ALWAYS: SURRENDER

Charles Fisher mobilizes the simple concept of antagonistic pleiotropy in a manner that clarifies the issues we have raised with respect to HWC in Bhutan, and—more importantly—deepens those issues with an enriched set of questions. Antagonistic pleiotropy, in its original formulation, describes any situation where a gene expresses both adaptive and maladaptive traits. Or, in the case of a single expressed trait, antagonistic pleiotropy describes the situation where one aspect promotes welfare while another undermines it (2007, 258). The maladaptive traits can be thought of in three distinct forms: as undesirable side effects that express or accumulate over time; as undesirable counter-effects that occur simultaneously; and as crucial structures that, through sheer momentum, "create vulnerabilities for species" as environments evolve (Fisher 2007, 77). A classic example of the first form would be testosterone in human males, which contributes to early reproductive fitness but "increases the chance of prostate cancer later in life" (2007, 400). A classic example of the second form would be the larynx, which enables, at once, speech and choking (2007, 204). An even more illustrative example would be terror, a "product of mimetic and narrative mind" that allows us to connect past and future, but also, in the present, "triggers diarrhea, a hindrance when survival is at stake" (2007, 304). But the most wondrous examples accompany the third form, because such examples invoke the foundational ironies that characterize human experience, as demonstrated so intriguingly by the Bhutanese experience of managing HWC. Fisher is at his finest when describing human imagination and abstract intelligence, which drive us towards innovation, creation, prediction, and management, all of which begin to undermine themselves as "evolving intelligence" begins to "antagonize itself" (2007, 368). In Fisher's compelling argument, we improve our survival odds with our developed minds—especially our ability to navigate, manipulate, and predict the material world through abstraction and categorization—but at the cost of increased fantasy and projected desire, both individual and collective. The cascade of unexpected consequences include the ability to destroy from a distance, the possibility of self-annihilation, and, beneath it all, the persistent knowledge of inevitable death and its attendant drive towards accelerated manipulation of the world, not to mention the exaggerated delusion of control (2007, 368). In a further irony, the delusion of full control—in its predictive mode—is perpetually undermined by the type of antagonistic pleiotropy where evolutionary adaptations are laid down in what, retrospectively, appears to be the wrong order. Fisher offers the example of mammalian eyes, which have "blind spots created by veins in front of the retina, and the exit of the optical nerve through it" (2007, 77). Time: the iridescence of interdependence.

Human intelligence—as a rich set of adaptive traits—is the most compelling issue in Fisher's version of antagonistic pleiotropy, and the most relevant to our discussion of Bhutanese policy, prediction, and management. Paul Errington—the trapper, naturalist, and zoologist famous for his studies of predation—wrote in 1967 that "life selects for what works, irrespective of our efforts to define and classify" (1967,

219). At the time, Errington's claim was noteworthy in the context
of predation studies, which tended to ignore the complex dynamics
of environments. One can appreciate Errington's insight that human
adaptations for definition and categorization have a more limited range
than we imagine. What Errington did not fully articulate is that human
drives to define and classify "life" are actually part of life, and have
"worked," indeed, have been selected for. Fisher locates human intelligence *within* the very evolutionary processes it tries to understand, and
notes the resultant ironies with more nuance than does Errington (in
this regard, we could connect Fisher to a panoply of otherwise uncon-
nected philosophers, everyone from Ludwig Wittgenstein to Donna
Haraway to Richard Rorty to Slavoj Žižek).

Bracketing the philosophical and neurobiological questions that
attend "consciousness," Fisher itemizes the possible adaptive functions
that would accompany the ability to define and categorize: connecting causes and effects; sequencing; evaluating, correcting, and reor-
ganizing; planning; forecasting possible outcomes; learning from the
past (2007, 202–3). All these features, obviously, give humans agency
within, and power over, their immediate local environments. (Indeed,
these are the very features that give rise to scholarship of ecology
and interdependence, and, eventually, to schools and institutes of
sustainability ...) But then the antagonistic pleiotropy kicks in. With
respect to human intelligence, innovation, and management, "evolution overdid its job," and the "very success of the mind in terms of
fitness is now our liability" (2007, 203). For Fisher, this iteration of
antagonistic pleiotropy is due to an intriguing collision of unexpected
consequences: communication is vital to rational thought, because it
ensures that knowledge is shared, refined, and adjusted; however, communicative abilities evolved largely out of emotion (2007, 203). For
Fisher, the maladaptive features that result from this collision sound
all too familiar: definition, classification, and prediction are accompanied by baseline agitation, endless planning and rehearsal, the inability to stop comparing outcomes, and a generalized, incessant mental
chatter (2007, 203). In Buddhist traditions, it's often called *monkey
mind*. Extrapolating from Fisher, the collective downside to human
intelligence is, depending on cultural factors, its attendant emotional
state: a meliorist and instrumentalist impulse that sometimes drives us
to compulsively re-engineer our failures in the name of an ill-defined

"progress" and to romanticize the impulse itself as uniquely human. Rage, rage against the dying of the light. Fisher notes another, ironic iteration of antagonistic pleiotropy: it is often difficult for us to slow down and deliberately assess the current value of our *monkey mind* gadgeteering impulses, since one feature of our highly developed minds is the ability to transfer what we have learned into habituated pattern, freeing up attention for immediate focus (2007, 282). But given that evolution, too, is relentless, *what we have learned* may have, in many cases, changed its environment and its adaptive value. Indeed, the perspectives from which to assess the value may have evolved. The patterns remain. Their advantages are lost. We have forgotten.

These unpredictables literally mean that human manipulation of the world—whatever its impressive and transient charms—can never be accompanied by thoroughgoing prediction and mastery, no matter how often we declare the opposite and cite whatever list of progressive innovations is currently trending. Antagonistic pleiotropy becomes a metaphor that is literally true, extended and extensive, a world view. It intensifies the truths of probability, non-linearity, and emergence, but also the impossibility of giving a full account for all of them, since that account would be embedded in the very environments it seeks to manage, and part of the very process for which it seeks to account. Who can name with confidence the processes that gave rise to the human insight of antagonistic pleiotropy, or the translation of that insight into symbols, or the effects such an insight might foster in a given environment or throughout the networks of linked environments? Who can account for the multiple perspectives—in both time and space—from which to evaluate the effects of that insight, and determine the appropriate responses? Evaluations must perpetually give way to re-evaluations, since whatever "effect" is in question will evolve in an evolving context under the scrutiny of evolving perspectives. And so—at least at finer levels of scale—that effect will change its relationships with its own relationships. Fisher frames the world view: "nature is like a fine sculptor chiselling out elegantly featured forms but then forgetting functional details" (2007, 125). As predictors, tinkerers, and manipulators, humans perpetually work on forms. In Bhutan, they do so with an openness and humility rarely seen elsewhere. But as part of nature, most of us who call ourselves "human" also forget that there are always more details: unknown unknowns that will predate, accompany, and arise from the form and the elegance. Time and agency are the iridescence of interdependence, always. To compound the issue, we perpetually forget that we are forgetting. And as any Buddhist will tell you, forgetting what we are is the fundamental problem for the human species.

Notes

Part of this work was carried out with the aid of a grant from the International Development Research Centre, Ottawa, Canada. Information on the Centre is available on the web at idrc.ca.

1 We use "substantialism" loosely to describe any position that attributes "substance" to actualities, or that assumes "things" have essences independent of— ontologically prior to—their own actions, qualities, and relations. The actual history of substance theory is, of course, much more complex and qualified and spans many epochs, cultures, and wisdom traditions.

2 In order to explore how GNH frames Bhutan's response to the unexpected interdependencies of HWC, semi-structured interviews were undertaken with ninety-six Bhutanese policy actors at the *gewog* (local), *dzongkhag* (district) and central government levels. Purposive sampling was used to select respondents from relevant central ministries as well as from 4 of Bhutan's 20 *dzongkhags* and 19 of its 205 *gewogs*.

References

Braidotti, Rosi. 2006. *Transpositions*. Cambridge: Polity Press.

———. 2013. *The Posthuman*. Cambridge: Polity Press.

Capra, Fritjof. 1996. *The Web of Life: A New Scientific Understanding of Living Systems*. New York: Anchor Books.

Code, Lorraine. 2006. *Ecological Thinking: The Politics of Epistemic Location*. Oxford: Oxford University Press.

Errington, Paul L. 1967. *Of Predation and Life*. Ames: Iowa State University Press.

Fisher, Charles. 2007. *Dismantling Discontent: Buddha's Way through Darwin's World*. Santa Rosa: Elite.

GNH Commission. 2009. *Tenth Five Year Plan 2008–2013*, vol. 1: *Main Document*. Thimphu.

———. 2011. *Eleventh Round Table Meeting. Turning Vision into Reality: The Challenges Confronting Bhutan*. Thimphu.

GNH Commission and UNDP. 2011. *Bhutan National Human Development Report 2011*. Thimphu.

Hewavitharana, Buddhadasa. 2004. "Framework for Operationalizing the Buddhist Concept of Gross National Happiness." In *Gross National Happiness and Development: Proceedings of the First International Seminar on Operationalizing Gross National Happiness*, edited by Karma Ura and Karma Galay, 496–531. Thimphu: Centre for Bhutan Studies.

Lokamitra, Dharmachari. 2004. "The Centrality of Buddhism and Education in Developing Gross National Happiness." In *Gross National Happiness and Development: Proceedings of the First International Seminar on Operationalizing Gross National Happiness*, edited by Karma Ura and Karma Galay, 472–82. Thimphu: Centre for Bhutan Studies.

Lopez, Jr., Donald S. 1999. *Prisoners of Shangri-La: Tibetan Buddhism and the West*. Chicago: University of Chicago Press.

Mathou, Thierry. 1999. "Political Reform in Bhutan: A Change in a Buddhist Monarchy." *Asian Survey* 39(4): 613–32.

McDonald, Ross. 2010. *Taking Happiness Seriously: Eleven Dialogues on Gross National Happiness*. Thimphu: Centre for Bhutan Studies.

McMahan, David L. 2008. *The Making of Buddhist Modernism*. Oxford: Oxford University Press.

MoA (Bhutan Ministry of Agriculture). 2009. *Biodiversity Action Plan*. Thimphu.

NCD (Nature Conservation Division [Bhutan]). 2008. *Bhutan National Human–Wildlife Conflicts Management Strategy*. Thimphu.

Planning Commission Secretariat [Bhutan]. 1999. *Bhutan 2020: A Vision for Peace, Prosperity, and Happiness. Part II*. Thimphu.

———. 2000. *Bhutan National Human Development Report 2000*. Thimphu.

———. 2002. *Ninth Five Year Plan (2002–2007)*. Thimphu.

RGoB (Royal Government of Bhutan). 1995. *Forest and Nature Conservation Act of Bhutan*. Thimphu.

———. 2005. *Bhutan National Human Development Report 2005*. Thimphu.

———. 2011. *The Third Annual Report of Lyonchhen Jigmi Yoeser Thinley to the Seventh Session of the First Parliament on the State of the Nation*. Thimphu.

Rinzin, Chhewang. 2006. *On the Middle Path: The Social Basis for Sustainable Development in Bhutan*. Netherlands Geographical Studies no. 352. Utrecht: Copernicus Institute for Sustainable Development and Innovation.

Schroeder, Randy, and Kent Schroeder. 2014. "Happy Environments: Bhutan, Interdependence, and the West." *Sustainability* 6: 3521–33. doi:10.3390/su6063521. Accessed 28 April 2015.

Thinley, Jigmi Y. 2007. "What Is Gross National Happiness?" In *Rethinking Development: Proceedings of Second International Conference on Gross National Happiness*, 3–11. Thimphu: Centre for Bhutan Studies.

Tideman, Sander G. 2011. "Gross National Happiness." In *Ethical Principles and Economic Transformation—a Buddhist Approach*, edited by László Zsolnai, 133–53. London: Springer. Issues in Business 33.

WCD (Wildlife Conservation Division). 2010. *Analysis of the Contributions of Protected Areas to the Social and Economic Development of Bhutan at National Level*. Thimphu: Wildlife Conservation Division, Royal Government of Bhutan.

CHAPTER 11

Effective Environmental Action in Canada: The German *Energiewende* as a Model of Public Agency

Mishka Lysack

> Climate is a function of biological activity on earth, and physics and chemistry in the sky. It is the prevalent weather conditions over time ... The goal is to come into alignment with the impact we are having on climate by addressing the human causes of global warming and bringing carbon back home.
> —Paul Hawken, *Drawdown: The Most Comprehensive Plan Ever Proposed to Reverse Global Warming* (2017, xiii)

Many policy pathways that address climate change so as to initiate a transition to a zero-carbon economy in Canada focus on reducing greenhouse gas emissions through carbon pricing. These pathways are agentive examples of what Daniel Gustave Anderson suggests are operational activations of theory that can lead to meaningful systems change: "If the practice of ecocriticism is understood as an activist knowledge and its object is ecology, then the theoretical frameworks appropriate to ecocriticism should directly inform ecological political action. Most straightforwardly, knowledge about the world should orient critics toward appropriate sites of intervention" (2014, n.p.). Among two of the current political parties on the federal level in Canada, the policy instruments used to price carbon include either a carbon price (Liberals) or cap-and-trade (New Democratic Party). Some provincial jurisdictions rely primarily on carbon pricing for addressing climate change—British Columbia, for example, with its carbon tax. Other iterations of energy policy, such as the Climate Change Strategy announced by Premier Rachel Notley of Alberta in

December 2015, blend different policy strands such as mandating an increase in renewable electricity and a phase-out of coal as well as the pricing of carbon. Other provinces, such as Ontario and Quebec, are also taking policy hybrid approaches to their energy transitions in order to address climate change—for example, a cap-and-trade schema of pricing carbon.

Countries in the European Union also employ integrated approaches to future-oriented environmental policy, including similar tools centred on pricing carbon through the EU's Emissions Trading Scheme (ETS). However, there is considerable disagreement in Europe regarding the effectiveness of the ETS as a climate protection tool, not only because of the low price on carbon, but also because of considerable "carbon leakage"—that is, the large number of allowances and exemptions given to industry that seriously undermine the policy's effectiveness in addressing climate change (Kemfert 2013). The many carbon "allowances" distributed to a significant number of European corporations allow these firms to continue with business as usual and ignore a significant quantity of GHG emissions, which further limits the effectiveness of the ETS.

In the EU, Germany and Denmark have provided significant international leadership by focusing on a pathway other than carbon pricing. Both are steadily transitioning to an energy system and economy based on 100 percent renewable energy, thus developing, in the words of Nancy Doubleday in this volume's final chapter, "alternative scenarios for the future that support equitable social-ecological resilience" (2018). Although Germany continues to participate as an EU member in the ETS, it is best known for its remarkable agentive transition to renewable energy known as the *Energiewende*, literally an energy pivot or historical turning point based on the substantive replacement of fossil fuels with an ambitious build-out of megawatts of renewable energy. Strongly resembling Bhutan's Gross National Happiness (GNH) pillars, which "include equitable socio-economic development, environmental sustainability, cultural preservation and promotion, and good governance" (Schroeder and Schroeder 2018, 7), the four key pillars of the *Energiewende* in Germany are as follows: (1) decarbonizing the economy and energy system through the Renewable Energy Law of 2000, which empowers German citizens as agents in the emerging renewable energy economy; (2) decentralizing energy production; (3) diversifying the energy market; and (4) phasing out nuclear energy. Besides these, a suite of other policy tools promote *Klimaschutz*, or climate protection.

For world leaders like Germany and Denmark, the strategic pathway in a world constrained by climate change leads towards (1) building up *MW of renewable electricity* as well as (2) increasing the *energy efficiency* of buildings and the renewable energy used for heating/cooling the built environment and (3) developing a *sustainable energy transportation grid*, leveraged from a renewable energy infrastructure platform. These endpoints, developed by an assemblage of agencies, resulted in public commitment and consensus that in turn led to concrete results. As of 2015, renewable electricity constituted 32.6 percent of the electricity in Germany's energy market, with wind, solar, and biomass generating most of the total renewable electricity mix (Morris and Hockenos 2016a). In 2014 this renewable energy transition achieved key milestones, such as preventing the emission of 147.869 million tons of CO_2 equivalents (Umweltbundesamt 2015).

With effective public leadership, Canada could follow in the footsteps of Germany and Denmark as it builds its own renewable energy economy and enacts climate protection. Ascribing to this more solution-focused model, Canada could focus on the positive pathway of stronger constructive action (renewable energy, energy efficiency, sustainable transportation) while limiting and diminishing negative action through the reduction of GHG emissions by way of regulations, public accountability, and carbon pricing. Stewart Elgie, Chair of Smart Prosperity, has described a climate change strategy that would include both "push" and "pull" policy tools (Elgie 2015). Within such a strategy, regulation would diminish fossil fuel use even as carbon pricing pulls the economy in a less carbon-intensive direction. At the same time, governments and society would invest in communities and infrastructure in order to drive decarbonization through an ambitious build-out of renewable energy across sectors.

THREE KEY BUILDING BLOCKS FOR GROWING CANADA'S RENEWABLE ENERGY ECONOMY

(1) Renewable Energy as a Tool for Job Creation, Economic Development, and Climate/Environmental Protection

The *Energiewende* has shown itself to be a key tool for achieving climate change mitigation and building out megawatts of renewable energy. Besides this, the energy transition in Germany and other European countries has generated substantial co-benefits. Significant and steady gains in energy efficiency and the building out of renewable energy have generated large numbers of new long-term jobs in Europe; it is estimated that two million people are currently working in renewable industry and related secondary employment (Federal Ministry 2015). Europe is now the world's second-largest employer in the renewable-energy sector (IRENA 2014); 40 percent of all employees in Europe's energy sector are working in renewable energy (CEPS 2014; Federal Ministry 2015).

Trend lines indicate that in Europe, renewable energy will continue to generate jobs over the next three decades. The total number of jobs in renewable energy will increase relative to the total of employees in Europe's overall energy sector. For instance, in Germany, 354,000 jobs in renewable energy were created by 2014, a figure far higher than for the coal mining and other fossil fuel sectors (Morris and Hockenos 2016b).

Evidence is accumulating that the renewable energy and energy efficiency sectors generate more jobs per unit of generation capacity than fossil-fuel technologies. US research indicates that the number of people who work for one year to generate 10 gigawatt hours of electricity is higher for renewable energy (solar: 5.5 persons; small hydro: 2.7; geothermal: 2.5, biomass: 2.1, wind: 1.7) than it is for fossil fuels (coal: 1.1 persons; gas: 1.1) (Wei et al. 2010). Economic modelling completed by American researchers like Dr. Robert Pollin, Director of the Political Economy Institute at the University of Massachusetts (Pollin et al. 2009a; see also Pollin 2008; 2009b) reveals similar results. For each $1 million spent on energy production, gas and oil generate 3.7 direct and indirect jobs; a significantly higher number of jobs are created by renewable electricity powered by wind (9.5 jobs per $1M) and solar (9.8 jobs per $1M) as well as by biomass (12.4 jobs per $1M).

Increased energy efficiency in the construction and transportation sectors also generates long-term jobs: these are found in smart grid construction and maintenance (8.9 jobs per $1M), retrofits (11.9 jobs per $1M), and mass transit/freight rail (15.9 jobs per $1M) (Pollin et al. 2009a, 28). Energy efficiency creates local capacity and generates jobs in local economies; research indicates seventeen jobs generated for every million euros invested (Urge-Vorsatz 2010). A European Commission impact assessment found that a 30 percent reduction in energy demand with a renewable energy target of 30 percent would generate 568,000 additional jobs in the EU (European Commission 2014).

Through similar economic initiatives in partnership with provinces and cities, Canada's federal government could enable the rapid expansion of the renewable energy sector in electricity, heating/cooling, and sustainable transportation. Building on the experience of Germany, Denmark, and other regions, Canada would thus be demonstrating how the development of renewable energy creates value in local communities, rural areas, and Indigenous communities, all while strengthening the Canadian economy and addressing climate change. By partnering with the provinces in infrastructure initiatives, Ottawa could foster after-manufacturing jobs in the renewable energy sector, such as those created by installation, maintenance, grid management, infrastructure development, cross-sectorial integration, and energy storage capacity.

(2) Setting Canada's Renewable Energy and Climate Protection Targets and Timelines

Within the framework of federal and provincial powers, Canada's federal government could provide important leadership for provincial, regional, and local governments by setting effective targets and timelines for renewable energy and energy efficiency in electricity, heating/cooling, and sustainable transportation. In setting these targets and timelines, Ottawa could use renewable energy and energy efficiency as a tool for economic development and job creation as well as for climate and environmental protection, including for accomplishing its GHG reduction targets.

Once there was an overall target reflecting Ottawa's objectives and timeframe, Canadian federal and provincial/territorial leaders could together develop a national vision, framework, and implementation plan. Here it would be established which concrete policy instruments were most suitable for specific constituencies and most effective in reducing GHG emissions and accelerating the transition to renewable energy. In countries like Germany and Denmark, governments on all levels are empowered to foster conditions that encourage and enable action by jurisdictions and citizen stakeholders, in addition to national and state efforts.

Target setting that employs clear timelines with five-year incremental stages fosters optimal conditions for energy transition (Leidreiter 2015; Couture and Leidreiter 2014, 10). Target setting also promotes agentive pathways in terms of the following environmental policy outcomes: (a) it demonstrates and communicates political will, effective leadership, a boldness of vision, and a clear public commitment to accountability and transparency; (b) it drives energy transition by providing a clear, public mandate for the mobilization of Canadians as stakeholders across sectors; (c) it ensures a more effective channelling of technical, administrative, and financial resources from both the government and the public; (d) it creates investment security and economic stability, as observed in Germany with its rise of interested investors in the period after national targets were announced (Fleck 2014); (e) it provides a firm foundation for deciding which policy instruments (e.g., carbon pricing) are the right tools to facilitate the Canadian *Energiewende*, and how these tools could best be implemented; and, finally, (f) it links different policy areas across sectors, such as climate and environmental protection, energy, infrastructure development, finance and fiscal development, economic development, industry, agriculture, human health, innovation and research, job training, and education. Ontario's then premier, Kathleen Wynne, emphasized the latter in her 2014 Mandate Letter for that province's Minister for the Environment and Climate Change, the Hon. Glen Murray, when she charged him to work with other Ontario ministers, "including the ministers of Finance, Energy, Transportation, Municipal Affairs and Housing, Economic Development, Employment and Infrastructure, Agriculture, Food and Rural Affairs, Research and Innovation, and Natural Resources and Forestry" (Wynne 2014). In this crucial period when environmental agencies still have the potential to attenuate tipping points, collaboration across conventional boundaries, such as those found in traditional governmental portfolios, will ensure that target setting works as it must.

Targets in themselves are ineffective without substantive governmental policy and legislative action reinforced by MRVs (measurement, reporting, verification) to drive change. That said, targets are indispensable to a climate change strategy that facilitates action. Targets, for instance, link public policy to research findings that indicate serious climate change tipping points towards irreversible environmental damage (Lenton 2011). Here and now, it remains instructive to examine the seven national targets of Germany's *Energiewende* and *Klimaschutz* program, which are as follows:

- GHG emission reduction by 40% by 2020, 55% by 2030, 70% by 2040 and by 80 to 95% by 2050, compared to the reference year 1990;

- primary energy consumption of 20% reduction by 2020 and 50% by 2050;

- energy efficiency increase of 2.1% per year as compared to final energy consumption;

- reduced electricity consumption by 10% by 2020 and by 25% by 2050, relative to 2008;

- heat demand reduction in buildings by 20% by 2020, with primary energy demand set to fall by 80% by 2050 (compared to 2008);

- renewable energies increase of 18% of gross final energy consumption by 2020, 30% by 2030, 45% by 2040 and 60% by 2050;

- renewable electricity increase of 35% by 2020 in gross electricity consumption, a 50% share by 2030, 65% by 2040 and 80% by 2050. (Leidreiter 2015)

Several features of Germany's target plans outlined above should be highlighted, as they indicate best practices regarding the metrics of energy policy development. First, the targets span different sectors of the economy and society, providing a broad view of key markers that demonstrate progress. Second, the country's GHG emission reductions themselves are ambitious (80 to 95 percent reduction by 2050, compared to reference year 1990) relative to those of most other countries, but these targets are both *necessary* when considered in relation to current scientific findings regarding the accelerating pace of climate change, and *achievable*, given that Germany has surpassed its targets in the past. Third, targets of expanding renewable energy production include electricity, building cooling/heating, and transportation. Fourth, a preponderance of targets focus on reducing energy consumption in different areas to ensure a diversified metric of collective human agency during this crucial time period: measuring energy efficiency, primary energy consumption, electricity consumption, and heat demand together. And fifth, all targets are stepped in increments of ten years. Indeed, with COP 21, best practice is now moving towards time increments of five years; this facilitates adaptation to the growing scientific knowledge regarding climate change acceleration and tipping points (Lenton 2011).

In Canadian energy and climate policy, three key targets are also time-sensitive: (1) Canada will need 100 percent renewable electricity, including transcontinental smart grid connections, by 2035 (Potvin et al. 2015); (2) its GHG emissions must be reduced by at least 26 to 28 percent (Potvin et al. 2015) by 2025; and (3) further GHG emissions need to decline by a minimum of 80 percent (Potvin et al. 2015) by 2050. The latter two targets measure from the reference year 1990. With respect to a more ambitious target of achieving 100 percent renewable electricity in Canada by 2035, Sustainable Canada Dialogues (http://www.sustainablecanadadialogues.ca/en/scd), a group of more than 70 Canadian scientists, social scientists, and engineers from every province in Canada (Potvin et al. 2015),[1] argues that this target is feasible given the considerable hydro assets already embedded in the nation's electricity grid in BC, Manitoba, Ontario, and Quebec. These hydro assets could be enhanced by additional east–west electricity linkages between provinces in strategic pairings. In its most recent report, *Re-Energizing Canada: Pathways to a Low Carbon Future*, Sustainable Canada Dialogues "examines how Canada can decarbonize its economy while remaining globally competitive" (2017, 5). The report concludes that to "succeed in the energy transition, it will be necessary to move beyond the general objectives of the [Pan-Canadian] Framework and adopt appropriate, specific policy tools and regulatory measures based on evidence and best practices. The current ambition will not allow us to reach our destination—a world that will have avoided a global temperature increment greater" than 2 degrees Celsius (Sustainable Canada Dialogues 2017, 4). Time is of the essence. Indeed, Climate Action Network Canada envisions 100 percent GHG reductions by 2050 based on growing scientific evidence that a 1.5 degrees Celsius average global temperature rise constitutes a necessary limit—a perspective shared by many of the world's developing nations (Climate Action Network Canada 2015; Comeau 2015).[2]

(3) Engaging Canadians as Empowered Agents in the Renewable Energy Economy

One of the most striking dimensions of the *Energiewende* in Germany is how many individual Germans are agents in the generation of renewable energy. The ownership distribution of installed renewable energy production for power production in Germany is not concentrated in a small number of corporations, but widely distributed among German citizens through diverse business models. In 2013, 35 percent of renewable energy capacity was in the hands of individuals, with an additional 11 percent held by farmers, for a total of 46 percent. The "big four" power providers in Germany hold only 5 percent capacity, a notable contrast (Leidreiter 2015). Moreover, the number of energy cooperatives in Germany has increased dramatically, from 66 in 2001 to 888 in 2013 (Klaus Novy Institut 2013). These statistics demonstrate how ownership of renewable energy is distributed and how it continues to proliferate in various models such as cooperatives, turning many Germans citizens into *de facto* clean energy citizens. I myself have visited two regions in Germany, each using a different business model for community ownership of renewable energy. In Rhein-Hunsrueck in southwestern Germany, the citizens and region lease lands to renewable energy corporations, with the profits being directed to both the regional government and individual landowners. In other regions, such as Steinfurt/Saerbeck in northwestern Germany, citizens become direct shareholders in their local Energy Transition Park (*Energiewendeparc*) by virtue of residing in that region, which itself owns the renewable energy with the assistance of both local and national banks.

Through an innovative and inclusive policy approach, Canadian federal and provincial governments and municipalities can strengthen the transition to Canada's new economy and energy system by empowering all Canadians to participate as citizen stakeholders and shareholders. Instead of being mere passive consumers of energy, Canadians can become agents of renewable energy as well as the drivers of a societal transition towards efficient buildings and sustainable transportation. By enabling citizens, farmers, regional cooperatives, and municipally owned utilities to invest in renewable energy and to receive a return on their investments, countries like Germany and Denmark have demonstrated that individuals and communities can transform a nation into an energy democracy. Policy can generate new business models, including energy and agricultural cooperatives, private–public partnerships between citizens and utilities, and municipally owned utilities. In Canada, these sorts of inclusive and dynamic business partnerships would share economic benefits from renewable energy, thus strengthening Canadian cities and communities, revitalizing rural and remote regions, and empowering Indigenous communities to become energy self-sufficient.

Innovative business models in Canada, such as the ComFIT (Community Feed-In Tariff) program in Nova Scotia (Nova Scotia 2014), which links renewable energy with community participation and local ownership, and the program of T'Sou-ke First Nation on Vancouver Island, show how ordinary Canadians can become empowered agents in the renewable energy economy. In September 2011, Nova Scotia launched the ComFIT program to foster community-based small- and medium-sized projects of 2 to 4 megawatts; these generated renewable electricity, besides offering stable and substantive revenue per kilowatt hour for the projects. The ComFIT program was directed at cooperatives, municipalities, universities, community groups, First Nations, combined heat and power (CHP) plants, and community economic development funds (CEDIFs). In these community-based projects, the stakeholders retain 51 percent share of the ownership, with the private sector able to own up to 49 percent.

Originally, ComFIT was envisioned as growing to generate 100 megawatts of renewable electricity. But between September 2011 and January 2014, a total of 89 projects generating 200 megawatts of energy (9 percent of Nova Scotia's total electricity generation) were constructed. Community Economic Development Funds (CEDIFs) comprise the majority of ComFIT projects, generating more than $10 million. CEDIFs strengthen local economies and create jobs in rural regions through local economic initiatives as new RE projects are built and continue to operate in these communities. ComFIT projects also increase energy security and economic diversification for communities as renewable electricity sources are geographically dispersed; they also deepen public support for renewable energy. Finally, projects developed through the ComFIT and CEDIF programs create added economic and social value in the long term.

Similarly, the T'Sou-ke First Nation on Vancouver Island, under Chief Gordon Planes, has provided leadership by integrating solar energy, retrofitting, sustainable agriculture, and land stewardship based on Indigenous knowledges and traditions (McKenna 2014). For Planes, this initiative makes perfect sense, given that "First Nations have lived for thousands of years on this continent without fossil fuels. It is appropriate that First Nations lead the way out of dependence and addiction on fossil fuels and rely on the power of the elements, the sun, the wind, and the sea once more" (T'Sou-ke Nation 2014). In partnership with Timberwest Forest Corp and EDP Renewables Canada, the T'Sou-ke Nation is developing large wind energy projects through a $750 million investment, having already completed its own community solar project of 75 kilowatts in 2009, which is distinguished by the Salish design of its solar array. The four threads of the T'sou-ke Nation's community development are (1) energy autonomy, (2) food self-sufficiency, (3) cultural renaissance, and (4) economic development (T'Sou-ke Nation 2014). T'Sou-ke Nation's project is just one of many similar initiatives among First Nations in Canada, which are weaving these four threads of development into one tapestry (Henderson 2013).

100 PERCENT RENEWABLE ENERGY: FROM HYPOTHETICAL IDEA TO PRACTICAL REALITY

Scientific and Technological Feasibility of 100 Percent Renewable Energy

For many people, the idea of 100 percent renewable energy seems wildly impractical, one that belongs in science fiction rather than in public policy. Yet a steadily broadening stream of scientific research is examining the technical feasibility of just such a goal, one that is delegitimizing and strategically relativizing the current hegemonic carbon-intensive energy systems that dominate economies today. Agents of the latter have forfeited their social authority in this time of climate destabilization and environmental degradation. The new research is visionary in that it expands the possibilities of existing innovations and implementations of sustainable energy in three key areas: electricity, heating/cooling, and transportation.

The history of this research is longer than one would think. It began with the revolutionary research project "Solar Sweden: An Outline to a Renewable Energy System" (1977), which postulated the feasibility of a 100 percent renewable energy system in Sweden by 2015 (Scheer 2007, 50–51). This was followed a year later by a report from France's La groupe de Bellevue titled "ALTER—a Study of a Long-Term Energy Future for France based on 100% Renewable Energies," which detailed how 100 percent renewable energy could constitute an energy system by 2050. Shortly after this, in 1980 in the United States, the Union of Concerned Scientists launched its own research project, "Energy Strategies: Toward a Solar Future," which demonstrated the same potential outcome: 100 percent renewable energy for the United States by 2050. This was followed by a similar study in 1982 for Western Europe by 2100 (Scheer 2007, 50–51).

In the years since, many comparable studies have followed (see the list in Scheer 2012, 28–33), with researchers in Germany playing a significant role. In 2002 the German Bundestag (federal parliament) asked Dr. Harry Lehmann to complete a study suggesting how a 95 percent renewable energy system could be constructed by 2050. In 2007, attention became focused in part at the municipal and regional levels—a EUROSOLAR scenario demonstrated how the German state of Hessen would be able to reconfigure its energy system on a renewable energy platform. Two years later, in 2009, a book focusing on 100 percent RE in municipalities, regions, and rural districts appeared, with profiles of many cities and regions that either already had completed a transition to 100 percent RE or were making significant progress towards this goal (Droege 2009).

In 2010, research into the technical feasibility of 100 percent renewable energy became more sophisticated and detailed through the work of scientists and technicians in Europe, especially in Germany. Three research groups led the way: (1) the German Advisory Council on the Environment (SRU), (2) the German Renewable Energy

Research Association (FVEE), and (3) the German Federal Environment Agency. Not to be outdone, Greenpeace commissioned its own study, Energy (R)evolution (2010), which posited that projected global energy needs in 2050 of 13.2 terawatts could be provided by a 94.6 percent renewable energy system (Scheer 2012). Since then, the research has become increasingly sophisticated and visionary.

MARK JACOBSON AND THE SOLUTIONS PROJECT

Research into the technical feasibility of 100 percent RE has not been limited to Europe. It is emerging in the United States as well, led by Mark Jacobson, an atmospheric scientist and professor of civil and environmental engineering at Stanford University. Combining attention to scientific detail with skills in bringing together research teams, Jacobson first surfaced in the public domain in 2009 with a co-authored article in *Scientific American* contending that all energy could be derived from 100 percent RE based on wind, water, and solar (WWS) sources by 2030 (Jacobson and Delucchi 2009). More detailed scientific research followed that examined the technologies, energy resources, and quantities and areas of infrastructure for 100 percent RE (Jacobson & Delucchi 2011) and provided reliability estimates, system and transmission costs, and enabling policies (Delucchi and Jacobson 2011). Besides calling for the electricity grid to be reconfigured with renewable energy, Jacobson extended his research to the heating and cooling of buildings as well as to sustainable transportation and mobility, thus offering an integrated tripartite model of sustainable energy systems. In this model, Jacobson envisioned the electrification of all energy sources, with renewable electricity providing the needed technological platform for the two other major sectors of the energy economy: the heating/cooling of buildings and built environment in cities, and the transportation sector.

Jacobson has recently asked his research team to take a nuanced and comprehensive look at the feasibility of converting the energy systems of New York State into a resilient and reliable infrastructure based on wind, water, and solar (WWS) (Jacobson et al. 2013). The transition to 100 percent RE based on WWS would reduce health costs and climate damage, generate jobs and economic development, and strengthen NYS's energy self-sufficiency. End-use demand for energy would be reduced by 37 percent, and job creation would significantly exceed job losses. Of particular interest is the expanding perspective that Jacobson and his colleagues developed by including significant co-benefits. For instance, deaths in NYS would drop dramatically, with a reduction of 4,000 deaths/year related to air pollution. In addition, health costs would drop by $33 billion per year (3 percent of the GDP of NYS in 2010), an amount that could finance the installation of 271 gigawatts of required 100 percent RE power infrastructure within seventeen years. This study concluded that the decrease in emissions would reduce projected US climate change costs for 2050 (damage from extreme weather, erosion of coasts, etc.) by $3.2 billion per year. By linking renewable energy technical studies with inquiries into significant health improvements and the reduction of health costs, Jacobson and colleagues signalled a significant shift in the template for future research.

100 Percent Renewable Energy Regions in Germany: Translating Scientific Research into Pragmatic Solutions

Research by Jacobson, Lehmann, and others into the scientific feasibility of 100 percent RE is crucial. These studies, besides mapping the specific technical requirements for transitioning the energy infrastructure to 100 percent RE, have solidified an economic alternative to traditional hydrocarbon-based energy. Future research will have to be translated into practical solutions, and concrete applications will have to be calibrated for specific ecological and cultural settings. Also, it will continue to be crucial to communicate this research so as to enable education and community engagement, both of which can transform energy infrastructures. But are there any actual working models of regions or communities that have already made the transition to 100 percent RE and that could provide concrete examples and templates for this energy transition?

The answer is yes—and it can be found in the 100 percent renewable energy regions (*Institut dezentrale Energietechnologien*), a network of 136 regional governments, cities, rural districts, and local communities and economies in Germany, which includes 21.2 million citizens or 26 percent of the population and which spreads over a combined area of almost 106,000 square kilometres (29.8 percent of Germany's land mass). This vast network was supported financially from

2007 to 2014 by the Germany's Environment Ministry, with technical advice and scientific support provided by the Federal Environment Agency. The project was centred in the IdE Institute of Decentralized Energy Technologies, based in Kassel, and was carried forward by staff at the University of Kassel led by Dr. Peter Moser.

As of April 2013, the 136 municipalities and districts in this network include (a) 73 regions that are already certified as utilizing 100 percent renewable energy, (b) 60 "starter regions" that have been assessed and accepted into the certification process, and (c) three urban regions with detailed action plans for achieving 100 percent renewable energy (Frankfurt, Osnabrueck, and Rostock). Some municipalities and regions in this network have been included in a European program that intertwines programs with 14 partners from 10 European countries in a collaborative implementation of their sustainable energy action plans (SEAPs) (Moser 2013). For several years, the network also organized an annual convention of 800 participants, facilitating the exchange of technical knowledge for transitioning to 100 percent RE and strengthening partnerships across economic, political, and geographic boundaries.

The contribution of this network in Germany cannot be underestimated. Combined with the research of Jacobson, Lehmann, and other scientists into 100 percent RE systems and infrastructure (providing both the technical resources required for transition and policy pathways for how this transition can be accomplished), it makes it clear that a transformation of energy systems to 100 percent RE is limited neither by science nor costs but only by political will and public support.

CANADA'S CLIMATE CHANGE AND RENEWABLE ENERGY STRATEGY

Based on the ethical imperative of limits (Lysack 2015b; 2008), what would be the most effective policy mix for Canada to adopt as a climate change and renewable energy strategy? Here are some key building blocks for an effective renewable energy and environment/climate protection policy strategy for Canada:

1 Establish 1.5 degrees Celsius as the target for average global temperature rise from pre-industrial levels below which humanity needs

to aim, rather than the target of 2 degrees Celsius. A consensus has been growing among scientists (Rogelj et al. 2015) as well as many political leaders, especially in the developing world and emerging economies, that 1.5 degrees needs to be the target for the earth, in accordance with the aspirations of the Paris climate change commitments. Using Lenton's (2011) scientific mega-analysis of global tipping points as well as other scientific findings one sees evidence that several elements of the planetary system are already heading towards or crossing dangerous and irreversible tipping points (Spratt 2015). For instance, retaining more than 10 percent of coral reefs while avoiding rapid increases in permafrost melt will require a global temperature rise limit of 1.5 degrees Celsius. While science has established that the planet has a limited carbon budget (Meinshausen et al. 2009), research by investment analysts and economists highlights that humanity can burn only a limited amount of fossil fuels, making upwards of 80 percent of current fossil fuel reserves stranded assets (Leaton 2012; McGlade and Ekins 2015).

2 Develop strategies of environment and climate protection to protect Canada's land, water and oceans, climate, and ecological communities. Protection of Canada's forests, water quality, land and marine ecosystems, and biodiversity is both an intrinsic valuing of these planetary life-support systems and a purposive addressing of climate change, one that involves using the planet's natural systems for carbon sequestration and pollution diminishment (Potvin et al. 2015, 43–46).

3 Advance a national strategy to transition to 100 percent renewable electricity in Canada by 2035 (Potvin et al. 2015).

4 Set a long-term goal to transition to 100 percent renewable energy or full decarbonization by at least 35 percent below 2005 levels by 2025, and 100 percent by 2050 (Comeau 2015; Climate Action Network Canada 2015).

5 Draw on Germany's *Energiewende* as a model for developing an integrated strategy to transition the three sectors of energy (renewable electricity, energy efficiency and heating/cooling of buildings and cities, and sustainable transportation) to renewable energy and zero emissions status (Leidreiter 2015; Couture and Leidreiter 2014).

6 Set five-year incremental commitment periods with opportunities for ratcheting up climate protection and energy transition plans, setting

the first target date as 2025 (Comeau 2015; Climate Action Network Canada 2015).

7 Develop frequent energy and emissions assessments, with robust MRVs (measurement, reporting, verification), of Canada's progress towards renewable energy (Comeau 2015).

8 Initiate robust stimulus incentives to encourage immediate reductions in GHG emissions so that Canada accelerates its transition and contributes its fair share in the world community (Comeau 2015; Climate Action Network Canada 2015).

9 Introduce and pass federal legislation for rapidly phasing out all coal-generated electricity and energy across Canada, including especially Alberta, Saskatchewan, New Brunswick, and Nova Scotia (Comeau 2015; Lysack 2015a).

10 Develop an approach to pricing carbon in Canada that includes a minimum floor price and schedule. The latter would include specific timelines for reducing GHG emissions (extending to 2025 and 2030) as well as a cap limit on Oil Sands emissions. This approach would need to be fair to all provinces (Comeau 2015) and congruent with Canada's carbon budget (Meinshausen et al. 2009).

11 Establish independent and transparent monitoring and verifiable measuring of toxins from energy generation with strict MRVs by an independent body embedded in the Office of the Auditor General of Canada, modelling the independent body of the respected Climate Change Committee in the UK (https://www.theccc.org.uk).

12 Eliminate subsidies for fossil fuels as soon as possible and redirect funding towards fostering renewable energy and infrastructure, environment and climate protection, and clean tech.

13 Over the next five years, contribute $4 billion in international trade support for developing countries for climate change mitigation and adaptation (Climate Action Network Canada 2015).

14 Drive innovation and research in the Canadian economy to support energy transition and environment/climate protection through the federal budget using Green Bonds, taxation, infrastructure spending, and trust funds (Comeau 2015; CANC 2015).

CONCLUSIONS

We have examined how both Germany and certain jurisdictions in Canada, such as the T'Sou-ke First Nation in BC and the province of Nova Scotia with its ComFIT initiative, have sought "to exercise agency on behalf of natural environments" (Trono and Boschman 2018, 5). But Canada needs leadership at the federal level if it is to pursue effective climate/environmental protection and a transition to a renewable energy economy and society. If the fourteen policy pillars and targets outlined above are enacted by political leadership in Canada, they will, based on extensive, documented research and on Germany's ongoing demonstrated policy experience, move Canada decisively towards fulfilling its COP 21 climate change commitments made in Paris in December 2015 as an enactment of future-oriented environmental right action.

Thus far, according to the most recent audit by Julie Gelfand, the Commissioner for the Environment and Sustainable Development (2017) in the Office of the Auditor General of Canada, in her "two reports, which focused on climate change ... the government's efforts to reduce greenhouse gas emissions have fallen short of its target and ... overall, it is not preparing to adapt to the impacts of climate change. Only five of 19 government organizations [have] fully assessed their climate change risks and acted to address them. The rest [have] taken little or no action to address risks that could prevent them from delivering programs and services to Canadians."

At this point, although thus far the Trudeau government seems focused more on style than on substance, and on rhetoric rather than effective action, there is still an opportunity for it to ratchet up its ambition and fulfill its Paris Climate Change Commitments. And, to repeat, time is of the essence. As Paul Huebener states in the opening chapter of this volume: "The environmental crisis is, in many ways, a crisis of time: from the apparent inability of societies to act with a clear sense of future responsibility, to the tensions that exist between accelerating resource use and the need for coherent ecosystems, to the anxious realization that time, for many humans and other members of the ecosphere, will soon run out, if it hasn't already" (2018, 1). It remains to be seen if it will be the Trudeau government, or another, that will set forth this exemplary and urgently needed leadership. Much is at stake.

Notes

1 As a social scientist, I am a member of Sustainable Canada Dialogues.
2 Potvin's group argues for a minimum floor of 80 percent GHG reductions by 2050 (2015, 8).

References

Anderson, Daniel Gustave. 2014. "Matter Matters: The Significance and Problems of the Heidegger Debate in *ISLE.*" *ISLE: Interdisciplinary Studies in Literature and Environment* 21(3): 600–605.

CEPS. 2014. *Impact of the Decarbonisation of the Energy Industry on Employment in Europe*. CEPS Special Report no. 82. Brussels.

Climate Action Network Canada. 2015. "Three Big Moves toward a 100% Renewable Energy System for Canada." https://climateactionnetwork.ca/2015/11/18/three-big-moves-toward-a-100-renewable-energy-system-for-canada/.

Comeau, Louise. 2015. "Alberta climate plan needs federal support and leadership so we do our fair share." http://climateactionnetwork.ca/2015/11/22/alberta-climate-plan-needs-federal-support-and-leadership-so-we-do-our-fair-share.

Commissioner of the Environment and Sustainable Development. 2017. Fall Reports of the Commissioner of the Environment and Sustainable Development to the Parliament of Canada. http://www.oag-bvg.gc.ca/internet/English/mr_20171003_e_42600.html.

Couture, Toby, and Anna Leidreiter. 2014. *Policy Handbook: How to Achieve 100% Renewable Energy*. Hamburg: World Future Council.

Delucchi, Mark, and Mark Jacobson. 2011. "Providing All Global Energy with Wind, Water, and Solar Power, Part II: Reliability, System and Transmission Costs, and Policies." *Energy Policy* 39: 1170–90.

Doubleday, Nancy C. 2019. "Culture as Vector: Agency for Social-Ecological Systems Change." In *On Active Grounds*, edited by Robert Boschman and Mario Trono. Waterloo: Wilfrid Laurier University Press.

Droege, Peter. 2009. *100% Renewable*. New York: EarthScan.

Elgie, Stewart. 2015. Comments in panel presentation in Ottawa Climate Talks, 23 November.

European Commission. 2014. *A Policy Framework for Climate and Energy in the Period from 2020 to 2030. Impact Assessment*. SWD 15. Brussels.

Federal Ministry for Economic Affairs and Energy, Government of the Federal Republic of Germany. 2015. *Power Upgrade 2030: The New Economic Rationale for an Ambitious EU Climate and Energy Framework*. Bonn.

Fleck, Bertram. 2014). Personal conversation. http://www.ucalgary.ca/oikos/node/252.

Hawken, Paul, ed. 2017. *Drawdown: The Most Comprehensive Plan Ever Proposed to Reverse Global Warming*. New York: Penguin.

Henderson, Chris. 2013. *Aboriginal Power: Clean Energy and the Future of Canada's First Nations*. Erin: Rainforest Editions.

Huebener, Paul. 2019. "The clock's wound up": Ecocritical Time Studies as a Response to Social Acceleration and Ecological Collapse." In *On Active Grounds*, edited by Robert Boschman and Mario Trono. Waterloo: Wilfrid Laurier University Press.

Institut dezentrale Energietechnologien. "100% Erneuerbare-Energie-Regionen (100% renewable energy regions)." http://www.100-ee.de.

IRENA. 2014. *Renewable Energy and Jobs: Annual Review 2014*. Abu Dhabi.

Jacobson, Mark, and Mark Delucchi. 2009. "A Path to Sustainable Energy by 2030." *Scientific American* 301(5): 58–65.

———. 2011. "Providing All Global Energy with Wind, Water, and Solar Power, Part I: Technologies, Energy Resources, Quantities, and Areas of Infrastructure and Materials." *Energy Policy* 39: 1154–69.

Jacobson, Mark, et al. 2013. "Examining the Feasibility of Converting New York State's All Purpose Energy Infrastructure to One Using Wind, Water, and Sunlight." *Energy Policy* 57: 585–601.

Kemfert, Claudia. 2013. *The Battle about Electricity: Myths, Power, and Monopolies*. Hamburg: Murmann.

Klaus Novy Institut. 2014. "Energy cooperatives in Germany." www.renewables -in-germany.com.

Leaton, James. 2012. *Unburnable Carbon: Are the World's Financial Markets Carrying a Financial Bubble*. March 2012. http://www.carbontracker.org/ report/carbon-bubble.

Leidreiter, Anna. 2015. *Achieving an Optimal Policy Framework for 100% Renewables*. Hamburg: World Future Council.

Lenton, Tim. 2011. "Early Warning of Climate Tipping Points." *Nature Climate Change* 1: 201–9.

Lysack, Mishka. 2008. "Global Warming as a Moral Issue: Ethics and Economics of Reducing Carbon Emissions." *Interdisciplinary Environmental Review* 10(1–2): 95–109.

———. 2015a. "Effective Policy Influencing and Environmental Advocacy: Health, Climate Change, and Phasing Out Coal." *International Social Work* 58(3): 435–47.

———. 2015b. "The Ethical Imperative of Limits." *Policy Options*, January–February, 23–26.

McGlade, Chris and Paul Ekins. 2015. "The Geographical Distribution of Fossil Fuels Unused When Limiting Global Warming to 2 Degrees C." *Nature* 517, 8 January, 187–90. doi:10.1038/nature14016.

McKenna, Cara. 2014. "T'Sou-ke First Nation Turns to Wasabi in Renewable Energy Push." *Globe and Mail*, 24 August. https://www.theglobeandmail.com/news/ british-columbia/tsou-ke-first-nation-turns-to-wasabi-in-renewable-energy -push/article20187542/.

Meinshausen, Malte, et al. 2009. "Greenhouse-Gas Emission Targets for Limiting Global Warming to 2°C." *Nature* 458, 30 April 2009. doi:10.1038/ nature08017.

Morris, Craig, and Paul Hockenos. 2016a. "Energy Transition." https://book
.energytransition.org/sites/default/files/2016-12/ET_Germany%20reaches%20
30%20percent%20renewable%20power%20in%202015-.png.

Morris, Craig, and Paul Hockenos. 2016b. "Energy Transition." https://book.energy
transition.org/sites/default/files/2016-12/ET_Renewables%20create%20
more%20jobs%20than%20coal%20power%20does-.png.

Moser, Peter. 2013. "100ee Regionen in Deutschland, Europa und der Welt. http://
www.100-ee.de/fileadmin/redaktion/100ee/Downloads/broschuere/Good
-Practice_Broschuere_Inhalt_Web.pdf.

Nova Scotia Department of Energy. 2014. COMFIT Program. http://energy.nova
scotia.ca/renewables/programs-and-projects/comfit.

Pollin, Robert, et al. 2008. "Green Recovery: A Program to Create Good Jobs
and Start Building a Low-Carbon Economy." Center for American Progress,
Washington, DC.

———. 2009a. "The Economic Benefits of Investing in Clean Energy." Political
Economy Research Institute, University of Massachusetts, Amherst.

———. 2009b. "Green Prosperity: How Clean-Energy Policies can Fight Poverty
and Raise Living Standards in the United States." Political Economy Research
Institute, University of Massachusetts, Amherst.

Potvin, Catherine, et al. 2015. *Acting on Climate Change: Solutions from Canadian Scholars.* http://www.sustainablecanadadialogues.ca/en/scd/endorsement.

Rogelj, Joeri, Gunnar Luderer, Robert C. Pietzcker, Elmar Kriegler, Michiel Schaeffer, Volker Krey, and Keywan Riahi. 2015. "Energy System Transformations
for Limiting End-of-Century Warming to Below 1.5°C." *Nature Climate
Change* 5: 519–27. doi:10.1038/nclimate2572.

Scheer, Hermann. 2007. *Energy Autonomy: The Economic, Social, and Technological Case for Renewable Energy.* Bath: Bath Press.

———. 2012. *The Energy Imperative: 100 Per Cent Renewable Now.* New York:
EarthScan.

Schroeder, Randy, and Kent Schroeder. 2019. "Declarations of Interdependence:
Unexpected Human–Animal Conflict and Bhutanese Non-Linear Policy." In
On Active Grounds, edited by Robert Boschman and Mario Trono. Waterloo:
Wilfrid Laurier University Press.

Spratt, David. 2015. "It's Time to Do the Math Again." http://reneweconomy
.com.au/2015/time-to-do-the-math-again-allowing-2c-global-warming-is-mad
ness-56033.

Sustainable Canada Dialogues. 2017. "Re-Energizing Canada: Pathways to a Low
Carbon Future." http://www.sustainablecanadadialogues.ca/en/scd/energy.

T'Sou-ke Nation. 2014. "T'Sou-ke Going Green … Really Green." http://www
.tsoukenation.com/tsou-ke-going-green-really-green.

Umweltbundesamt. 2015. "Environmental Trends in Germany: Data on the
Environment 2015." 36. https://www.umweltbundesamt.de/sites/default/files/
medien/378/publikationen/data_on_the_environment_2015.pdf.

Urge-Vorsatz. 2010. *Employment Impacts of a Large-Scale Deep Building Energy Retrofit Programme in Hungary.* Budapest.

Wei, Max, Shana Patadia, and Daniel M. Kammen. 2010. "Putting Renewables and Energy Efficiency to Work: How Many Jobs can the Clean Energy Industry Generate in the US?" *Energy Policy* 38:2. Amsterdam: Elsevier.

Wynne, Premier Kathleen. 2014. "2014 Mandate letter: Environment and Climate Change: Premier's Instructions to the Minister on Priorities for the Year 2014." https://www.ontario.ca/page/2014-mandate-letter-environment-and-climate-change.

CHAPTER 12

Culture as Vector: (Re)Locating Agency in Social-Ecological Systems Change

Nancy C. Doubleday

WICKED PROBLEMS: A PROLOGUE

The health and orderly use of the oceans and their tributaries are not new concerns. Science, law, and policy have devoted centuries of human effort to ocean and riverine management, beginning with concerns about freedom of the seas and territorial claims with the advent of the age of Western exploration and imperialism. Decades of international meetings, and the agreements they have produced, have been endorsed. Yet still the situation appears fraught: whether we are concerned for peace, social justice, health, or sustainability, our oceans appear to be in decline. Not everywhere—yet—but in enough places and at sufficiently large a scale to imply that trouble may await.

One recent post from a social action site presents a stark summary of this concern:

Dear friends,

When seasoned sailor Ivan Macfadyen returned from his last Pacific crossing he raised an ominous alarm:

"I'm used to seeing turtles, dolphins, sharks and big flurries of feeding birds. But this time, for 3000 nautical miles there was nothing alive to be seen."

This once vibrant expanse of sea was hauntingly quiet, and covered with trash.

Experts are calling it the silent collapse. Although very few of us see it, we are causing it—overfishing, climate change, acidification, and pollution are devastating our oceans and wiping out entire species. It's not just the annihilation of millennia of wonder and beauty, it impacts our climate and all life on Earth.

(personal communication, Emma Ruby-Sachs, avaaz@avaaz.org, 26 March 2015)

The same might be said of our global forests, our climate, our fresh water, and of all the minutiae that compose the living systems that sustain us. To confront this reality is an act of courage. To position "Environmental Humanities" against the question of futures, as this volume does, is in itself an act of courage. In this chapter we parse "agency" (conceptually rather than grammatically) within a framework of self-efficacy or capacity, but as an emergent property of dynamic cultural systems and as a force of quite a different order, rather than as individualistic "agency," customarily situated within a static, conservative view of culture. Through the contemporary heterogeneity of diverse practices and competing beliefs, we glimpse the possibility of an opening, or cultural shift—in brief, the recognition of an emergent "culture vector" capable of facilitating trajectory change in social-ecological systems. What follows is an attempt to unfold, illustrate, and ground this notion.

INTRODUCTION

If we wish for the capacity to develop alternative scenarios for the future that support equitable social-ecological resilience, and seek to avoid "unthinkable catastrophes"[1] whenever possible by anticipating and adjusting course proactively, then arguably we are at a point where we need to advance our conceptual, technological, and communicative approaches and deepen our collaborative abilities across scales. This will require of us deeper listening (Zedalis and Gould, 294), from landscape harmonies to the droning and howling of the unfamiliar (see Feisst, Chapter 3). In terms of preparing for the future, regardless of scale, first it is necessary for us to envision our desires, as well as their consequences (see Rekow, Chapter 6), and to understand the potential for the existence of other desires, human and non-human, thoughtfully considered in the chapters by Banting (Chapter 8) and Armbruster (Chapter 9). Much contemporary effort in social-ecological systems research is directed towards issues of participatory scenario modelling, including considerations of access, asymmetries of power, and equitable outcomes (e.g., Kok, Biggs, and Zurek 2007; Prell et al. 2007; Peterson, Cumming, and Carpenter 2003).

Similarly, if we wish to aspire to new futures, such as those envisioned for energy by Lysack (Chapter 11), besides visioning those futures lucidly, we must learn to act in new ways. There is a growing social consensus[2] in support of recognizing that the peace–health–justice–sustainability nexus requires interdisciplinarity in order to flourish, for two reasons: first, to provide context for wicked problems across scales, both temporal and spatial; and second, as a source of community formation and validation necessary to support the development of agency in individuals and groups. Visioning is empowered by the integration of knowledge from diverse sources and in novel configurations. Acting to realize new and unfamiliar outcomes requires, at a minimum, self-efficacy, capacity, and a sense of possibility. There are indications of the coalescence of this desire for new ways of being, ways that provide greater opportunities for peace, justice, health, and sustainability, in movements of all kinds, across scales. Visioning and acting towards new futures requires access to collaborative community in some form (Gibson-Graham 2003, 2010), as a source of diversity, as a place for safe-fail trials, and as a source of validation and empowerment (Patterson, Cumming, and Carpenter 2013). Valuing novelty, and the means for creating it, matters. Trono (Chapter 2) offers a possible cinematic entry point for engagement in the co-creation of new imaginaries, including the montagist re-creation of "commons" from proprietary images. This is a novel, creative, and inherently validating concept.

Clearly it is important to ensure that the foundation for enabling collaboration and cooperation is adequate and diverse. This respect for diversity of collaborative spaces percolates through this volume, underscoring the understanding that territory itself has the capacity to be agentive (see Boschman, Chapter 4) and fluid—literally, in the case of slumping permafrost on the Arctic coast.

At times it appears that collectively we possess more than enough knowledge to inform our choices and decisions about complex problems but lack the capacity to put what we know together in a conceptually and ultimately politically compelling way. Quite possibly this is because our conceptual framing is not sufficiently rich and robust. In particular, individual citizens often struggle with a sense of impotency and inadequacy in the face of the behemoths of social, ecological, and cultural change, such as globalization, war, and monumental inequality and social and environmental injustice. The tried and true response is a call for collective action, and this has been shown to be effective in many contexts. Difficulties in achieving change through collective action seem to arise where disparities of power and asymmetries of impact and benefit align against those perceived as powerless in a given situation. But what of those situations described as "wicked problems," where forces with intractable, perverse, and inexorable impacts bear down on individuals, communities, peoples, and entire ecosystems?

What can we learn about how we think about these situations in order to change our approach to our human project as a whole, in order to redirect possible future outcomes by changing our trajectory? Can we acquire new skills for "vectoring," as described by Trono and Boschman in the introduction to this volume?

Is it a gradual failure to destroy northern human and ecological communities and to irreversibly contaminate northern waters? Is the death of a few hundred ducks in a waste water pool as significant as the slow, incremental, and inexorable removal of a vast area of habitat essential for all migratory birds of North, Central, and South America along the Western Flyway during their annual migration? Is it a slow catastrophe to undermine efforts to restore the watersheds of the Great Lakes, to further threaten biodiversity, to increase the toxic body burden of the majority of people living in Ontario, to cast an industrial shadow over (and remove from agricultural production) 7,000 hectares of field and woodland in Great Lakes Basin? Is it acceptable democratic practice to add to the toxic loads in a region already identified for remediation and clean-up? What do we need to learn if we fail to understand that all of these impacts are connected to the fate of the ocean, which is the ultimate integrator of all human activity through time?

SYSTEMS AND CONCEPTS: PATHS TO CHANGE

In terms of bringing about conceptual change, collective efforts clearly matter, and hence the importance of collaborative work, meetings, and writing. Earlier collaborative achievements in conceptual development related to resilience underscore this point. For example, we have recognized the linkages that exist, particularly in terms of scale and system dynamics (Gunderson, Holling, and Light 1995); and we have established incontestably the integrated character of social–cultural–ecological systems, and the need to address them as such (e.g., Berkes and Folke 1998). We have envisioned systems of systems as "panarchies" of systems and adaptive cycles, offering alternative metaphors for organization and change (Gunderson, Holling, and Peterson 2002). We have pressed forward in social learning by learning from experience at boundaries and across divides (Berkes et al. 2005). We have learned about how we reconstruct our world in the process of reconstructing ourselves and our behaviours, as reported by Susemihl, for example (page 155 of this volume), and elsewhere by Westley, Zimmerman, and Patton (2007) and Doubleday (2007). We have also learned that when we engage in a new synthesis of co-management and adaptive management, "the results of the convergence open up some additional opportunities" distinct from each precursor (Berkes, Armitage, and Doubleday 2007). More than this, we are beginning to see the emergence of systematic approaches to support adaptive management (Patterson, Smith, and Bellamy 2013.)

These are all fruitful hybrids yielding new opportunities through collaborative strategies. They also point to the next steps and new syntheses: the big questions about how collaboration occurs; how communication can be effective in co-creating collaboration; and how the ever-present challenge of reconceptualizing the human project may be met in order to acknowledge and accommodate new insights into thinking about the health and well-being of the social–cultural-ecological systems that comprise our world.

THE CULTURE VECTOR

This chapter takes on a small but arguably significant task in attempting to clarify the possibilities for understanding culture as *action* as well as *object*, thus as a *culture vector* possessing a capacity for agency within social-ecological systems (SESs). This reconception is employed as a device for recognizing that culture has multiple modes, playing diverse roles ranging from the conservation and transmission of human knowledge to the defining of collective purposes and behavioural norms. It stems from observations of multiple sites of engagement from diverse disciplinary domains (e.g., Doubleday, Mackenzie, and Dalby 2004), varying with cross-scale changes in social-ecological relations and occurring in different locations and at different times, yet implicitly linked both by specified and unspecified *a priori* connections.[3]

The idea that we can look across time and space (for example, as Huebener does in Chapter 1 of this volume) and see causal and/ or correlative relations among important issues of the day and events and conditions in other places and other times, leads directly to new possibilities for our capacity to anticipate and address slow changes in our own time, across scales that are both spatial and temporal. Such slow changes could include, for example, loss of biological and cultural diversity, violation of human rights, climate change, deforestation, child poverty, toxic pollution, privatization of public goods, and the economic exploitation of the vulnerable, all leading to loss of collective confidence and to cumulative social–cultural–ecological–economic disorder. The Bhutanese examples discussed by Schroeder and Schroeder (see Chapter 11) offer concrete illustrations of unintended consequences and, as such a reminder of the value of humility in charting changes. At some point, these slow changes may move beyond disorder and increasing entropy to produce chaotic conditions that could legitimately be called catastrophic.

By introducing the prospect of a dynamic approach to culture within social-ecological systems, we are not abandoning our understandings of the significance of the interrelation of identity, collectivity, and place. These relations have established their importance and vitality as creative forces in many ways, most significantly perhaps in the emergence of the ecological knowledge of Indigenous peoples. Just as Inuit have asserted that they do not wish to be understood as "pickled Eskimos,"[4] we adopt an adaptive stance here and advocate for its central role in the articulation of a *culture vector*. By problematizing slow change across scales, and invoking culture as an informed source both of relevant knowledge and of agency for action, from communication to diversification to collaboration to innovation to transformation, and all places in between, we do two things in relation to wicked problems: (1) we deepen our understanding as to how early warnings of slow catastrophe might appear in terms of social-ecological change; and (2) we look for those elements of collaborative processes and emergent strategies that might help us anticipate slow catastrophe, and perhaps, in the best of worlds, avert some of it.

By looking to emergent expert strategies for analysis and intervention, such as resilience thinking about systems of systems (panarchies), participatory modelling, and communicative planning, and coupling these with recognition of the value of inclusion and diversity in processes such as adaptive co-management, open source social innovation, and collaborative interventions, we may just engender alternatives for the diversification of existing patterns of crisis-to-catastrophe management, conventional command-and-control reactivity, and unilateral political responses.

PANARCHY AND THE ADAPTIVE CYCLE: HEURISTICS FOR MAKING CHANGE

In 2002, Holling and colleagues identified two important paradoxes that contribute ultimately to stasis in systems and resistance to integration: pathology in development and management at regional scales, and the paradox of the expert. The insertion of a *culture vector* creates space for new options in this connection that potentially hold alternatives to paradox: by taking culture as both object and action—that is, as a construct capable of holding a continuum of identities, places, events, traumas, and beliefs, not all of which are in mutual accord, yet all of which are inevitably rich in diversity. As action, culture is arguably capable of conveying an implicit possibility of change and transformation, and learning and self-reflection, operating across scales of human organization as well as across temporal and spatial scales, and accessible to us as individuals and as collectivities, to examine, challenge, and redeploy our capacities in the context of wicked problems.[5]

Drawing from earlier work by Gunderson, Holling, and Light (1995), Holling and colleagues proposed *panarchy* (as the antithesis of hierarchy) as a paradigm to better address integration. They then targeted cross-scale dynamics of system change and cross-discipline approaches to complex systems where interlinked processes determine outcomes (in Gunderson, Holling, and Peterson 2002). Pursuing this strategy in search of a theory of adaptive change generates opportunities to innovate. Taken as a starting point, three interrelated and emergent understandings create the potential for complexity and novelty: there is, first, the reality of the interlinked nature of our worlds (Barabási 2003, Csermely 2006); second, the acknowledgement that we construct these worlds from divergent perspectives (Holling et al. 2002); and third, the pressing need to rethink the consequences of our understanding of the conventional economy–equity–environment triad. The wider implication of taking a panarchical approach is that by stepping away from the standard orthodoxy imbedded in the hierarchical ordering of relationships and knowledge and power, we acquire greater freedom to integrate our silos of knowledge and experience, to effectively collaborate in planning for desired futures, to recognize gaps and blockages, and to better anticipate incipient crisis and potential catastrophe.

Building on the adaptive cycle metaphor (derived originally from studies of ecological succession and disturbance) is a useful strategy because when we speak of slow catastrophes, we are concerned with future states and gradual failures. The adaptive cycle metaphor also allows for thinking about system change through time and across scales and presumes that system change (but not its rate) is a constant. Thinking at multiple scales and at a range of time periods is necessary in order to characterize the scope and magnitude of large-scale change whether from development or from cumulative effects. For functional evidence of the value of a multi-scalar view of time, see Huebener (Chapter 1).

SOCIAL INNOVATION, INCLUSION, AND DIVERSITY: POWER AND PROCESS

Communicative planning is another informative innovation. It evolved in part from the participatory approaches of the 1970s and 1980s into a hybrid model based on collaboration and on interest-based behaviours, characterized by networks of influence and knowledge and by the formation of epistemic groups around issues (Hillier 2002). This "cultural turn" in planning enriches earlier spatial and rational traditions by diversifying the range of actors and factors considered, as well as extending the field of participation to be more inclusive of those considered to be "at the margins" (Hillier 2002). While an exhaustive review is well beyond the scope of this chapter, it must be noted that planning practitioners have also formulated critiques (e.g., McGuirk

2001), as a response to the inability of consensus alone to constrain power where interests are vested. Elsewhere, asymmetries of power have been seen as defeating co-management initiatives (e.g., Nadasdy 2007). Alternatively, from the standpoint of applying the metaphor of the adaptive cycle, asymmetries of power have also been understood as potential drivers of self-organization and system change that can be harnessed in adaptive co-management (Doubleday 2007). Resilience thinking offers a systems approach to complexity that allows for consideration of spatial and temporal change and adaptation (Gunderson, Holling, and Peterson 2002; Westley 2002) and of surprise (Janssen, 2002).

Here resilience thinking is invoked on two fronts: in order to better understand the processes of social-ecological change, and to extend the reach of communicative planning and other collaborative processes to include a proactive vision of the ecological and cultural domains as open systems. By establishing a socially innovative discourse modelled on open source technology that is relevant to communicative planning and related processes and by using the adaptive cycle metaphor of Holling, Gunderson, and Ludwig (2002, 41), it becomes possible to conceptualize processes for collaborative planning and decision-making for system change. This strategy differs from previous attempts to problematize social processes of decision-making in the context of system change, such as that of precaution (e.g., O'Brien 2000; and others), which transforms debates about science into questions of environmental justice and environmental assessment (e.g., Dietz and Stern 2008, and others), and which casts public participation as a more appropriate category for advancing democratic action than citizenship-based rights or collaboration through adaptive co-management. What is different here is the starting point: a focus on adaptive, collaborative practices and strategies applied to planning and systems design within a paradigm that both assumes system integration and change and aims for resilience, while giving standing to individual and collective participatory influences by interposing the concept of a culture vector at a boundary membrane in social-ecological systems.

From a communicative planning perspective (but not unique to it), critiques have pointed out the heterogeneity of distribution of power and access to resources (including information), and how the role of planners and other experts may impede consensus, and have suggested that alternatives to consensus-based models are needed. This concern mirrors the paradoxes of Gunderson and Holling referred to above. Here we might take lessons from open source innovations and adaptive co-management in order to develop communicative planning strategies for engaging constructively with difference and power. One useful example from open source literature is the constellation model, proposed by Surman and Surman (2008) and developed by the Canadian Partnership for Children's Health and the Environment. Here, instead of assuming homogeneity of interests where goals are shared, diversity was a starting point and active steps were taken to recognize and work with it, and to share power and control. This may be a way of managing the "consensus problem" raised by McGuirk: each interest group was free to advance its approach and position according to its relative ability, subject to shared goals of improving children's health and the environment. Similarly, adaptive co-management can be used to frame a practical vision of hybrid, flexible, negotiated governance, primarily viewed as bound to places and/or issues, and relevant at local or regional scales.

TOOLS FOR COLLABORATION: ADAPTIVE CO-PLANNING FOR MEANINGFUL CHANGE

There are many tools that have been applied to a range of management issues and that are relevant to wicked problems in social-ecological systems, notwithstanding the valid critiques others have made concerning their impact on the distribution of power and influence (e.g., McGuirk 2001). These tools include reframing, redefinition, visioning and learning, scenario development, and modelling, to flag some examples. One of the constraints is that, almost by definition, most of these participatory approaches operate at a local or regional scale and are restricted to the participation of designated "stakeholders" (e.g., Prell et al. 2003). While this is clearly appropriate in some circumstances, such as limited entry systems, our global situation lays claim to valuing more inclusive approaches, and we anticipate opportunities for complementary strategies.

The first step is to adopt a conscious position, aligning with the cross-scale systems approach of panarchy and acknowledging the heterogeneity of the decision landscape. This includes a vision that change is inevitable and that it may be possible to actively open doors to creative change that supports human-scale public goods, such as vibrant ecosystems, peace, viable livelihoods, healthy children, and robust elderly.

Theory is also a tool when coupled with activities of learning organizations. Important contributions in the context of participatory planning include recognition that free and transparent flows of information are essential and understanding that, in the spirit of the work of Innes, "many types of information count, other than 'objective' information" (Innes 1998). For example, communicative planning offers a valuable nexus for praxis, combining theory and practice. In this vein, Harrison's review of the literature (2002, 159) may be helpful when he identifies pragmatism as a key. He revisits "radical empiricism" in the context of pragmatism (see note 7, below) and points to Joas's 1998 reference to pragmatism as "a theory of the creative character of human action." Elsewhere, critical theory has assisted in creating reflexive understanding and in turn contributing to what Bandura terms "self-efficacy" (Pinkerton 1991, 2007; Doubleday 2007). Learning by doing, and by watching what others do, is fundamental to human development.

Regional planning has bravely attempted integrative approaches to megaproject planning and land-use planning processes before, notably in northern Canada, and particularly in the case of the Lancaster Sound Regional Study in 1979. In this instance, offshore drilling was proposed in Lancaster Sound, an important ecological area, significant for its wildlife both to conservationists and to local Inuit. Peter Jacobs, author of the resulting Green Paper in 1983, said that the key to development of a planning process for Lancaster Sound was the establishment of the principle that a project cannot be evaluated without an understanding of its context (CARC 1986). The Inuit at the time took the position that

> the process of development is more important than the nature of the development project. They accept the idea of development, but within two limits: first, a development project must not cause long-term or extensive damage to the environment; and second, a development project must not reduce their economic and lifestyle options, which range from a subsistence economy to a mixed

economy to a cash economy. (Jacobs, interview with the Canadian Arctic Resources Committee, 1986, n.p.)

This position provides a productive frame for futures planning. It is highly relevant to First Nations facing the production phase of the Oil Sands in northern Alberta and to farmers contemplating a new refinery and pipelines, and to lobster fishermen in inshore areas of the Atlantic that have been targeted for intensive fish farming. It is a position that is also a precursor to ideals of sustainable development as articulated by Brundtland in the *Report of the World Commission on Environment and Development* (1986). The novel idea is to begin with the commitment to preserve system integrity and then insert development so that desired conditions are maintained (instead of conducting environmental assessment aimed at showing that development will have no effects). This position shifts the onus for effects and the weight of development costs to proponents, while retaining a greater degree of control of decision-making within communities. New technologies—in particular, open source models—now available offer infrastructure capacity for social processes in ways never before possible. Organizational models for managing open source approaches are evolving (e.g., Surman and Surman 2008).

In addition, complex (wicked) problems can stimulate novel visions and strategies for large-scale change in unexpected ways, as one NGO notes:

> [T]he size of the tar sands issue can seem daunting, but in reality few issues have presented an opportunity for a social justice movement to truly articulate a different vision of organizing the world, [a vision] that has as many entry points, and can provide as large of an impact. The scale and scope of the tar sands is huge and has tremendously deep implications for the way we approach questions that span the social justice spectrum. With a coordinated response involving all sectors of North American social justice movements currently impacted by the largest industrial project in human history we have the possibility to change the course of human and ecological fate like nowhere else. (Oil-SandsTruth.org)

Elsewhere, pragmatic grassroots responses to wicked problems on a local scale are emerging. Just south of Chemical Valley outside Sarnia, Ontario, the Aamjiwnaang First Nation Environmental Committee formed the Aamjiwnaang Bucket Brigade to take air samples when industry and government failed to do so to their satisfaction, and in 2011 Ecojustice (https://ecojustice.ca) mounted a Charter challenge. One consequence of the success of the Bucket Brigade is that the government air-monitoring station has been moved to a location *downwind* of the industrial complex, where it can actually sample emissions (Scott 2008).

This chapter's working hypothesis has been that by viewing wicked problems as an emerging "system of systems" with the potential for multiple "stable states," through the twin lenses of adaptation and resilience, it is possible to contribute to formulating a public response to what is clearly a public policy void. To achieve this, however, requires a cultural shift—that we move from culture as a set of norms and conditions, or as an objective given, towards a dynamic understanding of culture as changing and dynamic and as capable of transforming our relationships with one another and with the social-ecological systems we inhabit. This concept of a *culture vector* is intended to convey a sense of *culture as action,* with cross-scale, systemic learning and evolutionary capacities.

By considering long-term spatial and temporal change and finding means to integrate actors who lack roles and authority in decision-making, we forge an inclusive collective and learning community. The significance of weak links (Barabási 2003; Csermely 2006) in large systems also needs to be recognized. Here, adaptive governance creates opportunities for reconceptualizing societal decision-making by diversifying both the array of voices heard and the alternative outcomes that are considered. When dealing with complex systems and a mid- to long-range time horizon, is change an inevitability or a surprise—or a matter of perspective? Would greater inclusivity and diversity have generated a range of divergent views that reduced the degree of surprise? In the mobilization of culture formation as an emergent property of social-ecological systems it may be possible to broaden the base for adaptive governance beyond conventional interests as we move towards the formation of an open and inclusive social-ecological system that acknowledges interconnectivity and interdependency through space and time.

Catastrophic events often have triggers and impact paths at multiple scales, both spatial and temporal. After the fact, they become subjects of exhaustive, reflexive analysis and sometimes criticism. In the best of worlds, they may become agents of change. As Boschman illustrates with his interrogation of the "vampiric Other" (Chapter 4), this may also be true in the worst of post 9/11 worlds. Too often the signals and warnings that precede crises are ignored, coming to the fore in post-crisis analysis and sometimes resulting in punitive and/or remedial action. Clearly catastrophic damage cannot often be undone, and the best hope is that learning can occur and that catastrophes in analogous situations in the future can be averted through preventative action based on this learning from past practice. It is this pragmatic aspect of future application of understandings gained from past and current practice that William James[6] termed "radical empiricism"[7] and that some planners have invoked as a step towards praxis, linking pragmatism to planning, particularly with Dewey's strand of pragmatism (e.g., Harrison 2002, 159, referred to above, citing Dewey 1931; and clarified in his 2002 conference paper).[8] Here we could also draw from the theory and methods of sciences concerned with long-term change, for example, paleolimnology, where current conditions serve as proxies for past environments, allowing the reconstruction of processes of environmental change (e.g., Smol 2005).

Why are early warnings in problematic situations not seized as opportunities for redirection and change? According to Holling and Sanderson (1996, 152),

> if change is episodic (that is patchy in time the way ecosystems are patchy in space), one's only hope for leverage is to understand where in the episodic cycle the system is, and to act according. If, in fact, the tendency of social systems is to lock in to a given set of goals, outputs, and working processes, can it be said that the stable system is locked into a trajectory of development that can't be altered until it generates a crisis? (qtd. in Pritchard and Sanderson [2002])

Reflecting on this quotation, and on the observations of Ivan Macfadyen, the sailor referred to in my prologue and on the experience of the Aamjiwnaang Bucket Brigade, we can see release and reorganization taking place constantly, everywhere, from mid-ocean to the aging facilities of Chemical Valley, both literally and metaphorically. We also see the lobster fishing community acting to conduct research on the recovery of marine ecosystems previously exposed to fish farm waste, as well as participating in regulatory hearing processes.

Through communication, dissemination, and agency as citizen scientists, as caring and concerned humans, cross-scale learning is also taking place:

> [L]earning is viewed as a process of detecting and correcting error, and occurs under two conditions. The first is when intentions match outcomes of action, and the second is when intentions and outcomes do not match. Single-loop learning occurs when matches happen, or when mismatches are corrected by changing one's strategy or behaviour while preserving basic values and norms. Double-loop learning occurs by correcting mismatches by first changing or supplementing existing values and norms, then changing strategies or behaviour ... Learning occurs at both individual and social levels, but individuals are the agents for social collectives. Therefore, social learning does not occur until individuals encode what they have learned in social memory. (Diduck et al., 2005: 271)

The message here is dual: if we wish to anticipate system failure and avert catastrophes in the making, we need to understand the state of the system and the forces and processes that are operating to maintain that system's course, but we also need to engage in double-loop learning that clarifies the values and norms of all of those seeking to influence outcomes. We need to rebalance power through self-organization and adaptive governance enterprises that are strengthened by inclusivity. This is the call for the nurture of *culture as a vector* composed of values, principles, and norms, but also of action that enables self-efficacy and releases agency through engagement and participation. The concept of a *culture vector* flows from the spirit of adaptation that informs scenario creation, learning communities, and responsive fields such as communicative planning. Resilience thinking has offered us a foundation: social-ecological systems are more than systems of one or the other alone; and all systems are in flux. Envisioning a cultural dynamic—an evolving culture committing to action for inclusivity, diversity, equity, social justice, peace, health, and sustainability across the interfaces of social-ecological systems and across scales of human organization—creates opportunity for visioning. Adaptive co-management offers a strategic approach and social innovations that are rich with potential for new forms for cross-scale engagement.

CONCLUSION

In terms of supporting "variability that maintains renewal capacity," the goal, as seen from social, cultural, or ecological perspectives, is the protection and nurture of the deep wellsprings of sustainable change: "[T]he key question for future work is how we can implement ways to expand human opportunity, sustain resilience, and facilitate human learning" (Holling, Gunderson, and Ludwig 2002, 416). We are engaged collectively in the transformation of the earth and its social-ecological systems on continental scales, with nodes of impacts in widely separated and very different locales yet linked across scales. Those concerned for health and well-being, for peace and social justice, for sustainability, as well as those affected in some local nodes of impact, have begun to self-organize and to achieve learning that is durable and encoded in memory. Perhaps most importantly there is evidence that conceptual innovation, including a vision of collaborative learning and planning for a future of sustainable change, is occurring.

This spirit of resilience in action is new in scale and magnitude, both in spatial and temporal terms. It is new because it is the consequence of a broad understanding of the nature and meaning of social-ecological change distributed across scales, across time and space. It has been called into being not only by the large forces of energy development and industrial economics, but also by the creative communications, dissemination, and agency of citizen scientists, peacemakers, and change agents.

For some, it is clear that we have crossed a conceptual threshold and moved from slow change to slow catastrophe. Others remain invested in a status quo of competing constructions and contemporary priorities. Reframing issues in order to bridge silos and open the doors to change is heavy conceptual and practical work. However, through technological innovation, double-loop learning, and strengthened collaborative linkages, and in new initiatives in design and planning for resilience, like those in evidence here, we begin to co-create alternative scenarios, redirect gradual failures, transform slow catastrophe, and chart paths from crisis towards the opportunities we envision, by planning for the unthinkable, dreaming new dreams, and planning with courage and hope. This is an emergent culture, promising a vector of change. It also sounds in the deep harmonics of the multivocal, environmental humanism resonant throughout this volume.

Notes

1 Goldstein, 2009. https://www.ecologyandsociety.org/vol14/iss2/art33/.
2 See International Studies Association, "Call for Proposals," 2016; and Nancy Doubleday, "Sustainability and Peaceful Coexistence: New Opportunities for, and from, Arctic Regimes?" International Studies Association, Atlanta, Georgia, 17 March 2016. See also Outward Bound, "Peace Camp." This holistic approach has been advocated elsewhere (Abuelaish and Doubleday 2011).
3 In addition to decades of field observations, we also adopt an *a priori* stance and follow Commoner's Laws of Ecology, Leonardo da Vinci's exhortation to "see," John Muir's wilderness sense of interconnectivity, and Barbarási's 2003 exegesis on complexity, among others.
4 Zebeedee Nungak, speaking during the First Ministers's Meeting on the Canadian Constitution, held in Ottawa in April 1985.
5 Commonwealth of Australia (2007), *Tackling Wicked Problems: A Public Policy Perspective*.
6 Sincere thanks to Sean Gould for pointing out the need to fine-tune the reference around Harrison's use of Dewey and "radical empiricism" to include William James.
7 I am indebted to Professor Nicholas Griffin for the following reference to Chapter 74 of Ralph Barton Perry's biography of William James: *The Thought and Character of William James* (Boston: Little, Brown, 1935), where at page 388 Perry notes the first use of the term "radical empiricism" by James, occurring in a letter to Professor Ferrari (the Italian translator of James's *The Principles of Psychology*) on 22 February 1905.
8 Philip Harrison, "Subverting Orthodoxy: A Re-Look at the 'Truths' of Post-Apartheid Planning," a paper prepared for the Planning Africa Conference, Durban, September 2002. School of Architecture and Planning, University of the Witwatersrand. http://citeseerx.ist.psu.edu/viewdoc/download?doi=10.1.1.200.2677&rep=rep1&type=pdf.

References

Abuelaish, Izzeldin, and Nancy C. Doubleday. 2011. "Holistic Frontiers in Peace and Health Research." *International Journal of Peace and Development Studies* 2(5): 156–61.

Armitage, Derek, Fikret Berkes, and Nancy Doubleday. 2007. *Adaptive Co-management: Collaboration, Learning and Multi-Level Governance*. Vancouver: UBC Press.

Barabási, Albert-László. 2003. *Linked: The New Science of Networks*. London: Penguin.

Berkes, Fikret, Derek Armitage, and Nancy Doubleday. 2007. "Synthesis: Adapting, Innovating, Evolving." In *Adaptive Co-management: Collaboration, Learning, and Multi-Level Governance*, edited by Derek Armitage, Fikret Berkes, and Nancy Doubleday, 308–27. Vancouver: UBC Press.

Berkes, F., and C. Folke, eds. 1998. *Linking Social and Ecological Systems: Management Practices and Social Mechanisms for Building Resilience*. New York: Cambridge University Press.

Berkes, Fikret, Rob Huebert, Helen Fast, Micheline Manseau, and Allan Diduck. 2005. *Breaking Ice: Renewable Resource and Ocean Management in the Canadian North*. Calgary: University of Calgary Press.

Canadian Arctic Resources Committee (CARC). 1986. An Interview with Peter Jacobs, "The 1979–1983 Lancaster Sound Regional Study." *Northern Perspectives* 14(3). http://carc.org/pubs/v14no3/3.htm. Accessed 8 July 2015.

Commoner, Barry. 1971. *The Closing Circle: Nature, Man, and Technology*. New York: Knopf.

Commonwealth of Australia. 2007. *Tackling Wicked Problems: A Public Policy Perspective*. https://www.apsc.gov.au/tackling-wicked-problems-public-policy-perspective.

Csermely, Peter. 2006. *Weak Links: Stabilizers of Complex Systems from Proteins to Social Networks*. Heidelberg: Springer Verlag.

da Vinci, Leonardo. *The Notebooks*. https://www.britannica.com/biography/Leonardo-da-Vinci, accessed 19 September 2018.

Diduck, Alan, Nigel Bankes, Douglas Clark, and Derek Armitage. 2005. "Unpacking Social Learning in Social-Ecological Systems: Case Studies of Polar Bear and Narwhal Management in Northern Canada." In *Breaking Ice: Renewable Resource and Ocean Management in the Canadian North*, edited by Fikret Berkes, Rob Huebert, Helen Fast, Micheline Manseau, and Alan Diduck, 269–90. Calgary: University of Calgary Press.

Dietz, Thomas, and Paul C. Stern. 2008. *Public Participation in Environmental Assessment and Decision Making*. Washington, DC: National Research Council. http://www.nap.edu/catalog/12434.html.

Doubleday, Nancy. 2007. "Culturing Adaptive Co-management: Finding 'Keys' to Resilience in Asymmetries of Power." In *Adaptive Co-management: Collaboration, Learning, and Multi-Level Governance*, edited by Derek Armitage, Fikret Berkes, and Nancy Doubleday, 228–48. Vancouver: UBC Press.

———. 2016. "Sustainability and Peaceful Coexistence: New Opportunities for, and from, Arctic Regimes?" International Studies Association, Atlanta, Georgia, 17 March 2016.

Doubleday, Nancy, A. Fiona D. Mackenzie, and Simon Dalby. 2004. "Reimagining Sustainable Cultures: Constitutions, Land, and Art." *Canadian Geographer / Le Géographe canadien* 48(4): 389–402.

Ecojustice. 2011. https://www.ecojustice.ca/case/defending-the-rights-of-chemical-valley-residents-charter-challenge/.

Gibson-Graham, J.K. 2003. "An Ethics of the Local.'" *Rethinking Marxism* 15: 49–74.

Gibson-Graham, J.K., and Gerda Roelvink. 2010. "An Economic Ethics for the Anthropocene." *Antipode* 41(1): 320–40.

Goldstein, B. 2009. Resilience to surprises through communicative planning. *Ecology and Society* 14(2): 33. http://www.ecologyandsociety.org/vol14/iss2/art33/.

Gunderson, Lance H., C.S. Holling, and Garry Peterson. 2002. "Surprises and Sustainability: Cycles of Renewal in the Everglades." In *Panarchy: Understanding Transformations in Human and Natural Systems*, edited by Lance H. Gunderson and C.S. Holling, 315–32. Washington, DC: Island Press.

Harrison, Philip. 2002. "A Pragmatic Attitude to Planning." In *Planning Futures: Nre Directions for Planning Theory*, edited by Phillip Allmendinger and Mark Tewdwr-Jones, 157–71. New York: Routledge.

———. 2002. "Subverting Orthodoxy: A Re-Look at the 'Truths' of Post-Apartheid Planning." Paper prepared for the Planning Africa Conference, Durban, September 2002. School of Architecture and Planning, University of the Witwatersrand. http://citeseerx.ist.psu.edu/viewdoc/download?doi=10.1.1.200.2677&rep=rep1&type=pdf.

Hillier, Jean. 2002. "Direct Action and Agonism in Democratic Planning Practice." In *Planning Futures: New Directions for Planning Theory*, edited by Phillip Allmendinger and Mark Tewdwr-Jones, 110–35. New York: Routledge.

Holling, C.S., S.R. Carpenter, W.A. Brock, and L.H. Gunderson. 2002. "Discoveries for Sustainable Futures." In *Panarchy: Understanding Transformations in Human and Natural Systems*, edited by L.H. Gunderson and C.S. Holling, 395–417. Washington, DC: Island Press.

Holling, C.S., Lance H. Gunderson, and Donald Ludwig. 2002. "In Quest of a Theory of Adaptive Change." In *Panarchy: Understanding Transformations in Human and Natural Systems*, edited by Lance H. Gunderson, C.S. Holling, 3–22. Washington, DC: Island Press.

Holling, C.S., and Steven Sanderson. 1996. "Dynamics of (Dis)harmony in Ecological and Social Systems." In *Rights to Nature: Ecological, Economic, Cultural, and Political Principles of Institutions for the Environment*, edited by Susan Hanna, Carl Folke, and Karl-Göran Mäler, 57–85. Washington, DC: Island Press.

Innes, Judith E. 1998. "Information in Communicative Planning." *Journal of the American Planning Association* 64(1): 52–63.

Jacobs, Peter. *The Lancaster Sound Regional Study*. By Canadian Arctic Resources Committee, 1979–83.

Janssen, Marco A. 2002. "A Future of Surprises." In *Panarchy: Understanding Transformations in Human and Natural Systems*, edited by Lance H. Gunderson and C.S. Holling, 241–60. Washington, DC: Island Press.

Joas, Hans. 1998. "The Autonomy of the Self: The Meadian Heritage and Its Post-modern Challenge." *European Journal of Social Theory* 1(1): 7–18. https://doi.org/10.1177/136843198001001002.

Kok, Kasper, Reinette (Oonsie) Biggs, and Monika Zurek. 2007. "Methods for Developing Multiscale Participatory Scenarios: Insights from Southern Africa and Europe." *Ecology and Society* 13(11). http://www.ecologyandsociety.org/vol12/iss1/art8.

McGuirk, P.M. 2001. "Situating Communicative Planning Theory: Context, Power, and Knowledge." *Environment and Planning* 33(2): 195–217.

Morden, Paul. 2008. "Lambton jumps on the green bandwagon." *Sarnia Observer*, 7 November. http://www.theobserver.ca/2008/11/07/lambton-jumps-on-the-green-bandwagon.

Muir, John. 1988. *My First Summer in the Sierra.* (1911). San Francisco: Sierra Club Books.

Nadasdy, Paul. 2003. *Hunters and Bureaucrats: Power, Knowledge, and Aboriginal–State Relations in the Southwest Yukon.* Vancouver: UBC Press.

———. 2007. "Adaptive Co-management and the Gospel of Resilience." In *Adaptive Co-management: Collaboration, Learning, and Multi-Level Governance*, edited by Derek Armitage, Fikret Berkes, and Nancy Doubleday, 208–27. Vancouver: UBC Press.

Nungak, Zebedee. 1985. First Ministers' Conference on the Rights of Aboriginal Peoples, Ottawa, April 1985. Pers. comm.

O'Brien, Mary. 2000. *Making Better Environmental Decisions: An Alternative to Risk Assessment.* Cambridge, MA: MIT Press.

Oil Sands Truth. "Everyone's Downstream." *Oil Sands Truth: Shut down the tar sands.* http://oilsandstruth.org/conf. Accessed 2 July 2015.

Patterson, James J., Carl Smith, and Jennifer Bellamy. 2013. "Understanding Enabling Capacities for Managing the 'Wicked Problem' of Nonpoint Source Water Pollution in Catchments: A Conceptual Framework." *Journal of Environmental Management* 128: 441–52.

Perry, Ralph Barton. 1935. *The Thought and Character of William James.* Boston: Little, Brown.

Peterson, Garry D., Graeme S. Cumming, and Stephen R. Carpenter. 2003. "Scenario Planning: A Tool for Conservation in an Uncertain World." *Conservation Biology.* 17(2): 358–66.

Pinkerton, Evelyn. 1991. "Locally Based Water Quality Planning: Contributions to Fish Habitat Protection." *Canadian Journal of Fisheries and Aquatic Sciences* 48(7): 1326–33.

Pinkerton, Evelyn. 2007. "Integrating Holism and Segmentalism: Overcoming Barriers to Adaptive Co-management between Management Agencies and Multi-Sector Bodies." In *Adaptive Co-Management: Collaboration, Learning, and Multi-Level Governance*, edited by Derek Armitage, Fikret Berkes, and Nancy Doubleday, 151–71. Vancouver: UBC Press.

Prell, Christina, Klaus Hubacek, Mark Reed, Claire Quinn, Nanlin Jin, Joe Holdern, Tim Burt, Mike Kirby, and Jan Sendzimir. 2007. "If You Have a Hammer Everything Looks Like a Nail: Traditional versus Participatory Model Building." *Interdisciplinary Science Reviews* 32(3): 263–82.

"Priority Environmental Site Map." *Environmental Commissioner of Ontario.* http://www.eco.on.ca/eng/index.php?page=priority-environmental-sitemap #STCR. Accessed 2 July 2014.

Pritchard, Lowell Jr., and Steven E. Sanderson. 2002. "The Dynamics of Political Discourse in Seeking Sustainability." In *Panarchy: Understanding Transformations in Human and Natural Systems,* edited by Lance H. Gunderson, C.S. Holling, 147–72. Washington, DC: Island Press.

Scott, Dayna Nadine. 2008. "Confronting Chronic Pollution: A Socio-Legal Analysis of Risk and Precaution." *Osgoode Hall Journal,* 46(2). Special Issue on Environmental Law. http://ssrn.com/abstract=1262791.

Smol, John. P. 2005. "Tracking Long-Term Environmental Changes in Arctic Lakes and Ponds: A Paleolimnological Perspective." *Arctic* 58(2): 227–29.

Surman, Tonya, and Mark Surman. 2008. "Open Sourcing Social Change: Inside the Constellation Model." *Open Source Business Resource.* http://www.osbr.ca/ojs/index.php/osbr/article/view/698/666.

Westley, Frances. 2002. In *Panarchy: Understanding Transformations in Human and Natural Systems,* edited by Lance H. Gunderson and C.S. Holling, 333–60. Washington, DC: Island Press.

Westley, Frances, Brenda Zimmerman, and Michael Quinn Patton. 2007. *Getting to Maybe: How the World Is Changed.* Toronto: Vintage Press.

World Commission on Environment and Development. 1987. *Our Common Future ("The Brundtland Report").* New York: Oxford University Press.

ABOUT THE CONTRIBUTORS

KARLA ARMBRUSTER is Professor of English at Webster University in St. Louis, where she teaches literature, professional writing, and interdisciplinary courses; she also directs the Sustainability Studies minor. She is co-editor of *Beyond Nature Writing: Expanding the Boundaries of Ecocriticism* and *The Bioregional Imagination: Literature, Ecology, and Place* as well as the author of various ecocritical and animal studies-oriented articles and essays.

PAMELA BANTING (Associate Professor, University of Calgary) founded and served as the inaugural President of the Association for Literature, the Environment, and Culture in Canada (ALECC). She is the editor of the literary anthology *Fresh Tracks: Writing the Western Landscape* and the special issue of *Studies in Canadian Literature* on Canadian Literary Ecologies and the author of the essay "Ecocriticism in Canada" in the new *Oxford Handbook of Canadian Literature* as well as numerous other critical-theoretical articles. Her current research and teaching are in the areas of energy in literature/petrocultural studies, literature and culture in the Anthropocene, psychogeography, decolonization and sense of place, and animality.

ROBERT BOSCHMAN is Professor of American Literature and Ecocriticism in the Department of English, Languages, and Cultures at Mount Royal University in Calgary, Alberta. His monograph, *In the Way of Nature: Ecology and Westward Movement in the Poetry of Anne Bradstreet, Elizabeth Bishop, and Amy Clampitt*, was published by McFarland in 2009. *Found in Alberta: Environmental Themes for the Anthropocene* (co-edited with Mario Trono) was published in 2014 by Wilfrid Laurier University Press. Boschman's environmental memoir, published by the University of Regina Press, is forthcoming.

NANCY C. DOUBLEDAY is Hope Chair in Peace and Health, Department of Philosophy, at McMaster University in Hamilton, Ontario. Her work is transdisciplinary and integrative and has traversed ecology, contaminants, climate, environmental ethics, Indigenous rights, and social-cultural-ecological systems change. She works collaboratively with colleagues in Ocean Canada, Participedia, and Global Water Futures to develop methodologies for polycentric governance, health, environment, and justice.

SABINE FEISST is Professor of Musicology and Senior Sustainability Scholar at Arizona State University's School of Music and Global Institute of Sustainability. Focusing on twentieth- and twenty-first-century music studies, she published the monographs *Der Begriff 'Improvisation' in der neuen Musik* (Studio 1997) and *Schoenberg's New World: The American Years* (Oxford, 2011 and 2017). With Ethan Haimo, she authored *Schoenberg's Early Correspondence* (Oxford, 2016). Author of more than eighty articles in anthologies, journals, and reference works and US editor of *Contemporary Music Review*, she is currently writing a monograph on music inspired by deserts in the American Southwest and editing the *Oxford Handbook of Ecomusicology.*

SEAN GOULD, a native Idahoan, received his doctorate in moral psychology from Tilburg University. He is currently a raft guide and ski instructor in central Idaho.

PAUL HUEBENER is Associate Professor of English in the Centre for Humanities at Athabasca University. His book *Timing Canada: The Shifting Politics of Time in Canadian Literary Culture* (McGill–Queen's University Press) was a finalist for the 2015 Gabrielle Roy Prize. He is also a co-editor of *Time, Globalization, and Human Experience* (Routledge, 2017).

MISHKA LYSACK is a Full Professor in the Faculty of Social Work at the University of Calgary; he is also an Adjunct Assistant Professor in the Faculty of Medicine and has taught in the Faculty of Environmental Design since 2008. In 2014, Dr. Lysack established a knowledge mobilization and research partnership with the German Embassy in Ottawa focusing on the best practices in Germany in their leadership on their renewable energy transition (*Energiewende*) and climate protection (*Klimaschutz*). Through knowledge mobilization grants from both the Canadian (SSHRC) and German governments, Dr. Lysack and the German Embassy have brought a series of leaders in Germany to Canada to consult with their counterparts in three provinces (Ontario, Alberta, and British Columbia). Mishka Lysack was awarded the Queen's Diamond Jubilee Medal in 2013 for outstanding community leadership both in Alberta and on the national level in Canada regarding climate change and environmental protection. Canada West Foundation selected him as one of "40 Extraordinary Canadians" for its book *An Extraordinary West* because of his innovations in environmental education and leadership. More recently, Dr. Lysack was awarded both the Peak Scholar Award in 2018 for "Innovative Leadership for Renewable Energy Economy and Climate Protection" for "Excellence in Entrepreneurship, Innovation and Knowledge Engagement," as well as the 2018 Award in Faculty Leadership in Sustainability Education and Research for "outstanding achievement and excellence in sustainability at the University of Calgary."

LEA REKOW, PhD, is co-lead and co-curator of BifrostOnline, an international, open-access project promoting climate change awareness. Lea is also founder of Green My Favela (GMF), an urban restoration project based in the *favelas* of Rio de Janeiro. Lea has directed several mid-scale cultural institutions, was formerly special envoy for Open & Agile Smart Cities, an advisor to the European Urban IxD program, a fellow at the Center for Art and the Environment, and a consultant for GlobalCAD. She has sat on advisory panels including for Amnesty International (media arm), the MacArthur award, and the Lower Manhattan Cultural Council. Lea's research focuses on transdisciplinary practices for reclaiming degraded space in areas where people are living under socio-environmental stress. She has worked with ethnic minorities and marginalized communities in conflict zones in Burma, on large-scale reclamation projects on the Navajo Nation, and through GMF helped establish one of the largest urban, organic food security projects in Latin America.

KENT SCHROEDER is Director of International Development Projects at Humber College. He recently received the prestigious Certificate of Appreciation Award from the Prime Minister of Bhutan, Tshering Tobgay, and is one of eight recipients worldwide who have contributed to the implementation and refinement of Bhutan's Gross National Happiness Index and policy.

RANDY SCHROEDER is Professor of English, Languages, and Cultures at Mount Royal University in Calgary. He is interested in contemporary applications of interdependence, an ancient notion back in currency.

GENEVIÈVE SUSEMIHL is Assistant Professor to the chair of Cultural and Media Studies at the English Department, University of Kiel, Germany. She has also been a Postdoctoral Research Fellow at Carleton University, Ottawa, and at the Queen's Centre for International Relations, Kingston, Ontario. Her research interests cover the consequences of migration and integration, the importance of cultural heritage for identity and community building, regional studies, film and media, and Indigenous studies. Susemihl has published extensively on Jewish immigration to North America, the construction of Native Americans in literature and culture, cultural heritage, and storytelling. Her publications include "... *and it became my home": Die Assimilation und Integration der deutsch-jüdischen Hitlerflüchtlinge in New York und Toronto* (The Assimilation and Integration of German-Jewish Hitler Refugees in New York and Toronto; 2004), and *Bären, Lachse, Totempfähle: Die kanadische Inselgruppe Haida Gwaii am Rand der Welt* (Bears, Salmons, Totem Poles: The Canadian Archipelago Haida Gwaii at the Edge of the World, 2016).

MARIO TRONO studies visual cultures from an environmental perspective. He has a monograph on ecocinema forthcoming from Wilfrid Laurier University Press, and he co-edited (with Robert Boschman) *Found in Alberta: Environmental Themes for the Anthropocene* (2014). Mario was a co-founder (with Boschman) of *Under Western Skies,* a biennial, interdisciplinary conference on the environment, and teaches at Mount Royal University.

MORGAN ZEDALIS, PhD, has worked with hunters, ranchers, and Nimi'ipuu (Nez Perce) tribal members on landscape and wildlife management issues that face communities in Idaho. She currently works with the Payette National Forest's Heritage Program in cultural resource management in McCall, Idaho, her hometown.

INDEX

30 Days of Night (film), 110, 119–20
30 Days of Night (graphic novel): agents of the law in, 122, 123, 124, 133n22; allusions to 9/11 terror attack, 126; altered world of, 116, 131n6; characters, 110, 111, 124–25; depiction of new reality, 111–12; depiction of the city of Barrow, 132n20; Environmental Vampire in, 112, 119–20, 122; figure of Stranger, 122; horror of post-human invasion, 114–15, 122–23, 124; images from, *108, 109, 111, 115, 120, 121, 123, 125, 127*; theme of lost territory, 127, 130; theme of sacrifice, 126; theme of the Other, 123, 130; title of, 110; vampire ecology, 124–26
30 Days of Night: Night, Again (graphic novel), 111, 126, 128, *129*

Aamjiwnaang Bucket Brigade, 339, 341
abandoned uranium mines (AUMs), 173, *174*, 175, 176–77
abstraction, 58
acoustic ecology, 89, 99
actants, notion of, 4, 59
action, concepts of, 6–7
Adams, John Luther, 89
adaptive co-management, 331, 336, 339, 342
adaptive cycle metaphor, 334, 335
advertising: ecocritical reading of, 43–44; escape theme in, 44–46; event time in, 42–43
Aeschylus, 194
agency: academic debates over, 6–7; agentivity *vs.*, 6; collective, 8, 329–30; of display, 140; distributive, 5, 9, 59, 63; human, 7–9, 10; machines and, 8; post-humanist views of, 7–8; in psychology, 6; receptive, 18; time and, 1, 4, 16–17, 38

"Agency at the Time of the Anthropocene" (Latour), 4
"agency dilemma," 5–6
agentive realism, 5
Äit-Touati, Frédérique, 14, 22, 194, 196, 198, 205
Alaimo, Stacy, 5, 210, 211
Alberta: beef production, 77; Climate Change Strategy, 305; Cowboy Trail, 153; garbage management practices, 236–37; tourist attractions, 157n11, 196–98; weather, 76–77
Alberta Badlands, 22, *197*, 198, 203
Alexander, Caroline, 226
Alien (film), 81
Alliès, Paul, 121
Allô, ici la terre (multimedia spectacle): description of, 90; music and voices in, 92; "objective" and "subjective" texts of, 91–92, 102; origin of, 90–91; reception of, 92–93; themes, 91, 92
Among Grizzlies: Living with Wild Bears in Alaska (Treadwell), 215
Anderson, Daniel Gustave, 305
Andrus, Cecil, 267
"Anecdote of the Jar" (Stevens), 48–49
animals: concept of parallel lives of, 214; definition of wild, 239; effect of fossil fuel industry on, 236, 237–38, 239; humans' relationship with, 22, 23, 212–13, 219, 220–21, 237, 255–56; look of, 213, 214; as predators, 209–10; roadkill, 245–46; songs of, 97; as sources of energy, 237; as stakeholders of ecosystems, 256. *See also* bears; wolves
"anim-oils," 256
antagonistic pleiotropy, 290–91, 299–300, 301, 302
Anthropocene, 4, 33, 200

Books in the Environmental Humanities Series
Published by Wilfrid Laurier University Press

Animal Subjects: An Ethical Reader in a Posthuman World | Jodey Castricano, editor | 2008 | ISBN 978-0-88920-512-3

Open Wide a Wilderness: Canadian Nature Poems | Nancy Holmes, editor | 2009 | ISBN 978-1-55458-033-0

Technonatures: Environments, Technologies, Spaces, and Places in the Twenty-first Century | Damian F. White and Chris Wilbert, editors | 2009 | ISBN 978-1-55458-150-4

Writing in Dust: Reading the Prairie Environmentally | Jenny Kerber | 2010 | ISBN 978-1-55458-218-1

Ecologies of Affect: Placing Nostalgia, Desire, and Hope | Tonya K. Davidson, Ondine Park, and Rob Shields, editors | 2011 | ISBN 978-1-55458-258-7

Ornithologies of Desire: Ecocritical Essays, Avian Poetics, and Don McKay | Travis V. Mason | 2013 | ISBN 978-1-55458-630-1

Ecologies of the Moving Image: Cinema, Affect, Nature | Adrian J. Ivakhiv | 2013 | ISBN 978-1-55458-905-0

Avatar and Nature Spirituality | Bron Taylor, editor | 2013 | ISBN 978-1-55458-843-5

Moving Environments: Affect, Emotion, Ecology, and Film | Alexa Weik von Mossner | 2014 | ISBN 978-1-77112-002-9

Found in Alberta: Environmental Themes for the Anthropocene | Robert Boschman and Mario Trono, editors | 2014 | ISBN 978-1-55458-959-3

Sustaining the West: Cultural Responses to Canadian Environments | Liza Piper and Lisa Szabo-Jones, editors | 2015 | ISBN 978-1-55458-923-4

On Active Grounds: Agency and Time in the Environmental Humanities | Robert Boschman and Mario Trono, editors | 2019 | ISBN 978-1-77112-339-6